W9-BAX-068

New Transportation Fuels

New Transportation Fuels

A Strategic Approach to Technological Change

Daniel Sperling

UNIVERSITY OF CALIFORNIA PRESS
Berkeley • Los Angeles • London

University of California Press
Berkeley and Los Angeles, California

University of California Press, Ltd.
London, England

Library of Congress Cataloging-in-Publication Data

Sperling, Daniel.
New transportation fuels: a strategic approach to technological change/Daniel Sperling.
p. cm.
Bibliography: p.
Includes index.
ISBN 0-520-06087-3 (alk. paper)
1. Synthetic fuels industry. 2. Energy policy—Environmental aspects. 3. Transportation equipment industry—Technological innovations. I. Title.
HD9502.5.S962S67 1989
338.4'766266'097—dc 19 87-28805
CIP

Library of Congress Cataloging in Publication Data
Printed in the United States of America

1 2 3 4 5 6 7 8 9

Contents

Figures

TABLES

Acknowledgments

I am indebted to the many people who helped bring this book to fruition. First and foremost is my wife Tricia Davis, who has shown such confidence in me and has generously provided her considerable talents in helping organize the overall structure of the book and the contents of several key chapters.

Individuals who played an especially important role during the early stages of the book are Oreste Bevilacqua, who first attracted me to this topic; William Garrison, who provided the intellectual stimulation that inspired and motivated me; and Henry Bruck, my most diligent reviewer and enthusiastic supporter during the early drafts. During final preparation, Mark Christensen and Mark DeLuchi carefully reviewed the full manuscript and provided insightful and provocative comments, challenging me to go beyond conventional wisdom.

In addition, the following individuals enthusiastically reviewed one or more chapters for content, willingly sharing their considerable expertise and insight: Roberta Nichols, Tom Reed, Phil Patterson, Ken Koyama, Rob Motal, Eugene Ecklund, Sergio Trindade, Jeff Alson, and Barry McNutt.

Several students at the University of California, Davis, also made valuable contributions. Nancy Lindsay and John Bhend gathered and analyzed data for the two chapters on environmental impacts, Ken Kurani analyzed the diesel car experience, Mark Farman conducted the case study cost analysis in chapter 8, Quanlu Wang helped finalize and verify citations, and Mark DeLuchi made major contributions to chapters 12 and 15 and smaller contributions to other chapters. Special mention must be given to the library staff of the Institute of Transportation Studies of the University of California, especially Catherine Cortelyou and Michael Kleiber, who provided immeasurable assistance in finding and acquiring through interlibrary loans, new purchases, and painstaking searches the many references I used.

The manuscript was diligently and accurately typed by Lola Brocksen, Virginia Roy, and Barbara Sullivan, and the computer-generated graphics were prepared by multitalented and ever-creative Heidi Steiner. The manuscript benefited from editing by an outstanding freelance editor who generously gave of her time: my mother, Cary Sperling.

I am also grateful to many others who aided my research by providing information, ideas, and encouragement. Richard Shackson and the Mellon Institute provided an initial opportunity to address and interact with many industry people who subsequently provided invaluable information and insights. Others who provided information and assessments of recent developments include Andre Ghirardi, Phil Patterson, Barry McNutt, Robert A. Johnston, Ken Koyama, Sergio Trindade, Rob Motal, Margaret Singh, Marianne Millar, Alex Sapre, Paul Wuebben, Tom Cackette, Eugene Ecklund, Danilo Santini, Fred Potter, Jeff Alson, and many others too numerous to mention but not too numerous to remember.

The book was made possible by the considerate assignment of teaching and committee responsibilities by my department chairs, Gerald Orlob and Paul Sabatier, and by funding received from Argonne National Laboratory, the University of California Energy Research Group, and the University of California Institute of Transportation Studies.

I would like to thank all these people for making the preparation of this book a stimulating and enjoyable experience.

Several chapters in this book are taken from articles published elsewhere. Chapter 2 is excerpted from "Assessment of Technological Choices Using a Pathway Methodology," *Transportation Research* 18A:4 (1984): 343–354; chapter 4 is similar to "Brazil, Ethanol and the Process of System Change," published in *Energy* 12:1 (1987): 11–23; chapter 17 is a modified version of "Alcohol Fuels and the Chicken and Egg Syndrome," published in the proceedings of the Sixth International Symposium on Alcohol Fuel Technology (Ottawa, Ontario, 1984); and chapter 20 is an abbreviated version of "Testing the Validity of the Soft/Hard Energy Framework: Biomass Fuels for Transportation," *Transportation Research* 19A:3 (1985): 227–242.

PART I

Learning from the Past, Preparing for the Future

As of 1987 the vice president of the United States, the U.S. Environmental Protection Agency, Toyota, General Motors, Ford Motor Company, and the California Energy Commission were in agreement on one point: that methanol is the fuel of the future in the U.S.[1] But is it? What are the other options? Why have other countries, including Canada, embraced fuels other than methanol? What are the implications of following a methanol path? How do we branch onto a path? Those are the questions addressed by this book.

This book provides an analytical and conceptual framework to evaluate these many transportation energy options—to determine where and when each fuel option is most attractive, and where and when government should intervene to assist in the transition process. An overview of the organization of the book is provided at the end of this introduction. Before presenting that overview, however, a brief story is related to illustrate the type and magnitude of barriers facing new fuels and the difficulty of overcoming them.

A COUNTERFACTUAL STORY

In the early twentieth century, gasoline emerged as the dominant transportation fuel. It prevailed due to a particular set of circumstances: the ease of operating gasoline engines and the discovery of huge new petroleum reserves in Texas.

But what if circumstances had been different and petroleum had not been discovered in abundance? In that case, other fuels, probably ethanol in particular, would have prevailed. Decades later, when geologic exploration techniques and drilling equipment had been improved, petroleum would then have emerged as a viable alternative fuel. Would it have been accepted and gradually expanded its penetration of the transportation energy market? Let us briefly explore that hypothetical situation, for it provides insight into the exceptional barriers facing new transportation fuels.

In the early 1900s, ethanol was well known, having been manufactured by means of fermentation and distillation processes since antiquity. Ethanol had the advantage that it could be produced from a vast number of plentiful materials, including agricultural crops and food wastes.

In this counterfactual case, fuel production, fuel distribution, and fuel consumption activities would have evolved around alcohol—not petroleum. As demand for motor vehicles and transportation grew, ethanol production would have been expanded by growing more corn, sugar cane, sugar beets, sweet sorghum, and Jerusalem artichokes. Research efforts would have been aimed at making ethanol production more efficient and less expensive. Production processes would have been improved and modified over time so that they could utilize not only those crops (and food wastes that contained large amounts of sugar and starch) but also the vast amounts of otherwise unused and surplus cellulosic materials—including logging wastes, residues at lumber mills, and perhaps the cellulosic part of garbage. Eventually, large wood farms where trees are grown especially for energy production would have come into being. After several decades, in order to expand fuel production potential, production processes would have been developed to convert coal into alcohol. The price of these alcohol fuels would have been somewhat higher than gasoline prices actually experienced in the United States, and thus the proliferation of motor vehicles would have been somewhat slower than actually occurred with gasoline vehicles—perhaps at a rate closer to that which occurred in Europe.

Meanwhile, in this hypothetical situation, motor vehicles would have been slowly improved over time, just as occurred in the U.S., except that engines and vehicles would have been designed for alcohol instead of gasoline. Compared to gasoline cars, alcohol-powered cars would have higher compression ratios to take advantage of alcohol's higher octane, a smaller cooling system because of alcohol's cooler burning temperature, and other differences related to alcohol's different physical and chemical properties. The alcohol fuel distribution system would undoubtedly use the same transportation modes as those used for moving gasoline, but they would be utilized and deployed differently. Because biomass alcohol could be grown near where it was needed, trucks and railroads would have been the preferred transport modes; pipelines would be relatively unimportant until coal-derived alcohol production expanded.

Suppose that after seventy or eighty years of dependence on alcohol fuels, sophisticated new computer and seismic technology made possible the efficient exploration for and discovery of deeply buried petroleum and that improvements in drilling technology made possible its extraction. This new fuel, petroleum, would have had certain attributes that were superior to those of alcohol, such as being less corrosive, having more energy per unit volume, and having a lower ignition temperature. How would it fare?

WOULD GASOLINE BE ACCEPTED?

Petroleum's superior attributes would offer modest advantages. The higher energy density of gasoline would be appreciated because smaller fuel tanks could be used in motor vehicles and fewer barrels of fuel would be transported through the distribution system. The lower ignition temperature of gasoline would allow engines to start more easily in cold weather. Its lower corrosivity, however, would not be so important because all production plants, storage tanks, and motor vehicle engines would have been designed and built since long ago with materials that could not be corroded or damaged by alcohol.

But this new petroleum fuel would be perceived to have many disadvantages. It would face numerous technological, economic, regulatory, and logistical obstacles associated with fuel distribution and utilization. The new fuel would have a lower octane rating, higher toxicity, generally more dangerous and threatening air pollutants; would be more likely to explode and burn accidentally (for instance, in a crash or within a storage container); and if leaked or spilled, would be more threatening to the environment. Also, gum would form on surfaces where it was stored, and the fuel would leave carbon deposits in combustion chambers of engines.

From a fuel distribution perspective, the introduction of petroleum fuels would cause still more problems. Since petroleum would not be found in the same areas as coal, and since biomass alcohol would be marketed locally, no pipeline network would exist for carrying petroleum to end-user markets. An entirely new pipeline system would have to be built. Even in situations where there was spatial overlap of origin-destination shipment patterns of old and new fuels, owners of existing pipelines and storage tanks would be reluctant to accept this new fuel because it would leave gum deposits and might damage container walls—uncertainties and concerns that could only be resolved after many years of testing or experience.

A more profound problem than material compatibility would be the physical and chemical diversity of petroleum. Petroleum is composed of many different types of molecules ranging from light gaseous molecules with one carbon and four hydrogen atoms to dense, almost solid molecules containing thirty or more carbon atoms and a hundred or more hydrogen atoms. These petroleum molecules could be refined using physical and chemical processes to manufacture gasoline, but the refining process would also produce heavier molecules that are very unlike alcohol and would be completely unsuited to the ubiquitous alcohol-fueled spark ignition engine.

Perhaps in this counterfactual case the old, abandoned compression ignition (diesel) engine technology could be revived to burn some of the heavier liquids. But diesel technology was (and is) known to generate large quantities of unhealthy (and carcinogenic) pollutants. Moreover, even if the diesel engine were revived, it could only burn those petroleum products of medium density (middle distillates) that constitute about one-fourth or less of petroleum products. (In practice, the proportion of gasoline and diesel fuels that could be manufactured from petroleum would initially be very small because refining technology would be rudimentary. For instance, around the beginning of this century, petroleum refining processes could yield a maximum of only about 2 percent gasoline.) Thus half or more of the refined petroleum products would have no obvious market. If a market could not be found for the heavier petroleum components, then the total cost of exploring, extracting, and refining the petroleum (and disposing of and/or storing the unused unmarketable products) would have to be allocated to the small proportion of petroleum products (i.e., gasoline) actually used as an alcohol fuel alternative. Thus the price of gasoline would be very high initially.

Superimposed on these various technical and infrastructural barriers would be regulatory barriers. Highway safety regulators would be appalled at gasoline's volatility and flammability and would certainly oppose its use. Institutionalized air quality regulations designed specifically for alcohol would also impede the introduction of gasoline. Most states in the U.S. currently have vapor pressure limits; if these were based on alcohol's inherently low vapor pressure, alternative fuels such as gasoline, which have relatively high vapor pressures, would be precluded. Also, the various rules for emissions testing would be based on alcohol's properties and would likely prove onerous and inappropriate for petroleum. Most forbidding would be the motor vehicle emission standards themselves. Gasoline combustion in modified alcohol engines would generate high levels of nitrogen oxides, reactive organic matter, and possibly carbon monoxide. Emission rates for these new gasoline fuels would be exceptionally high because all emission control technology would have been designed for alcohol, and therefore would not be effective at reducing emissions from gasoline.

Even if all these technical, logistical, economic, and regulatory problems were resolved, consumers would still be reluctant to use the new petroleum fuels. The first generation of gasoline cars might have some advantages over alcohol cars. They would have a longer driving range than comparable alcohol cars and would possibly start in cold weather as well or better than alcohol cars (although by then various modifications would have been made to improve the cold-start capability of alcohol).

But there would be major disadvantages, the most prominent being diminished acceleration. Because motor vehicle technologies would have been optimized over a period of many decades to run on alcohol's unique properties, the modified gasoline vehicle would be relatively inefficient and would have countless small problems regarding spark timing, engine knock, overheating, vapor lock, etc. Moreover, even if first introduced in small blend proportions, gasoline would be unattractive to consumers because it would seriously degrade the performance of made-for-alcohol cars and would probably foul the spark plugs and otherwise be looked upon as an unsafe, dirty, low-quality fuel. What consumer would want to retrofit a vehicle if it meant that the vehicle would be more likely to explode in an accident and that it would have less acceleration and power? And what consumer would purchase a gasoline or diesel vehicle if there were no fuel outlets selling gasoline or diesel fuel?

In addition, consumers would have to pay a premium for the vehicles because the vehicle manufacturers would not have the economies of scale necessary to reduce the unit price.

In this alcohol fuel economy, petroleum would have a difficult time penetrating the transportation fuel market. A middle distillate petroleum such as diesel fuel would have almost no chance because no efficient engine technology to burn that fuel would be in existence. Gasoline, a light petroleum distillate, would have a better chance, but it would have to overcome negative health, safety, and environmental features, and its poor performance in motor vehicles. In addition, these petroleum fuels would be viewed with skepticism because petroleum is a finite resource.

To be accepted, gasoline would have to be significantly *less* expensive than alcohol (on an energy basis). Most studies would probably indicate, however, that gasoline would never be price competitive in the short *or* long term. Short-term analysis would point to the high cost of building refineries and a new distribution system, the lack of a market for the nongasoline components of petroleum, and the cost of redesigning the power plant, cooling, and emission control systems of vehicles. Gasoline would be even more costly than one might expect because of the initial vehicles' relatively poor efficiency using gasoline and the relatively higher cost that would be associated with pioneer petroleum refineries. So not only would the cost of producing and marketing gasoline be high, but vehicle efficiencies would be low. Long-term studies might cite the finiteness of petroleum supplies, petroleum's contribution to the greenhouse effect, the large sunk investment in alcohol fuels, and the greater health and safety hazards that would be created by using gasoline instead of alcohol. Studies might also point out that shifting to a petroleum energy supply system would lead to a concentration of economic resources in fewer people and corporations. The argument against petroleum would be compelling.

BACK TO REALITY

In reality, petroleum fuels have dominated the transportation sector. Infrastructure, vehicles, fuel outlets, and government regulations have all been built up to support the dominant fuel, gasoline. So a transition to nonpetroleum transportation fuels faces exceptional obstacles, just as the hypothetical alcohol world also faced exceptional barriers in the transition to gasoline. As a result, with few exceptions, the transportation sector throughout the world has remained dependent on petroleum. At some point it is in the interest of each nation to begin a transition. When is that point? Which alternatives are most attractive? What is the most efficient and effective means of beginning the transition?

The answers are not simple or straightforward. We have a powerful inertia that locks us into our current path and discourages the search for alternatives. It is difficult to know what alternative deviations are advisable, and when and where they should be pursued. For almost a century, system inertia has been building along our current path. We can deviate from our current petroleum path by reclaiming overgrown and nearly forgotten paths or by blazing new trails. But it is difficult to quantify the benefits awaiting us, and therefore we cannot know what costs are worth incurring along the way. It is impossible to specify the optimal step-by-step path. We can, however, make informed choices about which direction of change is preferable and the urgency of branching onto new paths. That is the challenge addressed here.

PREVIEW

This book is organized as follows. In the following chapter, the concept of pathways is presented as a general organizing framework for energy planning and policy analysis and for the subsequent analysis and evaluation of energy options in this book. This approach emphasizes making choices in a long-term context—not in terms of pursuing speculative end states, but in terms of pursuing possible directions of change that are based on fundamental values and beliefs.

The three chapters of Part II provide a historical and political context for the examination of alternative transportation energy options. In chapter 3 the reader is introduced to the extraordinarily long history of alternative fuels in the U.S., a history that began well before 1973. Chapter 4 reviews Brazil's ambitious and aggressive commitment to ethanol fuel—the single largest commitment to a nonpetroleum transportation fuel in the world. Chapter 5 examines other major alternative fuel experiences around the world—coal liquids in South Africa, oil sands and compressed natural gas (CNG) in Canada, and CNG and synthetic gasoline in New Zealand.

Parts III, IV, and V are the heart of the book. The three key components of a transportation energy system—feedstock production, fuel distribution, and end-use activities—are analyzed for each energy option.

In Part VI environmental impacts are specified. It is concluded that gaseous fuels (and solar electricity) are potentially far superior to other options, including petroleum, from an environmental perspective. Hydrogen produced from water with solar electricity is the most attractive of all. It is much more difficult to determine the superiority of one liquid fuel over another, however, because of the very disparate nature of the impacts, and because the magnitude of the impacts will be due more to a society's resolve to control those impacts than to inherent pollution-generating features of the energy options.

Part VII explores the fuel transition process. In chapter 17 the chicken-and-egg metaphor is used to analyze the difficulty of introducing fuels dissimilar to petroleum. Chapter 18 presents a set of strategies for accomplishing an efficient transition.

Part VIII addresses the difficulty of choosing an energy option. Chapter 19 argues that conventional cost studies are not satisfactory for making long-term choices in a changing world and presents the concept of "price-determining." In chapter 20 the concept of hard/soft choices championed by Amory Lovins is considered as an evaluation framework but found wanting. Chapter 21 is a key chapter: it proposes five paths of energy choices. In chapter 22 those paths are reviewed in terms of events in the 1980s, and in Part IX a determination is made of which energy choices are most attractive and plausible.

ONE

Is There a Transportation Energy Problem?

This planet has been blessed with large quantities of easily accessible petroleum. Despite its threat to safety and air pollution, petroleum is a superior source of energy. It has a high energy density, is easily and inexpensively transported, and can be transformed into a large number of products at relatively low cost. And despite our rather profligate use of petroleum, the planet Earth still has tremendous quantities of petroleum stored away in its crust. Although the cost of finding and extracting additional petroleum will become increasingly expensive, there is no threat of imminent scarcity. So what is the problem?

One concern is that in twenty to forty years or so, most of the easily accessible high-quality petroleum will be depleted and the cost of developing new oil fields will become very expensive. That is a very real and valid concern—although that concern can be mitigated and pushed further into the future by continued improvements in exploration, extraction, and refining technologies. But there is a more important factor that creates the potential for serious problems in the very near future—the uneven distribution of petroleum resources around the world.

UNSTABLE AND UNPREDICTABLE PETROLEUM MARKET

Most of the world's conventional oil resources lie within the narrow confines of the Middle East (see fig. 1). About 300 billion barrels of discovered oil reserves are located there.[1] In contrast, North America, including Mexico, the U.S., and Canada, has only about 83 billion bar-

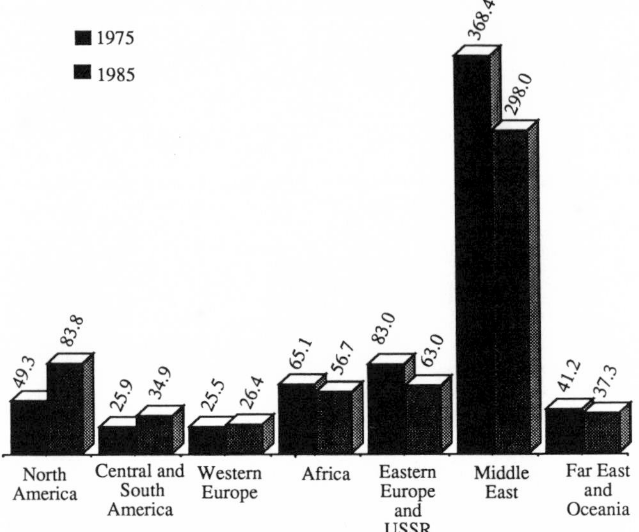

Figure 1. Crude Oil Reserves, 1975 and 1985, Billions of Barrels

Source: U.S. Energy Information Administration, *Annual Energy Review 1985* (Washington, D.C.: 1986).

TABLE 1 RESERVE/PRODUCTION RATIO FOR CRUDE OIL, LARGE OIL PRODUCERS, 1985

Country	Reserves (millions of barrels)	Production per year (millions of barrels)	Reserves÷ Annual Production
Kuwait	92,500	371	249
Saudi Arabia	171,500	1,237	139
Iraq	44,100	523	84
United Arab Emirates	33,000	435	76
Iran	47,900	803	60
Libya	21,300	387	55
Mexico	49,300	998	49
Venezuela	25,600	611	42
Norway	10,900	288	38
Algeria	8,800	235	37
Nigeria	16,600	537	31
China	18,400	905	20
Indonesia	8,500	459	19
U.S.S.R.	61,000	4,106	15
United Kingdom	13,600	911	15
Canada	6,500	535	12
Egypt	3,900	324	12
United States	28,400	3,274	9

SOURCE: U.S. Energy Information Administration, International Energy Annual 1985 (Washington, D.C.: EIA, 1986).

TABLE 2 WORLD PETROLEUM PRODUCTION, CONSUMPTION, AND NET EXPORTS FOR MAJOR OIL PRODUCERS IN MILLIONS OF BARRELS PER DAY, 1974 AND 1983

Country	Production		Consumption		Exports Minus Imports	
	1974	1983	1974	1983	1974	1983
U.S.S.R.	9.23	12.40	7.90	9.92	1.53	2.60
U.S.A.	8.70	8.60	12.1	11.60	-3.47	-3.20
Saudi Arabia	8.50	5.08	0.54	0.73	7.92	4.32
Mexico	0.60	2.79	0.60	1.17	negl.	1.61
U.K.	negl.	2.24	2.20	1.35	-2.24	0.90
China	1.31	2.13	1.23	1.83	0.08	0.30
Iran	6.06	2.48	0.66	0.59	6.40	1.83
Venezuela	3.14	1.91	1.28	0.89	1.86	1.03
Indonesia	1.36	1.33	0.36	0.50	1.09	1.04
Canada	1.66	1.34	1.69	1.32	-0.03	0.03
Nigeria	2.24	1.23	0.06	0.15	2.19	1.08
Iraq	1.95	0.94	0.04	0.17	1.83	0.77
Kuwait	2.59	1.08	0.34	0.50	2.24	0.57
United Arab Emirates	1.64	1.08	negl.	0.03	1.62	1.05
Libya	1.48	1.06	0.01	0.12	1.45	0.94
World	56.1	53.2	55.3	54.6	--	--

SOURCES: United Nations Dept. of International Economic and Social Affairs, 1979 Yearbook of World Energy Statistics (New York: 1981); U.N. Dept. of Int. Econ. and Social Affairs, 1983 Energy Statistics Yearbook (New York: 1985).

NOTES: Totals do not balance because of data accounting difficulties. Negl. = negligible (<0.005)

rels of discovered reserves and the Soviet Union only about 63 billion barrels. The ranking of countries by petroleum reserve/production ratios in table 1 further demonstrates that the Middle Eastern countries along the Persian Gulf (Saudi Arabia, Kuwait, Iraq, Iran, and the United Arab Emirates) have reserves far in excess of that needed to support current production levels. This uneven distribution of petroleum resources takes on great importance when viewed in light of the following factors.

First, North America and the Soviet Union consume nearly all the oil they produce, while the Middle East exports nearly all its production (see table 2). Thus the Middle Eastern countries, because of present and future surpluses, have a major influence on the world oil market.

Second, the cost of extracting petroleum in the Middle East is far lower than anywhere else. Much Middle Eastern oil can be produced for less than $2 per barrel while the rest of the world faces marginal production costs of $20 per barrel or more.[2] Thus not only do some

Middle Eastern producers have large surpluses, but their production costs are much lower than costs elsewhere. These large surpluses of cheap oil provide those countries with considerable discretion in how they price the oil, to whom they sell it, and when they sell it. As such, an inherent instability is created in the world petroleum market. For instance, in 1984 the five major oil producers in the Middle East (Saudi Arabia, Iran, Iraq, Kuwait, and the United Arab Emirates) produced only 10.3 million barrels per day (b/d), while just five years before they produced 20.7 million b/d. Saudi Arabia, with an installed capacity of about 12 million b/d, produced only 2 million b/d during much of 1985. The decrease in production had nothing to do with dwindling petroleum reserves—indeed, all have huge reserves—rather the cutbacks were due to market conditions and wars. In a matter of a month or so, these countries could increase production to 1979 levels or even higher and could sell these reserves for a profit at anywhere from about $3 per barrel and up.[3]

This instability in market structure is exacerbated by a third factor—the political instability of many Middle Eastern countries and their vulnerability to revolution, war, and sabotage. The 1978–1979 Iranian revolution resulted in production losses of 4 to 5 million b/d from that country. The war between Iran and Iraq resulted in large production reductions in Iraq. Saudi Arabia is also susceptible to revolution and sabotage; a radical government that is committed to a return to traditional society could come into power as in Iran. The result could be drastic cutbacks in exports.

The sensitivity of the world oil market to the actions of a few petroleum-producing countries is illustrated by recent history. Within a period of only thirteen years, several countries, serving their own special interests, were able to disrupt the world oil market three different times–twice by raising prices several hundred percent (in 1973–1974 and 1979–1980), and once (in 1985–1986) by collapsing prices. The problem facing oil-importing countries is unpredictability and vulnerability to supply and price disruptions—which is created not only by the possibility of intentional manipulation of the market but also by violence, sabotage, and revolution.

The U.S. or any oil-importing country may choose to respond to this unpredictability in the world petroleum market either by relying on market forces to make adjustments at the appropriate time or by using government intervention to anticipate and insure against sudden changes in oil prices and supplies. The risk with the latter strategy, government intervention, is diverting resources into projects and programs that turn out to be financial disasters because world oil prices do not rise as rapidly as expected.

The risk of the former strategy, relying on market forces, is being unprepared for precipitous future oil price increases, which result in turn in prolonged economic recession (and possibly social disruption). Also, by relying on the so-called market, oil importers may allow Middle Eastern countries to hold oil prices below the cost of alternative fuels for an extended time, and then to escalate prices rapidly. The oil importers would be unable to respond with substitute fuels because they would be unprepared.

If we were omniscient and prescient, we could predict the future development of the petroleum market and identify the correct or optimal strategy to follow. Since we are not, the notion of an optimal or ideal strategy is illusory. No sound scientific base of knowledge exists for assessing the likelihood of possible future events. No matter how intelligent and knowledgeable forecasters may be, the results are largely speculative. The pathetic performance of energy forecasters in the 1970s and 1980s is ample evidence of our inability to divine future prices and consumption of energy.[4] The same fundamental reality of unevenly distributed petroleum resources that led to unpredictability and instability in the world oil industry during the 1970s and 1980s will not go away.

THE PETROLEUM PROBLEM IS A TRANSPORTATION PROBLEM

Throughout most of the twentieth century, petroleum was steadily increasing its share of the energy market. Petroleum consumption in the U.S. increased from 0.1 million b/d in 1900 to 6 million b/d in 1949, and peaked at 18.85 million b/d in 1978. In 1900 petroleum contributed 2.4 percent of total energy consumed in the U.S.; by 1978 it contributed about 50 percent.[5] During the late 1970s an important reversal began: petroleum began to play a diminishing role in all sectors of the economy—except transportation. Petroleum's share of energy consumed by the industrial sector decreased from 32.4 percent in 1979 to 28.5 percent in 1985, in the residential and commercial sector from 16.6 percent in 1977 to 9.7 percent in 1985, and in the electric utility industry from 17.2 percent in 1977 to 4.1 percent in 1985. Altogether, fuel switching in these nontransport sectors resulted in a decrease in petroleum consumption of 36 percent between 1977 and 1985 relative to what it would have been if petroleum's share of each sector's energy consumption had remained at 1977 levels.[6] This pattern prevailed

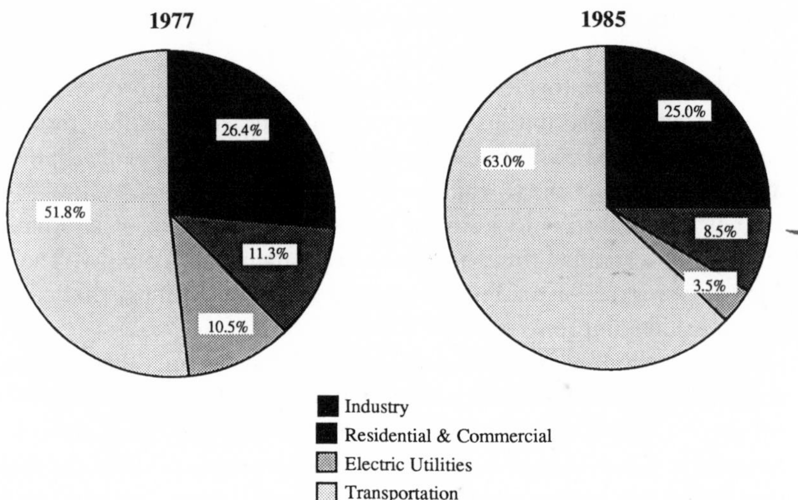

Figure 2. Petroleum Consumption by Sector in the U.S., 1977 and 1985

throughout the world (although it moderated after the 1986 drop in oil prices).[7] In Japan, for instance, consumption of residual oil for power plants and boilers fell 37 percent between 1979 and 1982, mostly because of fuel switching; this decrease amounted to 800,000 b/d, which was slightly greater than total gasoline consumption in the country.[8]

Fuel switching has not for the most part occurred in the transport sector. Petroleum has supplied between 95.4 percent and 97.3 percent of all transportation energy in the U.S. since 1958; it was 97.2 percent in 1977 and 97.2 percent again in 1985.[9] For highway vehicles, dependence on petroleum is even greater (about 98 percent in 1985 if ethanol and methanol additives are counted, about 99.5 percent if they are not). Thus, as other sectors shift away from petroleum, transportation accounts for an ever larger share of petroleum production. By 1985, transportation used 63 percent of all petroleum consumed in the U.S., an increase in market share of eleven percentage points in just eight years (see fig. 2).

The problem of dealing with petroleum dependency and petroleum imports is increasingly becoming a problem of the transportation sector. To a large extent, the petroleum problem is a transportation problem. Shifting the transportation sector to other energy sources may be the most intractable energy problem facing the United States and the rest of the world.

Transportation's continued dependence on petroleum fuels is a major concern because under almost any plausible scenario transportation energy demand will remain strong into the foreseeable future. Freight transportation worldwide has become increasingly reliant on trucks, which are relatively energy intensive, and passenger transportation has become increasingly reliant on automobiles for short trips and airplanes for long-distance travel. People have shown that they value the mobility of personal transportation (i.e., automobiles) very highly and are willing to pay a premium to retain it. Even if transportation fuel prices increase substantially to above $2.50 per gallon, the rate of auto ownership and the amount of auto usage will not be greatly affected (as long as income does not fall). Strong evidence indicates motorists are willing to pay well over $2 per gallon for fuel without making major changes in their transportation behavior. At those fuel prices, consumers may opt to purchase smaller, more efficient vehicles, but, significantly, they do not shift to mass transit.[10] For instance, even though fuel prices have been over $2 (U.S.) per gallon in Europe since the 1970s (and as high as $4 per gallon in some countries), the number of automobiles (and trucks) per capita has continued to increase. Since a large number of alternative transportation fuels, some from renewable resources, may be produced for $2.50 or less per gallon, there is good reason to believe that the auto-dominated transportation system of North America and Europe (and increasingly of Asia and South America) will continue into the foreseeable future.

The high value placed on transportation energy has important implications. It suggests that transportation users will continue to bid petroleum away from other users and that wealthier societies will bid petroleum away from poorer societies. These are important moral and strategic issues.

The major thrust of this book is twofold: to analyze in detail the technological, economic, environmental, and, to a lesser extent, political aspects of alternative fuels; and to synthesize this knowledge so as to formulate strategies for introducing alternative transportation fuels and to identify the implications of pursuing those strategies.

GROWING OIL IMPORTS

The urgency of dealing with the petroleum situation is illustrated by the unresponsiveness of oil production to higher oil prices. In 1972 the non-communist part of the world was consuming 47.6 million barrels of petroleum per day, representing roughly 56 percent of all commercial energy consumption in those countries.[11] The cost of finding and developing new oil fields to continue serving this demand was beginning to increase at that time. One might have expected that the jumps in oil prices, first in 1973–1974 and then in 1979–1980, would have been more than enough incentive to oil companies to replace what was being consumed. In some locations it was, but not in the U.S. or other oil-producing countries. In the U.S., production has outstripped new discoveries (i.e., proven additions to petroleum reserves) for every year since 1970.[12] Moreover, estimates of "undiscovered" oil in the U.S. by the Geological Survey have been severely reduced—by 75 percent in 1981 and still further in 1986.[13]

The cost of producing additional petroleum is increasing rapidly everywhere in the world except in some Middle Eastern oil fields. In the U.S. and many other countries, including most of the OPEC countries, almost all forecasts are for steady declines in oil production. As shown in figure 3, the U.S. Department of Energy (Energy Information Administration) forecasts that U.S. oil production, including natural gas liquids, will decrease from about 11 million barrels per day in 1986 to fewer than 8 million in 2000. At the same time, petroleum imports are forecasted to increase from about 5 million barrels per day in 1986 to almost 10 million in 2000. Thus, by 2000, imports would account for 50 to 60 percent of total oil consumption in the U.S. (These government forecasts are consistent with those of other major forecasting organizations and most oil companies and oil industry organizations, although past history indicates that forecasts of oil consumption and prices are not likely to be highly accurate.)[14]

The U.S. has already depleted about 84 percent of its proven recoverable petroleum reserves.[15] Stark reality is that the U.S. cannot and will not produce enough petroleum to meet projected demand, even at much higher energy prices. Most other countries are in the same situation. This situation became more critical in the 1980s as lower petroleum prices discouraged conservation and discouraged investments in petroleum exploration and development.

RESPONSES TO THE PROBLEM

The following strategies may be pursued to mitigate the volatility of petroleum markets and to reduce dependency on petroleum imports:

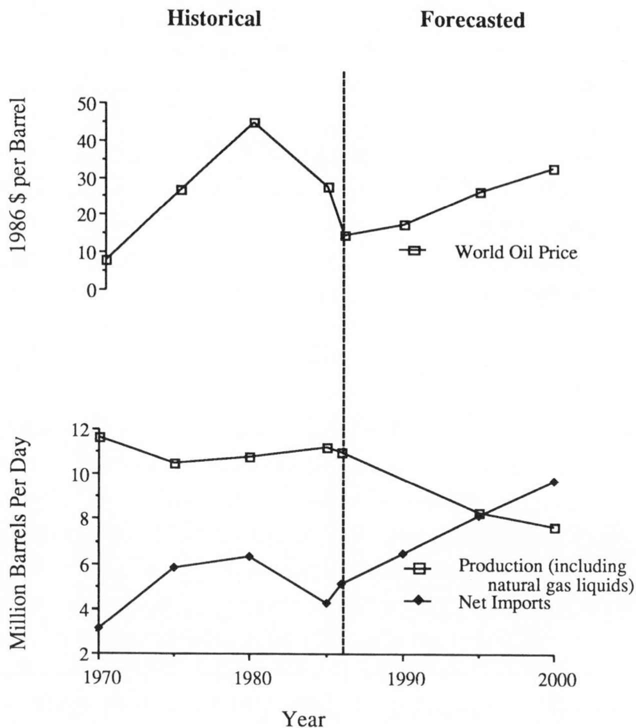

Figure 3. Forecasts of Petroleum Prices, Production, and Imports

Source: U.S. Energy Information Administration, *Annual Energy Outlook 1986* (Washington, D.C.: 1987), p. 37

Notes: Post-1986 numbers are base case forecasts.

Improved and intensified petroleum exploration and development,

Creation of storage buffers,

Petroleum conservation and interfuel substitution,

Expanded production and use of new energy sources.

The first option has been aggressively pursued through the normal workings of the marketplace. The rapid rise in petroleum prices in the 1970s gave the oil companies large cash surpluses that they used to improve exploration techniques, develop techniques to increase yields from existing fields (including heavy oil fields), and to expand investments in the exploration and development of less accessible oil fields, such as those located offshore and in arctic areas, and in complex geological formations such as the Overthrust Belt in the Rocky Mountain region. (The oil companies also invested cash surpluses in nonenergy businesses, but for most companies that was a short-lived and unsuccessful experience.)

Despite these greatly intensified exploration and development activities, the oil industry in the U.S. (and elsewhere) was barely able to sustain previous production levels. Meanwhile, the marginal cost of finding and recovering new oil (i.e., excluding refining, marketing, and distribution) has increased from an average of about $1 to $2 per barrel in the 1960s to $20 or more in the 1980s.[16] That marginal cost will continue to increase as oil development increasingly is directed at less accessible, lower quality, and smaller oil fields.

The creation of storage buffers, a second option, is another mechanism for dealing with future petroleum problems. The storage of large quantities of petroleum reduces vulnerability to abrupt supply cutoffs that might result from overt political decisions by Middle Eastern countries or from wars and revolutions that disrupt production and transport of petroleum. In 1975 the United States created a storage buffer called the Strategic Petroleum Reserve (SPR).[17] The initial storage goal for the SPR was 500 million barrels; in 1980 at President Carter's request the goal was increased to 1 billion barrels. During the first few years, filling the Reserve was slow, and was even suspended during and immediately after the 1979 petroleum shortages. In the early 1980s, the fill rate was increased to over 100,000 barrels per day, but that rate was expected to diminish in the late 1980s to about 35,000 b/d in response to federal budget deficits. There are almost 500 million barrels in five different underground salt domes in Louisiana and Texas. The total capacity of the five sites and one additional site under construction is 750 million barrels.[18] In an extreme situation in which all petroleum imports were stopped, the 500 million barrels could fully replace imports for over 100 days. The SPR represents one of the very few acts by the U.S. government to reduce vulnerability to the erratic petroleum market. The SPR does not, however, deal with the more fundamental long-term problem of increasing imports and increasing vulnerability.

Another approach with potential for more long-lasting effects is the reduction of petroleum consumption. One means for accomplishing this is to substitute other conventional energy materials for petroleum. As shown earlier, interfuel substitution has in fact resulted in substantial reductions in petroleum consumption in industrial, residential, commercial, and electric utility activities. Coal, natural gas, nuclear power, and more decentralized sources such as solar heat, wind power, geothermal energy, and small hydroelectric plants have supplanted petroleum in many activities, but not in transportation. In the transportation sector there has been practically no interfuel substitution.

Energy conservation in transportation is a highly rewarding strategy with great potential. Transportation energy consumption may be reduced by increasing the energy efficiency of transportation technologies, increasing occupancy rates and load factors, changing behavior so as to use technologies more efficiently, and by changing land-use patterns.

The technology-improving conservation strategy is a technical fix strategy and has already proven highly effective. By downsizing vehicles, substituting lighter materials, and improving engine efficiency, the energy efficiencies of domestic automobiles have been tremendously improved from an average of about 14 miles per gallon (mpg) for new vehicles in 1973 to about 26 mpg in the mid-1980s. The energy efficiency of commercial air transport (most of it due to efficiency improvements in jet planes) more than doubled between 1970 and 1985.[19] Improvements in energy efficiency of trucks have been more modest, increasing by about 30 percent on a vehicle-mile basis between 1976 and 1985. Even with these substantial improvements in transportation energy efficiency, the potential for still greater technology efficiencies exists, especially for automobiles and light trucks, which consume about 56 percent of all petroleum used in transportation.[20] Already, many cars average forty mpg or more, and some argue that efficiencies of 90 mpg by the year 2000 are feasible.[21]

Nontechnology strategies also have potential for significantly reducing transportation petroleum consumption. The most feasible is encouraging people to shift into carpools and vanpools. Doubling average auto occupancy in urban areas of the U.S. from the current 1.4 persons[22] to 2.8 persons would have reduced 1985 petroleum consumption in the U.S. by over 2 million b/d (the same effect as doubling automobile efficiency).

In practice, however, transportation energy consumption is not likely to decrease (see U. S government forecasts in table 3). Freight transport is not likely to reduce its energy consumption because trucks and air cargo, the least energy-efficient freight modes, are likely to expand their market share into the foreseeable future everywhere in the world. They provide a superior level of service for the transport of manufactured and high-value goods, the type of goods that tend to dominate as

TABLE 3 U.S. TRANSPORTATION ENERGY CONSUMPTION FORECASTS BY TYPE OF FUEL, MILLIONS OF BARRELS PER DAY

Transportation Fuel	Historical				Forecast		
	1970	1975	1980	1985	1990	1995	2000
Motor gasoline	5.06	5.90	5.85	6.06	5.96	5.56	5.48
Distillate (diesel)	0.75	1.00	1.32	1.49	1.63	1.78	2.00
Jet fuel	0.93	0.96	1.03	1.19	1.38	1.39	1.48
Residual fuel	0.35	0.34	0.66	0.38	0.45	0.51	0.58
Other[a]	0.49	0.41	0.43	0.37	0.36	0.36	0.37
Total (baseline forecast)	7.58	8.61	9.29	9.49	9.78	9.60	9.91
Total (high forecast)[b]	--	--	--	--	10.23	10.12	10.51
Total (low forecast)[c]	--	--	--	--	9.41	9.13	9.30

SOURCE: U.S. Energy Information Administraton, Annual Energy Outlook 1986 (Washington, D.C.: EIA, 1987), Table B2.

[a]Includes natural gas and electricity. Ethanol and methanol are included in motor gasoline. Quantities of nonpetroleum fuels are measured in oil-equivalent units.

[b]Based on scenario of low world oil prices and high economic growth rate.

[c]Based on scenario of high world oil prices and low economic growth rate.

societies become more affluent (i.e., consider the phenomenal growth of air express service in the 1980s). Nor is energy consumption in passenger travel likely to decrease much, if at all. As affluence continues to spread, the demand for personal transportation and transportation energy increases (see chap. 14).

Since fuel prices will probably never increase more than twofold or so from early 1980s levels (because alternative fuels would flow into the market at that price), it is unlikely that there will be significant pressure to make major improvements in the energy efficiency of personal vehicles. Some improvement in fuel consumption is likely as a result of increased ride sharing in those urban areas with significant traffic congestion and limited parking, but these behavioral changes are resisted by most people for various personal reasons that range from desire for flexibility to personality and scheduling conflicts among riders. Shifting people to mass transit has much less potential. In 1982 transit served only 2.5 percent of passenger miles in the U.S.[23] Even doubling this percentage would have a negligible effect on energy consumption, especially because the expanded transit service would tend to result in lower load factors, and therefore to be no more energy efficient than carpools. The last option, land-use densification, is the least plausible, since trends are in the opposite direction.

Government has played a small role in stimulating energy conservation. In 1975 the U.S. Congress passed legislation requiring automobile manufacturers to increase the average fuel efficiency of their automobiles to 27.5 mpg in 1985. Manufacturers reluctantly responded by downsizing vehicles and using lighter materials, but the sales of smaller and more energy-efficient vehicles did not increase significantly until the 1979–1980 petroleum price increase. The major effect of the fuel efficiency standards was to reduce uncertainty about the future—automakers knew they had to produce vehicles with a prescribed fuel efficiency. It is uncertain, however, what the ultimate effect of the mandated standards was—and to what extent the increase in gasoline price by itself would have induced greater fuel efficiency and conservation.[24] In any case, the resolve of government slowly dissipated when petroleum prices decreased in the 1980s. When Ford and General Motors claimed in 1984 that they could not meet the 1985 standard because consumers preferred larger (and less efficient) cars, the U.S. Department of Transportation relented and in mid-1985 reduced the standard to 26 mpg.

Another means of reducing energy consumption is to raise artificially the cost of energy use, either by raising gasoline and diesel fuel prices through taxes, or by taxing inefficient vehicles. Most countries have pursued both these options. The U.S. Congress regularly considered fuel surtaxes of up to 50 cents per gallon in the 1970s but ultimately rejected them each time. A proposal, known as a "gas guzzler" tax, to place a surcharge on less efficient automobiles and to provide a rebate to efficient automobiles was also considered. Eventually a watered-down tax only on very inefficient vehicles with no rebates to efficient vehicles was adopted.[25] Mostly only owners of luxury cars such as Rolls Royces have been forced to pay the gas guzzler tax.

In summary, petroleum conservation in the transportation sector has great potential. Implementation of ride sharing, vehicle efficiency improvements, and other conservation strategies should be pursued; those that are easiest to implement are vehicle efficiency improvements, because they change the technology without requiring behavioral changes by vehicle users. But conservation only delays the inevitable. And beyond some initial technical-fix strategies, changes are distasteful to consumers and therefore anathema to legislators. While conservation should be actively pursued, the need for alternative fuels is inevitable.

The fourth strategy is the subject of this book: the development and use of nonpetroleum resources as alternatives to petroleum. Two major studies of the early 1980s (by the Mellon Institute and the U.S. Congress Office of Technology Assessment) argued that, in the short term, greater benefit to society is gained by investing in fuel-efficiency than in alternative fuels.[26] Those studies found, however, that by the early 1990s, the benefits per dollar invested would be about the same for the two sets of options, and that thereafter alternative fuel investments would be preferred. A third study (by Purdue University researchers) determined that the optimal investment path is to invest first in easy-to-implement auto fuel-economy improvements, then in shale and coal liquids, and finally, after the shale and coal liquids options are fully developed, in more sophisticated (and costly) auto fuel-economy improvements.[27] The assumptions of those studies are outdated, but they support the general premise of this book that alternative fuels are inevitable and that their time is soon approaching.

TWO

A Conceptual and Analytical Framework

It is always wise to look ahead, but difficult to look further than you can see.

Attributed to
Winston S. Churchill

As new technologies proliferate and societal systems become more complex, the challenge of understanding the process of change and anticipating the future becomes increasingly important. But even as computers become more powerful and less expensive, discernment of the future remains elusive. While in the past, planning and forecasting relied principally on knowledge and intuition, the emphasis has now shifted toward the development and application of sophisticated mathematical modeling techniques. What was once recognized as an art is now cloaked in the vestments of science.

The combination of increased power and decreased cost of computing has tempted analysts and researchers to build more sophisticated models and to incorporate more variables and data into those models and analytical constructs. As a result, analytical capabilities have improved dramatically, even while our understanding of underlying phenomena and of relations between the many variables has not always kept pace. In many applications, models have been rich while data remain poverty stricken. A mismatch has developed between modeling sophistication and the knowledge used to build and calibrate those models. Obviously, this failing is not everywhere true. Where it is most true, however, is with large and complex systems.

Since the complex and multiobjective behavior of humans is highly unpredictable, one might suppose that as the presence of humans in these large, complex systems shrinks, quantitative evaluations and representations would be simpler, more accurate, and therefore more useful—and planners, forecasters, and policy analysts could proceed with greater confidence in their tasks.

But in a technology-based system, bringing the full weight of mathematical modeling to bear is not necessarily fruitful, even though human behavior is not at the heart of the analysis. Technologies do not develop or evolve in response to some set of universal laws. Technologies and technology-based systems evolve in response to goals, values, and beliefs of dominant social groups. Since this is the case, immutable laws of nature do not guide the evolution and development of transportation energy systems, and their future design and performance cannot necessarily be predicted by the use of more sophisticated and detailed quantitative treatments. There is little reason to believe that much progress in formulating laws of technology development is likely to occur.

One objective of this book is to suggest a more restrained approach to the evaluation and forecasting of future technologic options—one that relies on large doses of empiricism. The aim is to shift the emphasis from quantitative constructs amenable to high-speed numerical manipulation to a more robust analysis that addresses the fundamental factors that guide the deployment and development of technologies and technologic systems.

TECHNOLOGY EVOLUTION

There is a romanticized view that technology development comes about as a result of breakthroughs. Hopes for a more productive economy are often premised on technological breakthroughs and completely new technological systems. A more realistic view of technology development is that it is an evolutionary process. Consider the popular (and simplistic) notion of the development of the steam engine: that James Watt was in his mother's kitchen watching steam from a boiling kettle, realized the tremendous power of steam, and later invented the steam engine. The true origin of the steam engine is very different and much more interesting. It is a story of scientists, engineers, and machinists who over a period of centuries produced a series of modifications and improvements that eventually created the engine that changed the world. Contrary to scientific lore, the deployment of a steam engine was the cumulative result of many small innovations.[1]

The history of technological development is a history of many small cumulative innovations, not of a few large technological "breakthroughs." Thus we should not expect future changes to be the result of only a few major breakthroughs. The thinking process that views development in terms of breakthroughs is part of a larger paradigm that views change in snapshots (or end states) of time.

This end-state thinking is found in debates about transportation and energy; we argue whether hydrogen, alcohol, or coal-derived fuel is the superior choice, whether we should replace the auto and with what, whether "bullet" trains and magnetically levitated trains are answers to transportation and energy problems, and how to design the city of the future. These are all end-state images of the future drawn for us by technologists and futurists.

This preoccupation with end states may be explained by a limited understanding of the process of change. Technology forecasters attempt to predict the spread of innovations, but they do not specify or study the conditions and factors that lead to or encourage the diffusion of those changes. William Ascher, in an extensive review of the state of the art of forecasting, concluded that technological forecasting is quite erratic,[2] which is not surprising, considering the common perception that a primary requisite for good forecasting is "astute judgement and common sense on the part of the forecaster."[3] Ascher also asserted that the primary deficiency of technology forecasting is the high level of aggregation and the corresponding failure to develop a theoretical basis for the determinants of technological change.

Because of this lack of understanding of the determinants of technological change, people grasp for detailed images of the future at the expense of investigations of how those images can be attained or whether they are even attainable.

END-STATE THINKING

An end-state orientation manifests itself in a preoccupation with deployment of new technologies. Emphasis is placed on the attributes, impacts, and costs of a future end-state technology or technological system, rather than on how to reach a desired end state. Emphasis is on end-state scenarios, not paths of change. Deployment innovations are emphasized; system-changing innovations are overlooked.[4]

End-state thinking, which relies on images of future technologies, is not useful for several reasons:[5]

1. It does not help define societal goals and decision criteria regarding technological change;

2. It does not nurture and facilitate the process of consensus formation;

3. It is usually inaccurate.

Societal goals and decision criteria cannot be formulated or tested because specific attributes of new technologies are difficult to predict. Whether the effects are related to employment, noise, air quality, cost, or land use, not enough information is available for potentially affected parties to react in a meaningful manner. Neighbors of Kennedy Airport did not respond to the coming of the Concorde until the supersonic jet was already a reality.[6] Perceptions and reactions of people to the Concorde (and to the U.S. SST if it had been built) were not known until investment and deployment decisions had already been made. As further illustration of this failure to enlighten the process of goal and criteria definition, imagine going back to the 1940s, when national expressway (i.e., interstate highway) systems were being debated. Certainly a "technology assessment" was in order, but how likely is it that a clear, far-sighted, and sufficiently detailed picture of the future could have been drawn that would have led to national policies on land use, petroleum use, air pollutant emissions, noise, traffic safety, neighborhood preservation, and the many other conditions and concerns affected by widespread motorized transport? Not very likely. The same might be asked for nuclear energy and for genetic engineering; again the answer is negative. Human cognition and prescience is limited, and thus end-state descriptions do not help formulate societal goals and decision criteria regarding technological change.

A similar argument holds for the second point, which concerns the facilitation of consensus formation. Thinking about and specifying the future deployment of new technologies does not lead to consensus formation, because individuals and organizations do not (and perhaps cannot) fully perceive in advance gains and losses they will sustain. Consensus is more difficult to attain in a highly pluralist society such as the U.S., in which a diversity of values is set within a decentralized political and economic system,[7] than in more homogeneous societies such as Japan[8] and other countries with centralized political and economic systems. In most settings, speculations about the future do not facilitate the formation of consensus.

The third and most devastating criticism of end-state analysis and planning is that the specification of future technology scenarios is, as often as not, inaccurate. Ascher observed that there has been "little evidence, positive or negative, that technological forecasts [have] been successful,"[9] while another writer in exploring the basis of technology assessment observed that "our modes of thought are not yet adequate to the task before us."[10]

In more specific terms, a RAND corporation study pinpoints the problem of accurately specifying the attributes of future technologies. In studying pioneer chemical and energy process plants (e.g., synthetic fuel plants), they found that

> The extent of misestimation [of cost] remains severe even after excluding all external effects [e.g., inaccurate projections of inflation, unanticipated regulatory standards, labor strikes, bad weather] on plant costs. The 106 estimates analyzed [by RAND] range enormously in accuracy, from less than 20 percent of actual plant costs to as much as 10 percent above.[11]

The RAND study found that misestimations of plant performance were also severe, but less so than for cost misestimations. The RAND researchers concluded that

> severe underestimation of capital costs is the norm for *all* advanced technologies; the underestimation for energy process technologies mirrored that seen in major weapon systems acquisition, very large advanced construction projects, and major public works projects. . . . Because they lacked systematic understanding of these factors, planners dealing with pioneer process plants were severely handicapped. Their best options were either to *disregard early cost estimates* [emphasis added] for advanced technologies and rely on noneconomic criteria such as efficiency or environmental considerations, or to support costly design and engineering work to improve the estimates. Neither option could be very attractive to government and corporate managers, who need good early cost estimates to support planning decisions.[12]

The RAND researchers suggest that "the factors that account for poor cost estimates and poor performance [estimates] can largely be identified early in the development of the technology."[13] The problem is not only the inability to specify future costs and performance accurately but also the inability to specify the full range of important factors, including, in the case of alternative fuels, air quality impacts, soil erosion, toxic waste pollution, and site-specific effects related to the precise location of new process plants. I suggest that these factors could be identified and more accurately specified within a pathway context. An analysis of the imprecision of cost estimates is explored in chapters 8 and 19.

The concept of pathways provides an alternative approach. It places the emphasis on the "getting there" part of the story—on direction of change rather than on particular end states. Pathways present a framework that is well suited to policy analysis and planning. By shifting attention from end states to pathways, debates shift from disjointed discussions of some unfathomable and obscure future to discussions of how to generate the conditions that create and nurture the branching onto a new pathway. A pathway framework of analysis also allows and encourages an exploration of impacts and implications of technological choices. "[If] an individual senses a particular technological prospect as somehow offensive to the essence of humanity, the inchoate apprehension thus expressed ought to be treated not as immaturity to be overcome but as a potential source of wisdom to be explored."[14] By exploring alternative paths, these concerns can be incorporated.

The pathway concept has been dramatized by Amory Lovins as a framework for analyzing energy technologies.[15] He argued forcefully that evolving and new technologies be placed in a framework that would highlight underlying values and beliefs, and that would address the fundamental relationships between costs, performances, and the surrounding human and physical environments. But Lovins's work was advocatory and not rigorous (see chap. 20).

The concept of pathway-type analysis is not original. In recent years, for instance, policy analysts have begun addressing the disparity between analysis and implementation. This new emphasis on the implementation process has resulted in various efforts at introducing a perspective that examines the linkage between planning, decision making, and resulting changes.[16] In the discipline of economics, Richard Nelson and Sidney Winter have proposed an evolutionary theory of economics that has much in common with the pathway analysis proposed here.[17] Nelson and Winter argue that behavior is guided by a hierarchy of decision rules that aim at routinizing the decision process, and that decision rules are only modified when they become obviously inadequate or contradictory in producing desirable outcomes. Thus decision rules are updated periodically (sometimes implicitly, sometimes explicitly) as situations change; in other words, "the condition of the industry in each time period bears the seeds of the conditions in the following period."[18] This concept that decisions are made incrementally and are guided in a rather loose fashion toward desirable outcomes is the basis of a pathway perspective of system change.

Another area of intellectual endeavor that nurtures pathway concepts is strategic planning. Some theorists have observed that strategic planning should discard the end-state thinking that pervades planning; Robert Hayes, a Harvard Business School professor, argues,

> [T]urn the ends-ways-means logic on its head: means-ways-ends. A road map is useful if one is lost in a highway system, but not in a swamp whose topography is constantly changing. A simple compass that indicates the general direction . . . is far more valuable.[19]

In other words, the most effective approach is to choose the general direction of change that is most attractive, focusing on the process of change, not on some imagined end state.

PATHWAY ATTRIBUTES

Although a formal theory of pathway planning will not be constructed in this book, attributes of a pathway analysis that I consider to be most salient are presented below.

Pathways are useful for organizing the multitude of future options into a small set of streams of choices—by reducing vast interacting arrays of technologies, institutions, and environments into a form that does not overwhelm cognitive abilities (and, similarly, does not overwhelm the data-gathering efforts of mathematical modelers).

These pathway designs are useful for planning and analysis only if they are organized around sets of consistent values and beliefs. This feature of consistent values and beliefs is fundamental to pathway analysis. A path should be designed to represent a distinct and identifiable group of activities and actions. These actions and activities should derive from a coherent set of values and beliefs related, for instance, to environmental quality, centralization of decision-making authority, concentration of wealth, pricing and provision of public services, efficiency and equity of market systems, autarky, and national security. A summary of the values and beliefs underlying the five paths that are developed and evaluated later in this book is presented at the end of this chapter.

Another feature of pathway analysis is the specification of critical factors in a range of representative settings and in a temporal context; this increases the robustness of the analysis. Critical factors are those factors that most strongly influence—both positively and negatively—the implementation of technological change. Although in general theoretical analysis provides initial insights to focus the search for new and improved technologies, in practice theoretical underpinnings are weak. As a result, identification of critical factors, at least in the case of alternative fuels, must be based on extensive empirical analysis. Due to the incompleteness of data, lack of consensus on goals, values, and evaluation criteria, and uncertainty over future trends, final selection of critical factors requires considerable human judgment. Some of the critical factors (specified in later chapters) include diseconomies of scale in biomass feedstock collection, economies of scale in feedstock processing, structure of cost functions for major fuel transport modes, ability to share existing fuel distribution infrastructure, and uncertainty in energy prices. In Parts III, IV, and V, these critical factors are specified by examining relationships between key variables in different spatial and temporal settings—for optimum size of process plants by superimposing costs of collecting and processing feedstock and transporting fuel to end-use markets. A temporal analysis is particularly relevant in identifying and targeting market opportunities and anticipating obstacles and constraining factors that are likely to be encountered.

Placing emphasis on direction of change rather than on attaining specified targets encourages a more flexible and thoughtful approach to demand forecasting and market possibilities. This inductive mode of investigation simulates the real world in responding to shifting priorities, unpredictable events and innovations, and new information and knowledge. Emphasis is placed on viewing the process of change as one of constantly improving technologies and of constantly changing institutions.

MORE ON TECHNOLOGY INNOVATION AND DEVELOPMENT

Before proceeding to a discussion and analysis of the particular energy options, it is useful to address the general process of technology development. First, as suggested earlier, change occurs incrementally. The development of the steam engine was the product of many, mostly small, innovations over a period of many decades. The process by which new technologies and products are introduced and evolve may be characterized as follows. Initially they are relatively crude and flawed products; they are introduced into small niches and unique settings where their particular characteristics are highly valued. For instance, computing machines were valued by census takers, stationary steam engines by coal mine operators for pumping water, photovoltaic cells for space exploration and by remote marijuana farmers, and so on. In these market niches, nascent technologies and technological changes are nurtured—they are made more reliable and efficient and their performance is enhanced. Unit production costs are reduced as economies of scale are gained and as additional manufacturing improvements are introduced. Soon the products (technologies) are ready to compete in other niches where they have a smaller but still substantial advantage of some type. Eventually, if product and process innovations continue, the product is ready to compete in the general market. It will be shown that in this view the process of introducing nonpetroleum transportation fuels faces a special and formidable problem: the relative homogeneity and absence of niches in the transportation fuel market.

Another important observation is that there are no general laws that govern, explain, or predict the evolution of technologies. While existing transportation and energy systems have evolved to their present state over a very long period of time, their present format and structure is not premised on some immutable laws of technology and societal development. Rather, the existing format and structure of these systems is a response to the goals and purposes of society as codified in rules, regulations, and standards over time. Transportation energy systems are what Nobel laureate Herbert Simon labels artificial systems.[20] They are human built, designed in response to economic, political, and social goals and purposes. They are different from purely physical systems, such as ecological systems untouched by humans, which evolve in response to immutable laws of nature. Artificial (human-built) systems cannot, of course, violate laws of physics and nature, but their workings are shaped by societal goals and purposes, not by invariant laws.

Technological—or, more accurately, sociotechnical—systems are a function of conditions and perceptions of a previous time. Those previous perceptions and conditions prevail in the present in the form of rules, regulations, practices, and institutional structures. Such rules and practices are not necessarily based on current factor prices and current environmental and societal perceptions and they almost certainly are not related to future factor prices and perceptions and goals. In the case of the transportation sector, existing energy systems are based almost exclusively on the availability and use of two petroleum fuels: diesel (i.e., middle distillates) and gasoline.

Existing technologies and rules would need to be changed to accommodate a new fuel. The changes are not great, but they conflict with the specialized systems that have been created over time. The existing systems are so specialized that even small changes are resisted. This resistance to change is manifested in economic analyses that naively suggest that alternative fuels are inferior and will continue to remain inferior to petroleum on a cost basis into the foreseeable future. These cost analyses are used to justify inaction toward new fuels (and their supporting technologies). But to a large extent those cost analyses reflect past choices and decisions and accumulated experience. They are based on gasoline's short-term advantage of high efficiency of petroleum refining processes, gasoline engine operation, and petroleum transport, but these advantages do not necessarily hold up in the long term.

Another a priori advantage of petroleum, which also does not hold up in a long-term analysis, is years of experience in learning to adapt to petroleum's high fire hazard and polluting characteristics, and the absence of built-in institutional mechanisms for rewarding other fuels that are cleaner and less threatening. If gasoline and diesel fuel were new fuels (with no history) facing a feasibility analysis, the estimated costs and efficiencies for using those petroleum fuels would be far higher than the actual current costs and efficiencies. By having made a priori decisions about what is an acceptable level of pollution and toxicity, about the desirability of secure energy supplies, and about the distribution of financial resources, we have made "price-determining" choices. Decisions made a priori influence the economics of subsequent energy choices. We must recognize that energy choices cannot and should not be evaluated solely in terms of conventional "price-determined" economic analysis. It is in this sense that the subject of transportation energy choices is examined.

TRANSPORTATION ENERGY PATHS

In this book, transportation energy alternatives are organized into five pathways (see table 4). One idealized path is based on the production of biomass fuels in small and medium-sized plants and consumption near where the fuel is produced. Underlying the design of that path are values of self-sufficiency, widespread economic participation, local control, and world peace, and beliefs that biomass has the potential to supply a large proportion of transportation fuel needs at competitive costs, that it will not lead to increased soil erosion, and that it represents a permanent and sustainable solution.

A second path involves the conversion of coal, oil shale, and oil sands into petroleum-like liquids that can be transported in the existing

TABLE 4 TRANSPORTATION ENERGY PATHS

	Fuels	Feedstocks	Size and Cost of Individual Process Plants
Path I	Alcohols	Biomass	Small to medium
Path II	Petroleum-like liquids	Coal, oil shale, oil sands	Large
Path III	Methanol	Natural gas, coal	Large
Path IV	CNG, LNG	Natural gas, coal	Medium to large
Path V	Hydrogen, electricity	Solar energy	Small to medium

petroleum product distribution system and consumed in engines designed for gasoline and diesel fuel. This path would be supported by people who do not place much emphasis on environmental values and are more interested in national security, preserving existing investments, and generally maintaining the status quo.

In a third path, remote natural gas and later coal are converted into methanol. This path is motivated by a belief that the costs of modifying the fuel distribution system and motor vehicle technology are compensated for by the higher quality and lower pollution of methanol. The path reflects less emphasis on concerns for self-sufficiency and national security and more emphasis on air quality improvement. As does the second path, it reflects the faith that large organizations will act responsibly and vigorously.

The fourth and fifth paths rely on gaseous fuels. The fourth path is similar to the methanol-based third path in that first remote natural gas and then coal and unconventional gas are used as the feedstocks. In this case the end-use fuel is compressed or liquefied methane gas. This fourth path also is similar to the methanol path in that it is motivated (in the U.S.) by the opportunity to exploit large domestic reserves of coal and unconventional natural gas and the desire to reduce air pollution. The major difference between the third and fourth paths is the belief that the lower cost of gaseous fuels is more important to consumers than its greater dissimilarity to petroleum fuels and that the start-up barriers should not overshadow the large long-term benefits.

The fifth path is hydrogen-based. It would have to take root later than the other paths because the production technologies are less well developed. It promises to be a very clean and permanently sustainable fuel. Its most attractive feature may be that it contributes no carbon dioxide to the "greenhouse effect." An underlying belief is that low-cost solar technology will be developed as an energy source for converting water to hydrogen and that the environmental benefits of hydrogen will compensate for high production costs.

These five paths are developed and evaluated in detail later in this book. The attractiveness of each path is explored for different settings. The feasibility of the paths is tested by comparing them to actual energy activities and events in various parts of the world. Strategies for branching from the current petroleum path onto alternative paths are examined with the understanding that "like sailors who must rebuild their ship on the open sea without yet discerning [or ever being able to discern] its ideal design, we must simply do the best we can."[21]

PART II

Historical Review of Alternative Transportation Fuel Experiences

The next three chapters provide an overview of worldwide experiences with alternative fuels. They provide insight into decisions made by the public and private sectors and examine the influence of political and institutional forces in making transportation energy choices. Thus these three chapters provide a context with which to analyze the technological options and strategies developed later in the book.

In chapter 3 the intermittent bursts of enthusiasm in the United States for alternative transportation fuels are traced from the 1800s to the present. Coal conversion technologies are traced from their development principally in Germany to their application and continued improvement in the U.S.—some of these are now commercially ready, others are not. Similarly oil shale mining and conversion has been carried out in the U.S. and elsewhere for over 100 years, although processes for converting oil shale into oils that can be used as transportation fuels still have not been proven on a commercial scale, despite substantial investments. On the other hand, biomass fuels have powered motor vehicles since the beginning of the century, achieving some market penetration in the 1980s.

Chapter 4 is a review of Brazil's efforts to replace petroleum with fuels derived from biomass. By 1986 over half of all gasoline had been replaced by ethanol. Although Brazil chose a biomass fuel path because of limited domestic reserves of petroleum and extensive unused land, its commitment was motivated as much by agricultural interests as by energy concerns.

Chapter 5 is a review of alternative fuel initiatives in the Republic of South Africa, New Zealand, and Canada. South Africa, because of its vulnerability to international trade embargoes, exploited its domestic coal reserves using existing but inefficient coal conversion processes developed earlier in Germany. A large proportion of its petroleum demand has been satisfied by coal liquids. New Zealand, blessed with abundant natural gas but little petroleum, had the choice of using gas directly in vehicles as compressed natural gas (CNG), converting it to methanol, or converting it via methanol into synthetic gasoline. New Zealand chose the CNG and synthetic gasoline options; by 1986 synthetic gasoline replaced about one-third, and CNG about 10 percent, of petroleum-based gasoline use. Canada has been a worldwide leader in developing oil sands and, like New Zealand, has encouraged CNG use.

Note that the objective of these chapters (and this book) is *not* a comprehensive review of experiences with nonpetroleum sources of transportation energy. These chapters do not review, for instance, the innumerable experiments that have been conducted with electric vehicles, the use of CNG and LPG in Italy and elsewhere, or the many small efforts to produce and use various biomass fuels. Rather, the purpose of these chapters is to review those experiences that provide the most insight into the political and institutional factors that influence energy choices.

THREE

U.S. Flirtations with Biomass and Mineral Fuels

The United States has dabbled with nonpetroleum transportation fuels for most of this century. A major difference between the U.S. experience and the experiences of most other countries is that the U.S. had a large selection of resources to choose from, while countries such as Brazil, New Zealand, and South Africa had essentially just one major resource option. At various times the U.S. devoted substantial sums of money to converting oil shale, coal, and corn into liquid and gaseous fuels. Political and financial interest in these activities rose and declined several times during the twentieth century. But even as one misbegotten adventure after another disappeared from public view, groups of organizations and resource owners continued to nurture their dreams (and economic interests). Not until well into the 1980s did nonpetroleum fuels finally make a significant penetration into the transportation energy market. It is interesting that these fuels were not products of the megaprojects that received so much attention and scrutiny through the 1970s and early 1980s.

This chapter traces the history of nonpetroleum fuels produced in the U.S., first addressing fuels produced from oil shale and coal, and then fuels produced from biomass. Some lasting impressions remain from this historical review: the length and variety of those experiences, the substantial investments they attracted, and the number of times those investments were suspended or abandoned as economic, energy, and political circumstances changed.

PETROLEUM'S RISE TO SUPREMACY

In the twentieth century petroleum became the dominant and preferred source of energy for virtually every major energy-consuming activity. Petroleum use in the U.S. grew from 0.1 percent of total energy use in 1860 to 7.9 percent by 1915 and 40 percent by 1955. By 1920 the U.S. was already consuming almost a million barrels of petroleum per day as well as exporting a small amount. Despite, or possibly because of, this rapid expansion in consumption and production, fears were expressed periodically in the U.S. (and elsewhere) that petroleum supplies would not keep pace with growing demand, and that domestic production would be inadequate in time of war to meet military and domestic needs. As early as 1907 a Census Bureau report remarked that

> the importance of this measure [repeal of a tax on alcohol] may be realized when it is known that the supply of gasoline is limited, as petroleum yields only 2 percent gasoline; whereas the sources from which alcohol can be produced are inexhaustible.[1]

In 1921 a General Motors research scientist speaking before the American Chemical Society warned,

> For operating the motors of the country, an enormous and ever increasing amount of liquid fuel is required. This amount of fuel is so large that not only has great activity in the production of crude oil been necessary to meet the demand, but the reserves of crude oil are also being rapidly depleted. The yearly production of petroleum has become so large that exhaustion of reserves in the United States threatens to occur within a few years.[2]

Three decades later, in 1952, a report by the U.S. Bureau of Mines once again raised the issue of petroleum shortages, this time suggesting synthetic coal fuels as the solution:

> The increasing demand for gasoline and oil and the rising cost of finding new petroleum, coupled with America's growing dependence on imports and the unsettled international situation, have continued to emphasize the importance of [synthetic liquid fuel]. . . . Supplementing petroleum with synthetic liquid fuels will not only conserve this Nation's petroleum reserves, but also will bring into use its tremendous reserves of coal.[3]

These statements indicate recurring uneasiness about growing fuel demand and increased reliance on a finite fuel supply. However, the continuing discovery of new sources of oil both in the U.S. and abroad and the steadily decreasing cost to the consumer over time allayed these sporadic and scattered concerns and kept them isolated outside the mainstream political and industrial arenas.

TABLE 5 U.S. PETROLEUM CONSUMPTION AND PRODUCTION, 1950–1985, MILLIONS OF BARRELS PER DAY

	Crude Oil Production[a]	Consumption	Imports[a]	Consumption per capita (b/d)
1950	5.41	6.56	0.85	0.043
1955	6.81	8.46	1.25	0.051
1960	7.04	9.80	1.82	0.054
1965	7.80	11.51	2.47	0.059
1970	9.18	14.70	3.42	0.072
1973	8.78	17.31	6.26	0.082
1974	8.38	16.65	6.11	0.078
1975	8.01	16.32	6.06	0.076
1976	7.78	17.46	7.31	0.080
1977	7.88	18.43	8.79	0.084
1978	8.35	18.85	8.20	0.085
1979	8.18	18.52	8.39	0.082
1980	8.21	17.06	6.87	0.075
1981	8.18	16.06	5.74	0.070
1982	8.26	15.30	4.95	0.066
1983	8.69	15.23	4.82	0.065
1984	8.88	15.73	5.24	0.064
1985	8.97	15.73	5.07	0.066
1986[b]	8.67	16.17	6.06	0.067

SOURCE: American Petroleum Institute, Basic Petroleum Data Book (Washington, D.C.), published three times per year, Section VII, Table 2b.

[a]Imports do not equal production minus consumption because production excludes lease condensate, natural gas liquids, refinery processing losses and various accounting adjustments, and because exports are not included. Exports ranged between 169,000 and 815,000 b/d with values in the upper range beginning about 1982. Oil imported for Strategic Petroleum Reserve is excluded.

[b]Preliminary estimate.

The history of petroleum in the U.S. is one of continuing expansion and increasing dominance of energy markets.[4] Petroleum's growth did not necessarily mean that coal, its chief competitor during the first half of the century, was losing markets; rather, petroleum gained a disproportionate share of new markets. Coal's market share reached its peak in 1910 (at 77 percent), but production of domestic coal did not peak until 1947. Thus petroleum benefited from rapid increases in energy demand—a situation radically different from the stagnant market faced by alternatives in the 1980s.

By 1973, after many decades of cheap and plentiful oil, the U.S. became a nation with an extraordinarily high rate of petroleum consumption, a per capita rate roughly twice that of the industrialized countries in Western Europe. The U.S. was consuming over 17 million barrels per day, of which more than 6 million barrels were imported (see table 5). Despite increasing oil imports and forecasts of a doubling of oil consumption by the year 2000, the country was complacent and relatively unconcerned about the energy situation. For example, the Alaskan oil pipeline had been consistently and successfully opposed on environmental grounds for over three years, and there had been little effort to consolidate energy expertise and authority in the federal bureaucracy. The Arab embargo, beginning in October 1973, was a shock for which the nation and government were ill prepared.

SYNTHETIC FUELS FROM MINERAL RESOURCES

The term "synthetic fuels" has a vague and imprecise meaning, but it gained popular usage during the 1970s as a description of liquid and gaseous fuels produced from coal, oil shale, oil sands, and in some cases, heavy unconventional oil. Fuels derived from biomass were usually not called synthetic fuels. The term "synthetic fuels" will be used in this chapter to be consistent with the vernacular of the period being reviewed; in later chapters coal, oil shale, oil sands, and hydrocarbon feedstocks will be referred to as minerals, and plant matter will be referred to as biomass.

The production of liquid and gaseous fuels from coal, oil shale, and oil sands has an exceptionally long and varied history, many of the initial activities occurring outside the U.S. This section of the chapter on synthetic fuels traces coal-based and then oil shale activities through 1973. Then the "synthetic fuels" programs and endeavors of the 1970s and early 1980s are reviewed, followed by an overview of petrochemical methanol activities in the 1980s.

EARLY COAL CONVERSION ACTIVITIES (PRE-1973)

The first significant effort to convert solid coal into a more portable form dates back to 1792. A Scottish engineer, Murdoch, distilled coal in an iron retort and produced coal gas, which he used to light his home. By 1812 coal gas was used to light street lamps in London; shortly thereafter it spread to many other major cities of the world.[5] Gas distribution networks and gasworks for the manufacture of coal gas (known as "town gas") became common in large and middle-size cities, especially in Europe. By the end of the nineteenth century, coal gas was widely used for home heating and as a heat source for domestic and industrial application. In the latter part of the nineteenth century, a more advanced gasification technique, the forerunner of today's processes for converting coal into gas (and subsequently into methanol) was commercialized. These manufactured gases of the nineteenth century, however, were of a low energy density, about 130 to 500 Btu per cubic foot, and contained many impurities. The old gasifiers were too small, too uneconomic, and too polluting to be acceptable in the U.S. today.

Manufactured gas was replaced in the twentieth century by natural gas, which was being discovered in large quantities as a byproduct of petroleum exploration and development. Natural gas has a much higher energy density (about 1,000 Btu per cubic foot), lower production cost, and is relatively free of impurities. As natural gas pipeline networks expanded, local manufactured gas operations disappeared. By World War II, manufactured gases were virtually nonexistent.

The most important efforts in converting coal into liquid form took place in Germany after World War I.[6] Recognizing Germany's abundance of coal and dearth of petroleum, and sensitive to the claim that the winning side in the just-ended war had floated to victory on a sea of oil, the German government supported and encouraged chemists to accelerate the development of processes for converting coal into usable liquid products. Two separate processes were developed by the mid-1920s. One of these was by Friedrich Bergius, who received a Nobel prize for his pioneering work. In the Bergius process, a mixture of powdered coal, recycled coal oil, and a catalyst was forced under high temperatures into a high-pressure vessel filled with hydrogen. The resulting liquid product was then separated into three distinct components: a light gasoline, a medium-density diesel-like fuel, and a heavy dense oil. Most of the output was tars and heavy oil; in 1925 the process was producing only 6 percent gasoline.[7] This process was an early version of the liquefaction processes that were to receive so much attention during the 1970s.

A second process, developed by Franz Fischer and Hans Tropsch, followed in the tradition of the earlier coal gasification efforts. Fischer and Tropsch applied superheated steam to powdered coal, causing the coal to decompose into a gaseous mixture of hydrogen and carbon monoxide. After purification to remove sulfur compounds, this gas mixture was passed over a metal catalyst. A mixture of low-octane gasolines, high-grade diesel oils, and waxes was produced. Now known as the Fischer-Tropsch process, this process is the basis for the synthetic fuel industry in South Africa and is a forerunner of the more advanced indirect liquefaction processes that were tested in large-scale demonstration plants in the U.S. in the 1980s.

The Bergius and Fischer-Tropsch processes used high temperatures and pressures, requiring large capital investments to build the process plants. To carry on the development of these coal conversion processes, two major German companies (I. G. Farben and Ruhrchemie) bought the patents. The first hydrogenation plant (using the Bergius direct liquefaction process) was built in 1927. By 1939 fourteen hydrogenation and Fischer-Tropsch plants were in operation and six more were under construction. The hydrogenation and Fischer-Tropsch plants were reportedly producing about 25,000 and 14,000 barrels of liquid fuel per day respectively, contributing a total of about one-fourth to one-half of all liquid fuel production in Germany. Thereafter, the more technologically advanced and larger capacity direct liquefaction plants were favored. Peak production of coal liquids by wartime Germany was estimated to have been between 100,000 and 200,000 b/d of mostly aviation gasoline, with the largest plant producing about 17,000 b/d.[8] It was later estimated that production costs for these synthetic liquid fuels were between twenty and thirty cents per gallon—compared to about four to six cents per gallon of gasoline in the U.S.[9]

Efforts to produce liquids from minerals were undertaken elsewhere as well—in England, France, and Belgium, and in Manchuria and Korea with Japanese participation. All carried out advanced experiments and demonstrations before World War II. Except for that of Germany, the English effort was apparently the most advanced. It had a widely studied plant that produced between 750,000 and 1.1 million barrels per year (2,000–3,000 b/d) of gasoline from coal at a reported cost of about eighteen cents per gallon.[10]

Before World War II some sporadic small-scale research efforts on coal conversion were carried out by the U.S. Bureau of Mines (later part of the Department of Interior), but they were directed mostly at reviewing efforts elsewhere.[11] One original effort was a tiny plant, capable of processing 100 pounds of coal per day, set up in 1935. Based on that small plant, it was estimated that it would cost 19.2 to 22.6 cents per gallon to convert U.S. coal into gasoline, confirming the perception of U.S. oil companies that synthetic coal liquids were not a promising venture. Unlike European and most Asian countries, the U.S. had abundant petroleum supplies and little incentive to develop alternatives. A secret 1929 illegal agreement (renewed in 1939) between I. G. Farben and Standard Oil companies, in which Standard Oil promised to stay out of the world chemical business in return for I. G. Farben staying out of the world fuel industry, also played a role in restricting U.S. activities.[12]

The first major stimulus for synthetic fuels in the U.S. was created by the high fuel demands of its Allies in World War II. Despite mild opposition by the oil industry, based on fears that the government would enter the petroleum business, the U.S. Congress approved the Synthetic Liquid Fuels Act on April 5, 1944. The act provided an initial $30 million to the Bureau of Mines for research and construction and operation of demonstration plants. An additional $30 million and $27.6 million were appropriated in 1948 and 1950, respectively.

The coal conversion research and development program received a major boost from an extraordinarily efficient but unconventional version of technology transfer. In the latter part of the war, about thirty volunteer engineers and scientists from oil companies and the Bureau of the Mines received instant commissions as officers in the U.S. Army and together with other military personnel traveled to captured German synthetic fuel installations as soon as they were physically occupied. They interviewed German scientists, analyzed the installations, and shipped tons of industrial papers back to the U.S.

Around late 1945 the Bureau of Mines set up laboratories and pilot plants at six sites to study various synthetic fuel processes for converting oil shale, coal, and agricultural residues into liquid fuel. The oil shale plant was promising, producing gasoline for 7.5 to 9.5 cents per gallon, but the most ambitious were two coal plants built near St. Louis (in Louisiana, Missouri) that used the "borrowed" German technology. Seven cooperative German scientists were hired to help direct the design and operation of the plants. One plant, based on the Bergius patent, converted coal directly into liquids; the other used a modified version of the Fischer-Tropsch process to indirectly liquefy the coal. The first plant, which cost $10 million in addition to the $17 million already spent on the ammonia plant that it had absorbed, produced 200 to 300 b/d of liquids. The second installation was less expensive; the key gasification unit cost $5 million and produced about 80 to 90 b/d of liquids, of which more than half was gasoline.

These plants were vastly improved over the German plants, which had been built in a rushed wartime setting. For example, the average German hydrogenation plant was estimated to have a thermal efficiency of 28.9 percent, which was improved to 55 percent by the U.S. (and German) scientists and engineers. A study by an independent engineering and management firm estimated the cost of hypothetical, commercial-scale plants, using the hydrogenation process developed at the Louisiana, Missouri plant, to be 11.4 cents per gallon—which at that time was about 30 to 40 percent greater than the cost of producing gasoline from petroleum.[13]

Despite positive assessments of the projects, the plants were abandoned in 1953 when Congress abruptly cut off funding. Opposition to continued funding had come from the oil industry and some influential members of the Bureau of Mines; they argued that the knowledge gained was not worth the substantial cost, that theoretical and basic research had fallen far behind the engineering of the demonstration plants, and that the cost per gallon was too high. Two decades later, Bureau of Mines chief John O'Leary, overstating the case, called the closings "the most serious error in energy policy in postwar years."[14]

This abrupt abandonment of alternative fuels did not last long. A few years later renewed interest was motivated strictly by concern for the ailing coal industry, not for national security or shortfalls in petroleum production. Between 1947 and 1956 three thousand coal mines had closed, the number of mine workers had fallen by half, and coal's market share had dropped by 18 percent.[15] The depressed state of the coal industry is indicated by the fact that in 1953 the chemical industry spent thirty times as much for R&D as the entire coal industry earned in profits.[16] The coal industry was composed of thousands of mostly small operators who could not afford to support R&D programs or even an active trade association to lobby for government support.

At the behest of a few members of Congress from coal-rich but economically poor Appalachia, a coal research program was set up in the late 1950s and 1960s in the Interior Department to rejuvenate the coal industry. Out of confusion over the appropriate direction of this program—whether to focus on short-range research projects to gain immediate benefits or long-range research to develop new technologies and markets—there emerged a surprising commitment to synthetic fuel pilot plants. It was surprising because the benefits of such research were far off, while the coal industry was in need of immediate assistance.

By 1968 $53 million of the $60 million worth of contracts being managed by the Office of Coal Research were for synthetic fuel pilot plants; most of the remaining funds were used for studies of coal economics and marketing and coal mining and preparation.[17] Six pilot

TABLE 6 SYNTHETIC FUELS PILOT PLANTS FUNDED BY OFFICE OF COAL RESEARCH, 1962–1972

Project	Prime Contractor	Contract Duration	Government Funding Level	Location
Project Gasoline	Consolidation Coal Co.[a]	8/63- 3/72	$20,377,000	Cresap, WV
Project COED	FMC Corp.	5/62- 2/74	19,332,000	Princeton, NJ
CO_2 Acceptor	Consolidation Coal Co.	6/65- 9/73	16,606,000	Rapid City, SD
HY-GAS	Inst. of Gas Tech.	7/64-7/72	14,399,000 (2,384,000[d])	Chicago, IL
Low-Ash Coal	Spencer Chem. Co.[b]	8/62- 2/65	1,240,000	NA
Solvent Refined Coal (SRC)	Pittsburg and Midway Coal[b]	10/66- 4/72	7,640,000	Ft. Lewis, WA
BI-GAS	Bituminous Coal Research Inc.[c]	12/63- 2/71	3,438,000	Homer City, PA

SOURCE: Office of Coal Research, Annual Report, 1967 and 1972, in R. Vietor, Energy Policy in America Since 1945 (New York: Cambridge University Press, 1984), p. 170.

[a]Consolidation Coal was acquired by Continental Oil Co. in 1966.

[b]The Solvent Refined Coal Project started out as Low Ash Coal. Pittsburg and Midway Coal was a subsidiary of Spencer Chemical. In 1963 both firms were acquired by Gulf Oil Corp.

[c]BCR, Inc. is the research subsidiary of the National Coal Association.

[d]Contribution by American Gas Association (private trade association).

plants were constructed with these funds, three using coal gasification processes and three using direct coal liquefaction processes; all were built and operated by organizations affiliated with the coal industry (see table 6). The coal industry continued to founder while millions of dollars were invested in fuels research projects that provided no immediate benefits and that in the 1980s were still not providing returns.

The process technologies embedded in most of these plants would be retained and improved upon in the 1970s, but in the 1960s their future was bleak. The most prominent plant, "Project Gasoline," experienced huge cost overruns and seemingly interminable technical problems. It was allowed to die a quiet death in 1970 as the research program struggled to retain its dwindling budget. However, just as the coal synfuels program seemed about to die, the market conditions that had been causing it to atrophy began to shift. Petroleum prices and imports began to rise and shortages of natural gas seemed imminent; Congress began to expand coal research funding dramatically.

OIL SHALE AND OIL SANDS

A second major option for producing transportation fuels is the extraction of liquids from oil shale and oil sands. Oil shale is formed over hundreds of thousands and even millions of years by the accumulation and mixing of plankton plants and animals with sediment on the floors of lakes. Oil shale may be termed an incompletely developed crude oil.

Oil sand, sometimes called tar sand, is a mixture of sand grains, water, and bitumen. Oil sands are found in large deposits on every continent except Australia; the largest deposits are in Alberta (Canada) and Venezuela. A description of the Canadian oil sands experience is provided in chapter 5. Oil sand deposits in the U.S. are only a fraction of those in Canada, and little work has been done in the U.S. to exploit this resource. The U.S. does, however, have huge deposits of oil shale; those deposits are estimated to contain more recoverable oil than all of OPEC's crude oil reserves.

Oil shale production is not novel. It has been an ongoing commercial activity in many countries since the mid-1800s.[18] The earliest commercial activity was reportedly in France in 1838, where oil shale was distilled to make lamp fuel. In 1862 oil shale production was begun in Scotland and continued for about a hundred years, being shut down only when all the high-grade reserves had been depleted. At its peak, in 1913, the industry comprised six Scottish companies processing about 3 million tons of oil shale into over 4,000 b/d of liquid fuel. In Australia, oil shale production peaked at 3,000 b/d in 1947 and was abandoned in 1952 after the richest shales were exhausted.

Commercial oil shale industries are still operating in three countries outside the U.S. Brazil has reportedly had a small facility operating since 1862, and in 1970 built a 2,200 ton-per-day demonstration plant. In China a relatively large low-grade shale deposit is located at Funshun in Manchuria. It overlies one of the richest coal seams in the world. The Japanese recovered about 1.3 million barrels of oil from this shale deposit during their occupation in the 1930s and 1940s. After the war the Chinese modernized the Funshun operation and built a second complex; production reached approximately 50,000 barrels per day in the

1950s and is estimated to be still operating at that level.[19] The largest existing oil shale industry is in Estonia in the U.S.S.R. Production began in 1925 and has increased to about 31 million tons of shale in the 1980s. Most of this is burned directly for electricity production; about 20 percent, amounting to about 25,000 b/d, is converted into liquid fuel.[20]

The U.S. experience with oil shale dates back over a century. A sizable industry developed in the eastern U.S. in the 1840s and 1850s. Shale oil reportedly replaced wood as an industrial fuel and whale oil as a lamp fuel. In 1859, when Colonel Drake drilled the first commercial oil well in Pennsylvania, at least fifty small commercial oil shale facilities were operating on the east coast of the U.S.[21] Within a few years the industry collapsed. Very little is known about this brief episode.

In the western U.S., a few small businesses, mostly in California, Utah, and Nevada, experimented with mining and retorting of oil shale in the early years of the twentieth century. In 1915 the U.S. Bureau of Mines began initial research and testing of oil shale, and during World War I several experimental oil shale plants were built.

Interest in oil shale began to escalate shortly thereafter. The 1917 annual report by the Secretary of the Interior stated:

> As a result of the investigation of the western oil shales, it is believed that it is now commercially feasible to work selected deposits of shale in competition with the oil from oil wells, and that these oil-shale reserves can be considered of immediate importance to the oil industry and to the defense of the nation.[22]

This report, along with an enthusiastic National Geographic article, caused a rush to stake oil shale claims in a manner reminiscent of earlier gold and silver rushes.[23] In 1920 a law was passed that no longer allowed prospectors who found oil shale on federal land to take ownership of that land. After 1920 public land with oil shale on it could only be leased. But by then many thousands of square miles of shale land had already passed into private ownership. By 1922 at least 100 oil shale corporations had been founded, and by 1930, when the boom ended, another 150 had been established. Despite the investment of millions of dollars, total production during this period was less than 15,000 barrels. The only significant operation was in Nevada, where one company produced over 12,000 barrels.

The technology used to process oil shale was very crude. The shale was converted into a low-quality oil or burned directly to produce steam for electricity generation. The shale oil could not be used as a transportation fuel. The simple technology was a result of lack of interest and investment by major oil companies and of the lack of capital and technical knowledge by small operators. The increasing production and decreasing price of petroleum during the 1920s effectively doomed this second U.S. oil shale boom.

The third surge of interest in oil shale production began during World War II and continued afterward. The 1944 Synthetic Liquid Fuels Act funded the design and construction of a U.S. Bureau of Mines oil shale plant in Rifle, Colorado. This small experimental plant was closed in 1956 when funding was stopped on the recommendation of the National Petroleum Council, a trade industry group.[24] Other efforts at mining and processing oil shale during the 1950s and 1960s were carried out by Sinclair Oil and Gas (now part of Atlantic Richfield), Union Oil, Tosco (formerly The Oil Shale Corporation), Shell, and a consortium organized by Mobil.[25] All these operations were experimental and were eventually suspended. About 75,000 barrels of shale oil were produced from these ventures.

During the third oil shale episode, major advances were made in developing process technologies. Previously, in the U.S. and elsewhere, oil shale production had resulted in low-quality liquids that were often burned directly to generate steam for electricity. The liquids were unsuited for use as transportation fuels. Advances during the 1950s and 1960s made the use of shale oils for transportation energy more plausible.

After the plant in Rifle, Colorado, was closed in 1956, the U.S. government virtually ignored oil shale until the late 1960s. Only a few diehards in several energy companies and the U.S. Department of the Interior maintained interest.[26] In the 1960s new cost estimates suggested that oil shale might be competitive. Various estimates during the 1960s indicated that shale oil could be produced at costs ranging from $1.46 to about $5 per barrel.[27] In 1964 the Interior Department estimated that a 50,000 b/d plant would cost $60 to $142 million and could produce shale oil for $1.80 to $3.30 per barrel; in 1967 it estimated that a selling price of $4.34 per barrel was necessary to yield a 20 percent rate of return on a first commercial plant.[28] These estimates gained credibility when Tosco, Standard Oil of Ohio, and Cleveland-Cliffs Iron Co. formed a joint venture in 1964 to build a 1,000-ton-per-day prototype plant that was to be expanded by 1970 to a commercial-scale plant costing $130 million and producing 58,000 b/d.[29]

The U.S. government became actively involved again when Department of Interior Secretary Udall spearheaded an effort to rationalize the confused and complex land leases. About 80 percent of oil shale resources were thought to be on federal land, often mixed with private land holdings in small checkerboard patterns.[30] Considerable effort was made to create a coherent and long-range policy toward oil shale development by reducing uncertainty about accessibility to oil shale on public land. But when a test lease of oil shale land tracts in 1968 drew only three embarrassingly low bids, the entire effort was dropped.[31] Without a sense of urgency, the government was unable to develop a forceful and effective program. A sense of urgency was soon to arrive.

SYNFUEL POLICY AND POLITICS, 1973–1986

Even before the oil price shocks of 1973–1974, energy issues were beginning to attract attention. Domestic shortages of natural gas, and growing petroleum imports and prices during the early 1970s lent renewed credence to alternative fuels. In 1971 President Nixon called for "greater focus and urgency" in advancing synthetic fuel projects, and proposed a "cooperative program with industry to expand the number of pilot plants" and "the orderly formulation of a shale oil policy—not by any head-long rush toward development but rather by a well considered program in which both environmental protection and the recovery of a fair return to the Government are cardinal principles."[32]

The U.S. approached the watershed year of 1973 highly dependent on petroleum—an increasingly larger proportion of it from imports. Though there had been periodic concerns throughout the century that new discoveries would not keep pace with growing consumption, these fears proved groundless—until 1973. By 1973 the U.S. had succumbed to the low price of foreign oil and had increased imports until they accounted for 35 to 40 percent of total U.S. oil consumption. Alternative fuels became attractive in the 1970s as the U.S., Japan, and their European allies realized how vulnerable they were to oil supply disruptions and to increases in oil and gas prices. Interest in the development of a synthetic fuel industry in the U.S. came from electric utilities that feared diminished availability and higher costs of their oil and gas fuels, natural gas companies who sought substitutes for domestic gas, and most important of all, the vertically integrated petroleum companies. Coal companies were not participants because they were too small and/or too short of capital to consider major new investments.

In the next two years after 1973, industry and government revived their investment in oil shale and coal fuels, with the federal government investing $280 million and industry about $50 million. Construction began on three coal liquefaction pilot plants and five high-Btu gasification pilot plants, and additional research was initiated on a number of other new conversion processes. Several of the ill-fated projects of the Office of Coal Research from the 1960s were resurrected. At the same time the Interior Department began rebuilding its oil shale leasing program to encourage development of that resource. The department eased many conditions of the aborted 1968 lease, making them more favorable to industry, for instance, by allowing larger tracts, removing minimum production requirements, and eliminating restrictions on the licensing of patents resulting from work on the tract. It strengthened, however, environmental controls and monitoring requirements.[33] Environmental groups, which had grown rapidly in the late 1960s and early 1970s, were still displeased with the lease requirements and forced postponements of lease bids until January 8, 1974—seven days after OPEC raised the price of oil to $11.65 per barrel.

The initial results of the 1974 lease bids were deceptively encouraging for oil shale enthusiasts. Three consortia of oil companies made acceptable bids totaling $447 million on four tracts in Colorado and Utah. The extensive environmental and scientific profiles of the tracts required by the leases soon exposed some major problems, however. It was found that air quality rules to prevent significant deterioration would be violated (see chap. 15), that one tract did not have enough private land nearby to build a residential community for workers, that shale on another tract fractured in a manner that made underground mining difficult or impossible, that the title to one lease was unclear and likely to be contested, and so on.

But the most significant problem turned out to be uncertainty over cost and market conditions. During the two years of environmental and project planning, all three groups found that production costs would probably be two to three times greater than they had estimated—because of inflation in construction costs, unanticipated environmental control costs, and better knowledge resulting from more detailed mine and plant designs. Meanwhile, oil prices had stagnated. One investment group estimated that the production cost would be more than $20 per barrel (in 1975 dollars), considerably higher than the prevailing $12 per barrel price of oil.

In any case, the quadrupling of oil prices and the Arab oil embargo had finally secured the national security justification for alternative fuels. President Ford called on Congress in his 1975 State of the Union message to provide new incentives to spur production of synthetic fuels and to set a production goal of 1 million barrels per day of petroleum-equivalent fuel by 1985. The Administration believed that government should help develop this new industry.[34]

In September 1975 President Ford proposed an Energy Independence Authority to support synthetic fuels as well as other more conventional energy sources. It would be granted $25 billion plus an additional $75 billion in borrowing authority. The proposal was rejected by Congress. Liberals thought it was a handout to "big business" and a threat to the environment; conservatives were leery of governmental interference in the marketplace. Antipathy to oil companies and their expanding revenues and profits played a major role in this opposition. In 1976 Congress rejected by a one-vote margin in the House of Representatives another proposal: a scaled-back plan to provide $4 billion in loan guarantees and price supports for synthetic fuels.

The sense of urgency was dissipating. As oil prices stagnated and oil company revenues and profits expanded, public policy veered away from supporting large energy projects and began to reembrace oil price regulation, which of course was anathema to all investors in alternative fuels.

President Carter's 1977 energy plan rejected the large-scale supply-side approach of Nixon and Ford and emphasized energy conservation and solar energy. Alternative fuels were given a minor role in the plan—a few pilot and small demonstration plants were to be subsidized—but even these few projects faced strong opposition. These proposed projects were to produce gaseous fuels from coal, a response to the widespread natural gas shortages of the previous winter. The Carter Administration funded the preliminary design of five coal gasification projects with the intent of selecting the two most successful and of sharing funding of subsequent construction and operating costs on a fifty-fifty basis with the private participants.[35] Of the five groups, three were consortia of gas utilities, one was an oil company, and the other was a mixed consortium headed by W. R. Grace. The project designs by the three consortia of gas utilities grew out of earlier coal projects that had been funded by the government in the 1960s.

Shortly afterward, in 1978, the newly established U. S. Department of Energy (DOE) provided additional funding for the design of two coal liquefaction plants (SRC-I and SRC-II), again with the intent of providing subsequent funding for successful designs.[36] DOE promised $600 million for the SRC-II plant, which was to be built near Morgantown, West Virginia; the other participants in the proposed $1.6 billion demonstration plant were the governments of Japan and West Germany (25 percent each), a German and a Japanese company, and Gulf Oil. The SRC-I plant was a joint venture between the State of Kentucky and several nonoil companies and had similar financing arrangements.[37] Neither of these projects came into being, however, as one by one the participants withdrew after 1980.

By 1979 events elsewhere in the world had once again created a heightened sense of urgency. In February 1979 the Iranian oil industry was disrupted by the revolution and rise to power of Ayatollah Khomeini. Iran's oil exports dropped from 5.2 million b/d in 1978 to 1.1 million in early 1979. The combination of this supply cutback, the Soviet invasion of Afghanistan and its implied threat to the neighboring oil-producing countries of the Persian Gulf, and the growing world demand for oil allowed OPEC once again to raise prices dramatically. OPEC prices had ratcheted up from $12.70 per barrel in late 1978 to $30 in December 1979, and to a peak of $36 per barrel for "benchmark" oil in December 1980.

Carter's proposals for huge synthetic fuel subsidies became politically more attractive as a result of two phenomena. One was the dramatic increase in world oil prices during 1979. The second was a growing acceptance of the notion that the most effective force for conservation was higher energy prices. It was becoming widely recognized that suppressing domestic petroleum prices below world market levels to protect poor people was a terribly inefficient and costly policy—that a more efficient policy was to allow oil prices to rise to world levels, to tax the resulting "windfall profits" accruing to petroleum companies, and then to return some or all of those revenues to consumers. The key mechanism for making a synthetic fuel program politically viable proved to be the tax on those "windfall" profits.

In April 1979, before the second price shock, President Carter announced that he would gradually end price controls on domestic oil and proposed a tax to capture some of the "huge and undeserved windfall profits" that oil companies would gain from decontrol. By summer of 1979, oil prices were rising rapidly and spot shortages were developing. Motor vehicle drivers faced long lines at service stations in many parts of the country. Carter's standing in public-opinion polls dropped to its lowest level, and Carter himself had become frustrated by the difficulty of reducing energy demand.

On July 15, 1979, after a widely reported ten-day "domestic summit" at the presidential retreat, Carter emerged to make a televised address devoted to energy issues. He rejected his earlier emphasis on conservation in favor of a supply-side approach. Describing the nation as suffering from "a crisis of confidence," he said the energy crisis could be the issue through which the nation could generate "a rebirth of the American spirit."[38] Carter announced he was setting quotas on imported oil, but more important, he asked Congress to authorize "the most massive commitment of funds and resources in our nation's history to develop America's own alternative sources of fuel."[39] He proposed that a special Energy Security Corporation (later named the Synthetic Fuel Corporation), similar to one proposed by President Ford earlier, be established to spearhead this effort. He asked Congress to establish an Energy Mobilization Board, which he described in a July 1979 speech to Congress as an agency that "will have the responsibility and authority to cut through the red tape, the delays and the endless road blocks to completing key energy projects."[40] The three linked proposals—for a windfall profits tax, Energy Security Corporation, and Energy Mobilization Board—were sharply debated throughout 1979 and into 1980.

The Crude Oil Windfall Profits Tax Act (PL 96–223) was signed into law on April 2, 1980. This Act provided for the collection of a forecasted $178 to $277 billion in taxes over the next thirteen years, based on a formula that claimed up to 70 percent of the difference between the selling price and $12.81 per barrel of oil. The Act did not set up a dedicated trust fund for alternative fuels as originally proposed; rather, tax revenues were deposited in the general revenue fund to be allocated at the discretion of Congress.

In June 1980 the Energy Security Act (PL 96–294) was approved by a wide margin and signed into law. This act gave the President most of what he had proposed, but in a two-phase approach that slowed the pace of funding, reduced some of the risks, and sought to limit government's role to that of a banker. The principal provision of the act was the establishment of the United States Synthetic Fuel Corporation (SFC), which was provided with $20 billion and was to be run by a seven-member board of directors nominated by the President and confirmed by the Senate. This Board of Directors was relatively independent in that its members had seven-year terms and, unlike many presidential appointees, could not be fired by the President; members could be removed only for neglect of duty or malfeasance.

The corporation was to use price guarantees, purchase agreements, loan guarantees, and direct loans to encourage private industry to produce the equivalent of at least 500,000 b/d of crude oil by 1987 and 2 million b/d by 1992. If unable to spur private industry to produce synfuels, the corporation was empowered to enter into joint ventures with private firms or to hire contractors to build and operate as many as three government-owned synfuels plants. The $20 billion authorization included $4.475 billion for biomass fuels, solar energy, and conservation programs. The act also provided for (but did not specifically appropriate) an additional $68 billion that could be made available in four years, after the SFC had gained approval from Congress for a "comprehensive strategy" on how it planned to attain the production goals identified in the Act. The corporation could not obligate funds after September 30, 1992, and was to end operations by September 30, 1997.

The proposed Energy Mobilization Board initially had strong support in Congress, based on the perception that the need to develop new sources of energy was sufficiently compelling to merit special treatment outside the regular system of laws and procedures. This early support fragmented. Eventual rejection of the Board in the June 1980 Energy Security Act was due to the opposition of two disparate groups of legislators: conservatives who were concerned that the Board would override states' rights and prerogatives, and liberals and environmentalists who were concerned it would override environmental laws. It was feared that "a crash program guided by an Energy Mobilization Board might run roughshod over the body of regulatory, adjudicatory, and participatory controls that the environmental movement had cultivated for a decade. Thus, the program threatened not only the environment, but environmentalism."[41]

Although the Energy Mobilization Board was rejected, the synthetic fuel program still retained strong political support, generous funding, and enthusiastic interest in the private sector. In the fall of 1979, a government solicitation for proposals to do subsidized feasibility studies attracted 951 responses; over 100 of those initial proposals received funding. A subsequent solicitation in November 1980 for proposals to construct fuels projects attracted sixty-three applications from the private sector.[42] The U.S. Department of Energy (DOE), under orders to initiate the government's synfuel efforts in the interim until the Synthetic Fuel Corporation (SFC) was fully operational, granted subsidies to three of those projects in summer 1981: a $1.12 billion loan guarantee to Tosco Corporation for 75 percent of that company's 40 percent share of the large (47,000 b/d) Tosco-Exxon Colony oil shale project near Parachute, Colorado; a $400 million purchase agreement and price guarantee for a 10,400 b/d Union Oil Company oil shale project also near Parachute; and a $1.5 billion loan guarantee to a consortium of natural gas and other nonoil energy companies to build a 23,000 b/d (oil equivalent) coal gasification plant to produce high-Btu substitute natural gas (SNG) in Beulah, North Dakota. In January 1982 the SFC Board narrowed the list of remaining applicants to eleven additional projects. By the end of 1982, however, enthusiasm was dropping sharply. Only four projects were under construction: the Great Plains SNG and Union oil shale projects being supported by DOE, and two small coal gasification projects that were initiated without federal assistance. Exxon had withdrawn from the Colony oil shale project (with estimated losses of $1 billion), forcing the much smaller Tosco Corporation to cancel the project. Two other major oil shale projects (Cathedral Bluffs and Rio Blanco), already under construction using private funds, had been postponed indefinitely, again with huge losses to the sponsors. The major reasons given for these withdrawals were unanticipated stagnant oil prices and higher than expected costs.

The SFC followed with second, third, and fourth general solicitations and then targeted solicitations, in many cases asking sponsors to resubmit their proposals as the projects became better defined and more "mature." The intent of targeted solicitations was to encourage and support a set of projects with diverse feedstocks and process technologies.

In July 1983 the SFC, desperately seeking fundable projects, awarded a price guarantee of $120 million to the Coolwater coal gasification project, which had already been under construction for one-and-a-half years. The project was located in southern California; it used Texaco coal gasifiers to produce synthesis gas from high-quality coal. Electricity was generated in a combined cycle operation. Six organizations including the electric utility industry's research arm (EPRI), an electric utility, Texaco, a construction firm, General Electric, and a Japanese consortium participated in this joint venture. The plant was designed to produce the energy equivalent of 4,300 b/d of petroleum. It was completed in 1984 a month ahead of schedule, under budget at $260 million, and operated at full capacity within a few months.[43]

A similar coal gasification project was given a $620 million price guarantee in April 1984. It was designed to use lower-quality western coal (subbituminous and possibly lignite) and is exclusively owned by Dow Chemical Company, which developed its own gasifier and intends to license and market the technology after it is demonstrated. The project cost about $300 million, half of that being for the conventional electricity-generating portion of the plant and other support facilities such as coal-handling facilities that were already in place in 1984. Construction began in late 1985 and was completed in 1987. By mid-1987 the plant was operating near its design capacity of 5,170 b/d of oil-equivalent energy.[44] It was completed under budget, and Dow managers claimed it was exceeding performance expectations.

Two other awards were made by the SFC in late 1985. One was an additional $500 million in loan and price guarantees for the troubled Union oil shale plant and the other was $60 million in guarantees for a small heavy oil project in Texas.

One other notable synthetic fuel project of the early 1980s was a small plant built by Tennessee Eastman Company (a subsidiary of Eastman Kodak) to gasify coal and convert the gases into a range of chemicals including methanol. The project was proposed in the early 1970s, pilot plants were built in 1977, and the full-scale commercial plant began operation in early 1984. The company requested no assistance from the DOE and the SFC (though it did benefit from various tax incentives). Tennessee Eastman will release no cost or operational data on the project, but insists the plant was running at 90 percent of capacity within several months and as of 1984 considered the project an attractive investment.[45] Some of the methanol output was sold to the fuels market, but the principal purpose of the plant was to produce acetic anhydride and other chemicals for manufacturing film and other products.

DOWNFALL OF THE SFC

In the early 1980s, two sets of problems intervened to plague the synthetic fuels program: the economic problems of dropping oil prices and rising interest rates, and the political and management problems impeding the SFC's effectiveness. Around February 1981 world oil prices stabilized and soon afterward began a slow descent. The feasibility of synthetic fuel projects was based on assumed oil prices of $40 or so per barrel in the mid-1980s and steadily increasing prices thereafter. When prices did not increase, and in fact decreased, investors became nervous. When interest rates approached 20 percent around 1981–1982, they panicked. The estimated costs of the synthetic fuel projects were typically $2 to $4 billion each (and over $10 billion for one Canadian oil sands project), huge risks even for Exxon, the largest corporation in the world.

The political and managerial problems plaguing the SFC from beginning to end were mostly related to the Reagan Administration's general opposition to government support of alternative fuels (other than in the form of basic research), and its initial opposition to the SFC. When President Carter appointed seven members to the SFC board of directors in September 1980, two months before the presidential election, Senate Republicans and others blocked their confirmation in anticipation of Reagan's victory in November. President Reagan, after assuming office in January, appointed a board chairman, Edward Noble, at a salary of $135,000 per year and gave him permission to bring in a group of former business associates and longtime friends, most of whom had little knowledge of alternative fuels and had little experience working with government under close public scrutiny. There was opposition from Congress about salaries—eleven senior officials of the SFC received higher salaries than a U.S. senator's annual $72,600 salary. Congress was also concerned that putting Noble, who earlier had advised Reagan to eliminate the SFC, in charge of the SFC would be like assigning a fox to guard the hens. Moreover, a steady stream of news reports began to emerge of conflicts of interest and various business improprieties by board members. The corporation's original president and his replacement were both forced to resign.

By mid-1984 only two of seven directors remained, not enough to constitute a quorum, and top SFC managers were reportedly leaving en masse. At congressional hearings in 1983 and 1984 typical criticisms of the SFC by members of Congress included the following: "The people in charge are obviously incompetent and incapable of running the program . . . management is appalling . . . the biggest program of waste and abuse I have ever seen."[46]

Clearly management was not entirely responsible; the deteriorating environment for alternative fuels certainly created exceptional pressures. Throughout, conflict raged within and outside the SFC over its proper role: was its role to expedite the commercialization of synthetic fuel by expanding capacity, to protect the taxpayer by carefully scrutinizing the economics of projects, or was it to demonstrate the maximum number of different process technologies?

In late 1984 three new directors were appointed, but by that time the corporation had lost much of its credibility with both Congress and industry. The SFC's 1984 annual report suggested:

> The lack of a Board quorum weakened the Corporation's credibility with project sponsors, and the intense and raging debate in the Administration and Congress about the continued existence of the Corporation and the size of its program further undermined industry commitment to synthetic fuels. Sound business judgment precluded major investments in projects while the political climate remained so unsettled.[47]

Congress rescinded $7.2 billion of the corporation's appropriated funds in late 1984, leaving about $8 billion. In compliance with a law passed in December 1985, the SFC went out of business in April 1986.

The status of major nonbiomass synthetic fuel projects in the U.S. as of 1987 are listed in table 7. Five of the ten projects in the table are ex-

TABLE 7 COMMERCIAL-SCALE NONBIOMASS SYNTHETIC FUEL PLANTS IN THE U.S. (AS OF 1986)

Sponsors	Location	Conversion Route	Capacity (oil equivalent)	Status
American Natural Resources Co. et al. (Great Plains)	Beulah, N. Dakota	Lignite → SNG	21,000 b/d	Operational 1985, defaulted on loans
Union Oil	Parachute Creek, Colo.	Oil shale → liquids	10,000 b/d	Extended start-up 1984
Tosco and Exxon (Colony)	Parachute Creek, Colo.	Oil shale → liquids	50,000 b/d	Suspended 1982
Texaco, Southern Calif. Edison, Bechtel, GE, EPRI, Japanese consortium (Coolwater)	Daggett, Calif.	Coal → gas → elect.	4,300 b/d	Operational 1985
Dow Chemical	Baton Rouge, Louisiana	Coal → gas → elect.	5,170 b/d	Operational 1987
Tennessee Eastman	Kingsport, Tenn.	Coal → gas → methanol and chemicals	3,600 b/d	Operational 1984
Occidental, Tenneco (Cathedral Bluffs)	Rio Blanco, Colo.	Oil shale → liquids	14,100 b/d	Suspended 1982
Gulf, Amoco (Rio Blanco)	Rio Blanco, Colo.	Oil shale → liquids	10,000 b/d	Suspended 1982
Chevron (Clear Creek)	Rio Blanco, Colo.	Oil shale → liquids	50,000 b/d	Suspended 1985
Phillips, Sohio, Sun	White River, Utah	Oil shale → liquids	?	Suspended 1985

pected to be operational by 1987; the other five were suspended in 1982 through 1985 after construction was begun. At least one hundred million dollars were reportedly spent on each of the five suspended projects. The five suspended oil shale projects were originally designed with a capacity of about 100,000 b/d. Over time they were downsized to the levels indicated in table 7. No other coal or oil shale projects are being actively pursued.

Of the four projects completed by 1986, three proved to be technical successes. The Coolwater, the secretive Tennessee Eastman, and Great Plains projects did not suffer cost overruns or delayed start-ups and performed well after start-up. All were based on coal gasification processes. The only significant technical problem reported for any of the three plants was higher than expected air pollutant emissions from the Great Plains plant. The technical success of these demonstration plants suggests that the technology is ready to begin commercial production of methanol or high-Btu gas from coal.

The plants were not economically successful, however—not because they had higher than expected costs, but because energy prices were lower than had been forecast when the decision was made to construct the projects. In 1980–1981, for instance, it had been expected that the Great Plains plant would produce gas for about $10 to $12 per million Btu and that it would be able to make a profit at that level. But natural gas prices did not rise as expected. When the Great Plains plant began operations in 1985, its costs were about $10 per million Btu, about $5 for capital charges and $5 for operating costs. Thus, although the plant was operating as expected, it was uneconomic. The investors requested a price guarantee (to establish a price floor for the gas) from the SFC in 1983 but were turned down in 1985. As a result, the owners defaulted on their loans and abandoned the project. The U.S. government became liable for the loans (as a result of the $1.6 billion in loan guarantees) and DOE assumed ownership.

The DOE, intent on reselling the plant, attempted to improve operating efficiency and succeeded beyond all expectations. By early 1987 the plant was operating at slightly over 100 percent of design capacity and operating costs were down to almost $3.00 per million Btu, with expectations that costs would be reduced further in the following two years to about $2.75 per million Btu.[48] These remarkable cost reductions are indicative of the very steep learning curves associated with new production technologies (see chap. 8). As of 1987 DOE was attempting to sell the plant for about $300 to $450 million (which represents a significant but not a total financial loss to the government).

The Union oil shale project, unlike the three coal gasification plants, experienced a series of technical problems after construction was completed in late 1983. The company requested, and received in late 1985, $500 million in additional guarantees to help reengineer the plant. Unlike the coal gasification plants, the oil shale plant included a large number of new technologies that had never been built or tested anywhere except in small pilot plants. The Union project demonstrates the difficulty of scaling up new process technologies and the long time frame needed to commercialize new processes. It also demonstrates the less undeveloped state of oil shale production technology.

In any case, by 1985 it was clear that enthusiasm for alternative fuels had evaporated. Firms had delayed and canceled new synthetic fuel projects and legislators had withdrawn their support for SFC spending. The country was not even coming close to attaining the ambitious goals of 500,000 b/d oil equivalent by 1987 and 2 million b/d by 1992. Total production in 1987 was less than 50,000 b/d oil equivalent and probably will not be much more in 1992.

METHANOL FROM NATURAL GAS

Before moving on to biomass fuels, one other set of activities merits attention: the production of methanol from natural gas. It is addressed here and in much greater detail later in the book because most natural gas is located in areas where the local market for gas is not large enough to consume even a tiny proportion of the available reserves. Since natural gas cannot be effectively shipped across oceans by pipeline (or by any other transport mode), this remote natural gas must be either liquefied as LNG or synthesized chemically into methanol. Thus methanol from natural gas is a plausible future energy option for the U.S. and many other countries. A brief history of methanol production from natural gas follows.

The production of methanol from natural gas is essentially the same process that produces methanol from the gas products of coal and biomass gasification. As noted earlier, gasification technologies were first developed in the 1800s; methanol synthesis, the chemical joining of hydrogen and carbon monoxide molecules, was first described in a 1913 German patent. During the 1920s, commercial processes for synthesizing methanol from hydrogen and carbon monoxide gases were developed, but these early processes required very high pressures (3,000 to 5,500 pounds per square inch) and moderately high temperatures

(700° to 800° F). During the 1960s, Imperial Chemical Industries, Ltd. (ICI) developed catalysts that permitted methanol synthesis to take place with low pressure. A large worldwide methanol industry now exists that uses this low-pressure synthesis process to convert natural gas into methanol. Methanol has several industrial applications: as a solvent, a feedstock for formaldehyde, and numerous other chemical uses. Worldwide production capacity, using natural gas as the feedstock, was 350,000 b/d in 1983 and 545,000 b/d by 1986. U.S. methanol production capacity is about 125,000 b/d. In the 1980s oil refineries began using methanol (either directly, or indirectly as MTBE) as an octane-enhancing additive to gasoline, in large part to replace lead, which was being phased out. By 1985 about 16,000 b/d of methanol was being diverted to fuel use (about 70 percent as MTBE); in 1987 this increased to 24,000 b/d, although about 94 percent was in the form of MTBE. No government subsidies were provided. This important development is investigated further in later chapters.

GASOHOL AND BIOMASS FUELS

The history of biomass fuels in the United States is distinct from that of other alternative transportation fuels. The only significant biomass fuel has been ethyl alcohol (ethanol), which in the U.S. is produced mostly from corn but also from other materials such as grain sorghum, molasses, and food-processing wastes. During the twentieth century periodic campaigns to promote ethanol fuel have coincided with low agricultural prices and fears of inadequate petroleum supply. It is a history of unfulfilled promises and special-interest politics.

Sustained efforts to expand biomass fuel production were made in 1906, again after World War I, during the economic depression of the 1930s, and again in the late 1970s and 1980s. In each case farmers were seeking to expand the market for grain.[49] In 1906 farm leaders and the distillery industry, seeking to create a market for surplus crops, were successful in removing a federal tax of $2.08 per gallon that had been placed on beverage alcohol since 1861. About 1906, a few service stations reportedly sold ethanol fuel at thirty cents or more per gallon, but the price was considerably higher than gasoline, which sold for ten to twenty cents per gallon. Henry Ford, a lifelong sympathizer and promoter of agricultural development, supported the "farmer's fuel" by equipping some early Model T engines (actual number unknown) with adjustable carburetors that could be easily switched to operate with either gasoline or ethanol. Interest in ethanol dissipated, however, as awareness of the tremendous size of the newly discovered oil fields in Texas and Oklahoma spread (a 100,000 b/d gusher in 1901 at Spindletop, Texas, was the first major find).

With the outbreak of World War I, Henry Ford campaigned for the production of alcohol from grain as a means to offset gasoline shortages and to further his ambition of applying industrial science to agriculture. After World War I, this campaign continued as fears of petroleum shortages grew. The Standard Oil Company of New Jersey (now Exxon) marketed a blend of 25 percent alcohol and 75 percent gasoline in the Baltimore region during 1922–1923, a period of high gasoline and low alcohol prices. The alcohol used by Standard Oil had been produced during World War I for munitions production, but after the war it was unneeded and made available by the military at low prices in the Baltimore area. The marketing experiment soon ended, in part because of quality-control problems (principally phase separation caused by water intrusion and clogged fuel lines caused by ethanol's solvent properties).

Until the late 1970s the strongest push for ethanol occurred after the repeal of Prohibition during the 1930s, when agriculture (and the rest of the country) was in a period of economic depression. A lobbying campaign was mounted to require that gasoline contain ethanol and that tax benefits be provided for alcohol-gasoline blends. The effort was initially directed at Congress, but it soon shifted to state legislatures in Iowa, South Dakota, Illinois, and Minnesota. Arrayed on one side were farmers, corn products manufacturers, and beverage-alcohol distillers; on the other side were major oil companies, automobile manufacturers, and auto clubs. Both sides conducted wide-ranging but not always truthful campaigns. For proalcohol groups, the movement was viewed as a panacea for the nations' and farmers' economic problems (wheat was selling for less than 25 cents per bushel compared to $3 during World War I, and corn for as little as 7 cents per bushel). Advocates of alcohol fuel presented arguments of increased self-sufficiency, farm relief benefits, military importance of alcohol in the manufacture of munitions, and the superior power, antiknock, and fuel efficiency characteristics of alcohol fuel. Opponents countered by citing the high cost and economic impracticality of biomass alcohol as a farm relief program—and by pointing out the technical problems of phase separation, scavenging, and corrosion created by using alcohol in motor vehicles.

For critics it was an uneconomic gimmick that would be costly to consumers and at best only a temporary expedient for the farm community. The oil industry in particular was alarmed. They feared alcohol might be foisted on them. Indicative of this alarm is the following memo from a committee of the American Petroleum Institute, a trade association of the oil industry, to its members:

> The situation is so critical that action must be taken at once. Make contact immediately with automobile clubs, motor transport interests, and similar organizations in your territory. Acquaint them with the real nature of this scheme and enlist their support in opposing this legislation. Direct the attention of newspaper, trade paper, and automotive publication editors to alcohol blend legislation, and encourage them to consider the economic advisability. . . . Prompt action is necessary to counteract the aggressive and misleading propaganda of the proponents of the blending schemes.[50]

The oil industry's campaign in 1933 and 1934 to delay and defeat alcohol fuel legislation was successful.

Supporters did not give up the fight, however. A nonprofit foundation set up to foster the interests of industrial chemistry was enlisted in their campaign. The Chemical Foundation, funded by royalties from impounded German patents, was an advocate of agrarian reform. With support from Henry Ford and others, the Foundation loaned $116,000 in 1935 for the construction of an experimental 240-barrel-per-day plant (3.7 million gallons per year) in Atchison, Kansas, that was to ferment and distill ethanol fuel from Jerusalem artichokes, sweet potatoes, sorghum, corn, potatoes, and various grains. The plant operators had to overcome legal problems regarding denaturing (rendering alcohol unfit for human consumption) and logistical problems in securing a guaranteed and steady supply of feedstock. The final composition of the fuel (marketed as "Agrol Fluid") contained 78 percent anhydrous ethanol, 7 percent other alcohols, and 15 percent benzol produced from coal (the latter two components serving primarily as denaturants). The fuel was blended with gasoline in proportions of 5 percent to 17.5 percent; the most popular was "Agrol 10," which contained a blend of about 10 percent Agrol Fluid and 90 percent gasoline, and cost about one cent per gallon more than regular gasoline. By the spring of 1938, the fuel was reportedly sold in approximately 2,000 service stations in eight midwestern states. In 1938 plans were announced for a significantly expanded effort that included several additional plants and wider marketing.

Soon, however, unfavorable publicity instigated by the oil industry and continuing high costs began to dampen consumer demand. By early 1939 alcohol fuel production came to a halt. Lobbying efforts to exempt alcohol blends from the one cent per gallon federal gasoline tax were renewed but were unsuccessful in the 1939 and 1940 sessions of Congress. During World War II, the original Atchison plant was reopened and several new alcohol plants were built to supply alcohol for use in manufacturing synthetic rubber, gunpowder, and medicine, and as a petroleum extender and octane booster, mostly for aviation and submarine fuels. At the end of 1944 the U.S. was producing almost 60 million gallons of biomass-based ethanol annually (3,900 b/d)—nearly four times the nation's 1942 rate.[51] The ethanol was produced mainly from grain and molasses, but also from by-products of paper production.

After World War II grain prices rose rapidly and surpluses dwindled as food shipments were sent to Europe. The fermentation alcohol plants either closed or switched to beverage-alcohol production. Petroleum replaced biomass as the primary feedstock for the manufacture of industrial alcohol.

Other private efforts to promote biomass alcohol were made by corporations such as Seagrams, which attempted to increase ethanol yields from sweet potato and grain feedstocks, and International Harvester and Chrysler, which tried to build prototype alcohol-fueled tractors for use in developing countries. However, these were primarily minor research efforts and received little attention.[52]

In these early times, alcohol fuel was doomed by its high costs. By the 1930s oil was cheap, as low as ten cents per barrel for East Texas oil, and the cost of fermented alcohol was high. According to historical records, the lowest cost attained by the Atchison, Kansas, plant was twenty-five cents per gallon, which was five times the average cost of wholesale gasoline (at the refinery gate).[53] Even with increased efficiency, ethanol plants of that era could never have produced alcohol at a price competitive with gasoline.[54]

During the next few decades the fuel market option for agricultural products receded as large new oil discoveries in the Middle East and elsewhere continued to keep oil prices low and supplies abundant. Ethanol was not forgotten, however.

With the 1973 Arab oil embargo and the jump in oil prices, agricultural interest in biomass fuel rose again. Indicative of revived interest was a multiyear "two million mile Gasohol road test" in Nebraska, sponsored by a state-funded committee composed of farmers, oil industry officials, and Nebraska business leaders, that began in December 1974. Farm groups in the Midwest and other parts of the country began a lobbying campaign in state legislatures and in Congress to gain subsidies for ethanol fuel.

Meanwhile, a grass-roots alcohol fuel industry was beginning to form. Under growing political pressure from the many "backyard" enthusiasts and farm groups, the federal government (Treasury Department) in 1977 and 1978 greatly eased restrictions on the construction and operation of small "stills" (under 10,000-gallon-per-year capacity). By fall 1980 over 6,000 permits had been granted by the Treasury Department (although only a fraction were ever used to build and operate stills).[55]

Ethanol was once again promoted as a "home-grown" fuel—as a renewable fuel that would help attain energy independence and that would benefit the "little guys," not "big business." There was little opposition initially; as a small-scale effort it posed little threat to other interests.

It was quickly determined in the Nebraska tests (and elsewhere) that most vehicles could operate on gasoline containing up to about 10 percent ethanol—accordingly, that blend proportion became the standard. Initially these 10/90 blends were sold under the registered trademark "gasohol," but soon the word "gasohol" became a generic label.

After 1973 legislative activities involving ethanol fuel increased. In 1978 some Midwestern farm states began granting rebates and exemptions of state gasoline excise taxes for gasoline containing 10 percent ethanol. A landmark event was the passage of the national Energy Tax Act (PL 95–618) in November 1978, which allowed gasoline containing 10 percent ethanol (if produced from biomass) to be sold free of the four-cent-per-gallon federal excise tax until October 1, 1985. Other less important benefits granted in that Act and in the Crude Oil Windfall Profits Tax Act of 1980 included income tax credits for blending alcohol fuel, investment tax credits, participation by alcohol production in the crude oil entitlements program, and loan guarantees for alcohol pilot plants. The Windfall Profits Tax Act of 1980 extended the excise tax exemption to 1992. During this time federal R&D funding for alcohol fuel increased dramatically, from a minuscule $2.9 million in 1977 (fiscal year) to $24.9 million in 1980.[56] The 1980 Energy Security Act (PL 96-294), which set up the Synthetic Fuel Corporation, authorized $1.2 billion for use as loans, purchase agreements, and loan and price guarantees to promote the production of alcohol and other fuels from biomass. Extraordinarily ambitious goals were also set in that Act to produce 60,000 barrels per day of alcohol from biomass by the end of

1982 and to replace 10 percent of domestic gasoline consumption (500,000 to 650,000 b/d) by 1990. Indicative of enthusiasm for biomass alcohol during this period was the fact that eighty-two bills were introduced into the U.S. Congress in the 1979–1980 session (96th Congress) to support alcohol fuels, and that 10 percent of the bills were passed into law (compared to a reported average success rate of 2 percent).[57]

The new administration of Ronald Reagan, which took office in January 1981, was ideologically opposed to subsidies for building demonstration or commercial plants for alternative fuels and attempted to make substantial cuts in biomass fuel programs. Although Congress blocked those efforts, the resulting loan guarantee programs that were begrudgingly set up by the Administration at the Departments of Energy (DOE) and Agriculture (USDA) provided little stimulus to the nascent industry. Of a total of fifty-seven projects that applied to the DOE for assistance, only eleven plants representing a total annual capacity of 365 million gallons were offered loan guarantees.[58] As of mid-1985 only two of these were in operation: a 52-million-gallon-per-year (g/y) plant in Indiana and a 25-million g/y plant in Tennessee. Several were under construction and others had been canceled.

The USDA program had similarly unfruitful results. Loan guarantees were offered to twelve different ethanol fuel plants between 1980 and 1982 (one additional guarantee was offered in 1984 for an expansion of one of the plants).[59] The largest plant was to have a capacity of 60 million gallons per year and the smallest, 168,000 gallons per year. As of 1986 all but three were in various stages of liquidation.

Throughout the late 1970s and 1980s the single most important government incentive, indeed the only significant incentive, was the extraordinarily generous exemption of 10/90 (gasohol) blends from the fuel excise tax. Until 1983 the federal tax was four cents per gallon. Since blended gasoline contained only one-tenth alcohol, the subsidy to alcohol was forty cents per gallon (10 × 4 cents/gallon). With states granting additional, and in many cases even more generous benefits (Louisiana offered sixteen cents per gallon), the net subsidy for gasohol was over $1.00 per gallon in some states at a time when gasoline was selling for eighty to ninety cents per gallon.

Not surprisingly, the response to these generous subsidies was swift. Around the beginning of 1978 the first gasohol pump in the nation was opened. By 1979 there were about 2,200 outlets selling about 40 million gallons of ethanol fuel (2,600 b/d); sales of gasohol continued to rise swiftly and steadily from that time (see table 8). In 1980, when

TABLE 8 U.S. ETHANOL FUEL MARKET

Year	Ethanol Fuel Sales (b/d)[a]	Market Penetration of Blends	Number of Outlets Selling Blends	Number of States w/Excise or Sales Tax Exemptions	Retail Price, Unleaded Regular Gasoline[c]	Average Ethanol Fuel Whole-Sale Price[c]
1978	1,300	negl.	negl.	NA	$0.67	$1.45
1979	2,600	0.2%	2,200	8	$0.90	$1.76
1980	5,200	0.4%	9,000	25	$1.24	$1.75
1981	5,500	0.8%	NA	NA	$1.38	$1.75
1982	15,300	2.4%	NA	NA	$1.30	$1.66
1983	28,900	4.2%	NA	32	$1.24	$1.68
1984	37,000	5.4%	NA	33	$1.21	$1.54
1985	51,700	7.3%	NA	33	$1.20	$1.48
1986	52,100	7.1%	NA	31	$0.93	$1.05
1987[b]	55,400	7.1%	NA	28	NA	$0.95

SOURCE: Information Resources, Inc. (Washington, D.C.). Gasoline prices are from U.S. Energy Information Administration, Monthly Energy Review (Washington, D.C.: EIA, published monthly).

[a]Includes fuel ethanol imports from Brazil of about 260 b/d in 1981, 910 b/d in 1982, 3,600 b/d in 1983, 5350 b/d in 1984, and 6500 b/d in 1985 (official U.S. Department of Commerce statistics are somewhat at variance with these import estimates).

[b]Estimated as of April 1.

[c]Current year prices (not adjusted for inflation).

Congress extended the four-cent federal tax exemption to 1992, new large investments were made by the private sector in ethanol fuel production. Subsequent increases in the federal tax exemption to $0.05 and then $0.06 in 1983 (in concert with increases in the gasoline excise tax to $0.09 per gallon) further enhanced the attractiveness of ethanol fuels.

The reaction of the oil industry to biomass alcohol was mixed.[60] Initially the industry strongly opposed gasohol use. Companies banned the use of company credit cards for gasohol purchases, required separate storage and pumping facilities for gasohol sold at company outlets, and created other barriers and disincentives. By 1980, after a series of congressional hearings called for by leaders from farm states, this policy of opposition had shifted. Most companies reluctantly accepted gasohol use, and a few large oil companies even became participants, partly as a strategy to diversify and hedge their investments, and in some areas to protect their market from independents and cooperatives. Texaco, Cities Service, Ashland, Chevron, and Amoco (Standard Oil of Indiana) participated in joint ventures to produce fermentation ethanol to be used for blending with gasoline.

Encouraged by the simple nature of ethanol production technology, the feasibility of constructing very small plants that required only small amounts of capital, and of course by the large subsidies, investor interest in ethanol fuel production quickened. A national survey conducted in 1980 found that over 300 firms had already prepared to build ethanol fuel plants (with a planned gross production capacity of over 4 billion gallons per year, or 270,000 b/d); only five of the 300 plus firms were oil and gas companies, and none of those five was a major energy company.[61] It was, in fact, this grass-roots nature of the industry that appealed to the public.

Although the image of ethanol fuel activities—an image of many small investments by farmers, cooperatives, and small businesses—persisted into the 1980s, reality was very different. The ethanol fuel industry was, and continues to be, dominated by one very large corporation, Archer Daniel Midland Company (ADM). ADM is a major grain trading and processing firm, ranking seventy-fourth in the Fortune 500 list of largest companies in 1984, with annual sales of $4.9 billion.[62] ADM entered the ethanol fuel market in the late 1970s by converting corn processing plants to fuel production. ADM produced 54 million gallons of ethanol fuel in 1979, roughly 80 percent of all ethanol fuel produced that year. ADM continued to expand its capacity; by 1985 it was operating three large plants that were producing about 300 million gallons annually (19,600 b/d), which was about 60 to 70 percent of total ethanol fuel sales in the country. Other large plants were operated by smaller firms (sometimes with capital from participating oil companies), whose major business activities were beverage-alcohol production, grain trading, and corn processing. One firm (Midwest Solvents) used the old Atchison plant to produce ethanol fuel.

The almost instantaneous creation of an ethanol fuel industry in the U.S. did not represent a rash of investments in new fuel production facilities. On the contrary, just as in the early years of the Brazilian ethanol program, rapid expansion of production capacity occurred mostly through the addition of alcohol facilities to existing food processing plants.

In late 1981 fuel ethanol production dropped somewhat as a result of lower petroleum prices and determinations by fuel distributors that the costly long haul from midwestern production sites to the Northeast and elsewhere was not economically viable. A backlash against gasohol also developed; governmental support for ethanol began to be perceived as another farm support program and an ill-advised one at that. Some states, mostly outside the cornbelt region, reduced their tax incentives. By mid-1981 ethanol fuel marketers and promoters responded to adverse publicity by eliminating the label "gasohol." Thereafter, most fuel outlets rarely advertised their fuel as containing ethanol; occasionally the fuel was labeled "ethanol plus" or something similar, but often no mention was made. The "home-grown" sales pitch that had appealed to patriotism was abandoned. By late 1982 the combination of refocused marketing strategies and a drop in corn prices (by far the most important feedstock for ethanol) led to a recovery in ethanol sales. Ethanol fuel sales continued to rise steadily. By 1985 ethanol fuel accounted for about 0.6 percent of gasoline sales, which indicates that 6 percent of all gasoline contained ethanol. In some states, such as Iowa, Nebraska, Indiana, Kansas, and Kentucky, market penetration was much higher—between 20 and 35 percent.[63]

The continuing expansion in ethanol fuel production was due in part to the substantial subsidies, but also to its value as an octane-boosting gasoline additive. Until the mid-1980s, almost all ethanol was used as a gasoline "extender" by fuel marketers, although many drivers bought the fuel for its higher octane rating. Following the accelerated phaseout of lead beginning in late 1985, however, some refiners started making a substandard gasoline with a rating of 84 to 85 octane; fuel marketers downstream would then add 10 percent ethanol to get the standard 87 octane rating. The emerging use of ethanol as an octane enhancer rather than as a gasoline extender increased ethanol's value.

The high price of ethanol rendered the U.S. ethanol fuel industry vulnerable to the possibility of massive low-priced imports of Brazilian ethanol (which cost about eighty to eighty-five cents per gallon to produce in the early 1980s) (see chap. 4). For that reason domestic producers, especially ADM, lobbied hard and successfully for tariffs on imported ethanol. In 1981 a $0.10 per gallon tax was placed on ethanol imported for fuel purposes (PL 96-499). The legislation that established that tariff provided for a further increase to $0.20 per gallon in 1982 and to $0.40 from 1983 through 1992. Subsequent legislation increased the duty to $0.50 in April 1983 and to $0.60 in 1984. The latter increases were intended to match the previously mentioned increases in domestic fuel tax exemptions that had been granted to "gasohol"; this assured that imported alcohol could not benefit from the domestic subsidies (and that imported ethanol would not be price competitive with domestically produced ethanol).

The size of the subsidies required to keep the ethanol fuel industry viable continues to grow. A U.S. Department of Agriculture report estimated that twenty-nine states contributed $302.5 million to ethanol fuel in 1985 and that the federal gasoline tax exemption of six cents per gallon caused a loss of about $500 million per year from the Federal Highway Trust Fund.[64] These subsidies amounted to over $1 per gallon of ethanol fuel (over $1.30 per gasoline-equivalent gallon).

A DISTORTED BIOMASS FUEL INDUSTRY

Biomass fuels finally achieved significant penetration of the transportation energy market in the 1980s. But the actual form taken by the biomass fuel industry was, as will be shown later (chap. 6), not in the immediate interests of the country. It used corn, an expensive and energy-intensive crop; it received a subsidy of a dollar or so per gallon to be competitive; and it was dominated by a single large corporation. A biomass fuel industry need not take this form; indeed, several other biomass fuel options are far superior.

The distorted industry structure that now exists can be readily explained, however. Corn fermentation processes were used because corn was readily available, the process technology was well known, and some fermentation-distillation facilities were underutilized. The industry was highly concentrated because smaller firms, even with the large subsidies, could not compete with large experienced firms such as ADM—they did not have fuel marketing expertise, lacked adequate investment and operating capital, did not have enough volume to keep transportation costs down, and did not have the expertise and size to improve production efficiencies.[65]

SUMMARY

This historical review has shown that coal and oil shale have been utilized as resources for the production of liquid and gas energy for over a century. They are not exotic fuels, and many of the processes for converting them into liquids and gases are not based on exotic technologies. The question has always been and will continue to be: What conditions must prevail for these resources and processes to become economically attractive alternatives to petroleum? While that question will continue to be addressed throughout this book, several observations can be made at this point.

First, perceptions of technical readiness of fuel options and the reality of technical readiness are not always in agreement. This discrepancy was greatest for oil shale (for producing high-quality liquid fuels). A second observation is the precariousness of public support for energy options and the huge role public support, or the lack thereof, plays in the development of those new options. First the promise and then the disarray and ineffectiveness of the U.S. Synthetic Fuel Corporation undoubtedly stimulated and then truncated investments in coal and oil shale activities, since the huge cost of new coal and oil shale projects makes those projects risky investments even for the largest corporations. All the large coal and oil shale projects in the U.S. were either canceled or went into bankruptcy by 1985. Several smaller projects survived as demonstration plants. This fickleness of public support is explained in large part by the complexity and ambiguity of the issues. The public, including many policy makers, often does not have access to coherent and accurate information. Given the uncertainty about the future, and taking into consideration the decentralized nature of political and economic power as in the U.S., it is not surprising that a strong, lasting consensus of support for alternative fuels has not come into being.[66]

In any case, the U.S. has a rich history of alternative fuels. The coal and oil shale–based activities remain a history of unfulfilled promises. While biomass fuels finally reached large-scale commercial production in the 1980s, they did so in a distorted form. Natural gas–based methanol was the only nonpetroleum fuel to make a sizable contribution without needing government subsidies, although this methanol was aided by depressed prices caused by a worldwide supply glut.

Those fuels that actually entered commercial production were those that required the least investment and least technological change. Thus fuels produced from coal and oil shale were least successful because they depended on major new investments and major technological advances. Methanol from natural gas and ethanol from corn required little or no investment and little or no technological change. The process of change is incremental and often moves in a direction not consistent with the overall interests of a society, unless a society makes a deep and prolonged commitment.

FOUR

Ethanol in Brazil

The objective of this chapter is to describe and analyze the events, motivations, and forces that led to the introduction of ethanol in Brazil and to discern important lessons that might be relevant for other regions and countries as they contemplate a transition from petroleum transportation fuels. It is shown that the pursuit of ethanol fuel in Brazil was not based on long-term plans or deep-seated values, but was an ad hoc response to a particular set of circumstances: a large depressed sugar industry, an ambitious domestic auto industry, and mounting foreign debt. It is not the purpose of this chapter to conduct a definitive evaluation of the Brazilian ethanol program—indeed, because of the scarcity of good economic, financial, and environmental data, and the uncertainty of future petroleum prices, such an evaluation would be difficult. Nonetheless, to give some sense of the impact of the program on the Brazilian economy, society, and environment, an overview evaluation that addresses the major questions and issues is provided in the appendix to this chapter.

BACKGROUND

Brazil is among the largest nations in the world. It is fifth largest in land area and sixth largest in population (135 million in 1985). Brazil is a rapidly industrializing country with a per capita income of $1,880 (U.S.) in 1983, and is categorized by the World Bank as an upper middle–income country.

Brazil was governed by the military from 1964 until it transferred power to a democratically elected civilian government in 1985. The central government was, and continues to be, deeply involved in the economy; it regulates or holds partial or full financial interest in many economic activities. Brazil continues to have a politicized market economy in which state officials strongly influence investment patterns.

Brazil is blessed with many natural resources, but in late 1975 when Brazil was about to embark on its transition to alcohol transportation fuels, it was widely perceived that petroleum was not among those abundant resources.[1] As a result of booming economic growth, and a decision to favor highway development over railroads and to promote the automobile industry, Brazil became increasingly dependent on petroleum in the 1960s and 1970s. As trucks gained an increasing share of intercity freight shipments (accounting for about 70 percent of ton miles in the 1970s vs. less than 20 percent in the U.S.), and as domestic automotive output quadrupled between 1964 and 1974,[2] transportation fuel demand increased dramatically. By 1974 Brazil had 4 million cars and 1.4 million trucks and buses. Within the short period of seven years between 1967 and 1974, annual petroleum consumption nearly doubled to about 750,000 barrels per day (11.5 billion liters).[3] By 1974 about 42 percent of all domestic energy was supplied by petroleum and most of the remainder came from hydropower (20 percent) and firewood and other combustible biomass (34 percent).[4] Approximately 80 percent of petroleum was imported.[5]

The quadrupling of world petroleum prices in 1973–1974 had a chilling effect on the economy; the annual growth rate dropped from 10 percent in the late 1960s and early 1970s to 7 percent between 1974 and 1980 (and to less than zero in the early 1980s, although, of course, petroleum prices were not the only cause of this economic decline). In 1973, before oil prices quadrupled, Brazil was spending $600 million per year on petroleum.[6] By 1978, petroleum imports increased to over $4 billion and continued increasing into the early 1980s. Brazil's foreign debt began to rise in the 1970s from $6.2 billion in 1973 to almost $80 billion in 1982, the highest in the developing world. The importation of foreign oil became a critical issue.

Brazil had few readily available options for reducing petroleum imports, especially for the transportation sector. Domestic reserves were thought to be sparse; considering the growth of the economy, conservation would not be sufficient. The country had abundant hydropower, which it was continuing to develop and which could easily be substituted for petroleum in industrial, commercial, and residential uses, but hydroelectricity is unsuitable for most transportation uses. Short of fundamental structural shifts away from internal combustion engines and motor vehicles, biomass fuels appeared to be the one option that could make a large difference.

SUGAR POLITICS

The biomass feedstock that met the two criteria of being easily converted to liquid fuel and widely available was sugar cane. Although the imperative of reducing petroleum imports and developing an alternative fuel option was inspired by the worldwide petroleum crisis, the motivating and sustaining force was the political clout of the sugar cane industry and, to a lesser extent, agricultural and distillery equipment suppliers. The opportunity to bolster and stabilize the volatile sugar market by dampening price fluctuations and increasing the value of domestic sugar cane was tempting.

Diverting sugar cane to fuel was not a new idea. In 1931 the Brazilian government had decreed that ethanol must be blended into gasoline in a 5-to-100 proportion; the number of distilleries producing fuel-grade ethanol increased from one in 1933 to about fifty-four by 1945.[7] During the petroleum-scarce years of World War II and shortly thereafter, the average annual ethanol content in gasoline reportedly reached 40 percent in northeastern Brazil.[8] In 1966 a government rule was enacted to encourage the use of up to 25 percent ethanol in gasoline in order to reduce excessive ethanol supplies (otherwise used for beverage alcohol) and to dampen fluctuations in sugar prices. Thus by 1975 there was a long history of diverting sugar cane into ethanol fuel production, but it was primarily a device to support the sugar market.

Because of its dependence on sugar exports, Brazil was vulnerable to sugar price fluctuations. While not its major export, sugar accounts for about $200 million to $1 billion of Brazil's export earnings per year. This wide range in export earnings is due to the drastic fluctuation in world sugar prices, as shown in figure 4. The existence of a transportation market for alcohol gave the sugar industry the flexibility to divert sugar into alcohol production when sugar prices were low, partially protecting it from the vagaries of the volatile world market.

In the early 1970s Brazil had endeavored to modernize the sugar cane industry by encouraging higher productivity and by building larger, more efficient mills for extracting sugar from the cane.[9] But just

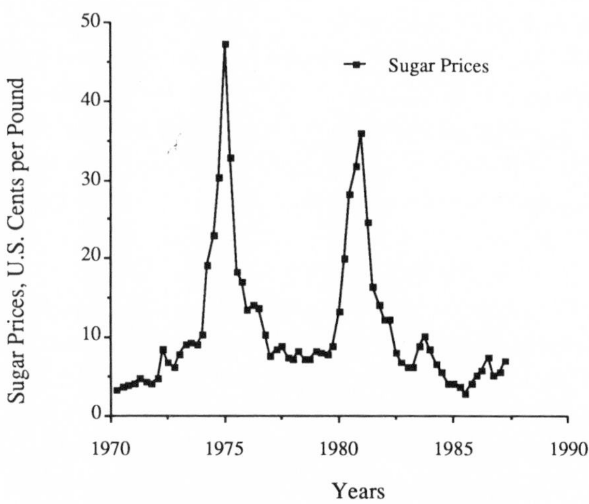

Figure 4. World Raw Sugar Prices

Source: U.S. Department of Agriculture, *Sugar and Sweetener, Situation and Outlook Yearbook* (Washington, D.C.: USDA, 1987, SSRV12N2)

as output began to boom, the world market once again collapsed. Sugar prices began a precipitous decline in 1974 (plummeting from highs of 70 cents per pound in 1974 to prices that eventually dropped to under 10 cents per pound in 1976). The stage was set.

PROALCOOL (BRAZIL'S NATIONAL ALCOHOL PROGRAM)

The emergence of a major alcohol fuel initiative in 1975 was a response to three phenomena threatening the Brazilian economy: a mounting foreign debt, low sugar prices, overcapacity in sugar production, and petroleum supply disruptions and price increases. As will be discussed, the alcohol fuel program was not a carefully planned or a well thought out initiative. It resulted from a peculiar (though not necessarily unique) set of political and technological circumstances. Over a period of just a few years and on the basis of little analysis, Brazil ventured boldly onto a promising but risky new path.

On November 14, 1975, the president of Brazil, Ernesto Geisel, officially established the National Alcohol Program (Proalcool). The principal stated goal of the program was to save foreign exchange by using alcohol as a supplement to the nation's gasoline supply (and secondarily as a chemical feedstock). The initial stated target was to produce and use 790 million gallons (3 billion liters) of biomass-based fuel annually by 1980, which would account for roughly 20 percent of anticipated gasoline demand. Over 90 percent of the fuel was to be ethanol derived from sugar cane, 8 percent was to be ethanol derived from cassava (also known as manioc and yucca), and 1 percent was to be vegetable oil derived from babacu palm as a substitute for diesel fuel.[10] A National Alcohol Commission was formed to develop and coordinate the program. Included in the Commission's mandate was a provision to establish a favorable credit program to stimulate production and to guarantee a market for all alcohol produced. The following social and economic objectives were also explicitly stated: (1) reduction of regional income disparities, (2) reduction of individual income disparities, especially between rural and urban sectors, and (3) expansion of employment opportunities and economic growth in agriculture and related industrial activities, including the manufacture of mills and distilleries as well as motor vehicles. These more socially oriented objectives did not receive much serious attention.[11]

By 1979 the targeted goal of 790 million gallons had been reached.

EARLY IMPLEMENTATION

Although Brazil had a strong central government, the introduction of alcohol fuel was not comprehensively planned or well managed. In the U.S. the introduction of new fuels is hindered in large part by the vast number of firms participating in fuel production, distribution, and end-use activities, by the reluctance of corporations to be initial risk takers, and by the reluctance of government to take a lead role. In Brazil the obstacles were different; they were related to conflicts amongst public and quasi-public organizations.[12]

President Geisel was an enthusiastic supporter of ethanol fuel, but even though he was at the head of a strong, centralized government, he was not able to, or in any case did not, direct the implementation of Proalcool along a well-planned, deliberate path. Although the state had considerable resources at its disposal and directly influenced much of the economy, it was not a monolithic entity. Each public corporation and government agency pursued its own interests. The Sugar and Alcohol Institute (IAA), which regulated the sugar cane industry and advocated sugar growers' interests, lobbied to make sugar cane the exclusive feedstock. Copersucar, a cooperative of large São Paulo sugar and ethanol producers, lobbied to procure subsidies and to oppose the establishment of large state-administered plantations. Petrobras, the huge state petroleum monopoly (and one of the fifty largest corporations in the world at the time), lobbied to gain control of the entire program and to protect its liquid fuels monopoly.

As with any major new government program that is superimposed on existing activities, responsibilities were dispersed among many organizations. The large size of the public sector in Brazil exacerbated this problem of dispersed responsibilities. The National Alcohol Council, comprising representatives of seven agencies, determined project selection criteria and approved distillery construction plans; the ministerial-level Council for Economic Development set fuel production targets and producer prices; Petrobras and the National Petroleum Council determined gasoline production levels and retail fuel prices; and the National Monetary Council, comprising representatives of the monetary and fiscal authorities, determined how much credit would be available to the alcohol program. An investor interested in a distillery would submit a detailed technical proposal to the IAA and National Alcohol Council. If the technical proposal was approved, a financial proposal was then submitted to the state-owned commercial Bank of Brazil to receive subsidized credit, and finally the Central Bank of Brazil would review the financial contract.

One of the early obstacles was the reluctance of these two financial institutions to provide credit. By October 1976, of thirty-two projects approved by the National Alcohol Council, only eight had received funding; by May 1977 the funding rate had barely increased to twenty-eight awards out of 112 approvals. Funding approvals by the banks in 1978 and 1979 continued to lag far behind the council's project approvals. The Bank of Brazil, even though state-owned, was reluctant to issue what it considered "bad" loans, not unreasonable given the questionable economics of the projects and the uncertain future of alcohol fuel; the Central Bank was reluctant to issue loans at below-market rates because of the inflationary effect on the economy.

The program remained fragmented in the early years as various public sector organizations pursued their own objectives, each in accordance with its own parochial interests and agency missions. Some were promoting alcohol, others were actively impeding it. One independent researcher observed that the ethanol program "suffer[ed] from weak program vision, vague institutional definition and organizational restrictions which impede effective leadership . . . [and that] there was little or no information, analysis or supporting research to guide decision making. In consequence, many decisions were made on an ad hoc basis."[13] The transition to ethanol proceeded haphazardly on the basis of little analysis and little overall direction.

Even so, by 1978 alcohol production had increased substantially. This early growth was easy in the sense that few technological or institutional changes were necessary. Sugar cane had been grown for centuries. Ethanol had been blended in gasoline for several decades. No changes were needed in motor vehicles since the ethanol content in gasoline was restricted to less than 20 percent, nor were changes required in fuel pumps at stations; and the existing fuel storage and distribution infrastructure could adequately handle the ethanol output. Also, distillery equipment was manufactured domestically and was readily available.

Another factor contributing to this easy early expansion was the availability of unutilized and underutilized distilleries, and the large numbers of existing sugar mills; relatively little new investment was required to expand alcohol production at existing distilleries and annex distilleries built adjacent to sugar mills. Moreover, the initial annex distilleries often used the molasses byproduct, not the cane itself, as the feedstock for ethanol production.

Autonomous distilleries are riskier investments. They require large investments in sugar cane mills as well as distilleries, use sugar cane directly as feedstock, cannot easily switch back to sugar production, and have lead times of up to four years (compared to one year or less for annex distilleries). Few autonomous distilleries were built during the 1976 to 1978 period;[14] in June 1979, 89 of the 104 operating fuel distilleries were annex distilleries attached to preexisting sugar mills.[15]

Investments in ethanol fuel by sugar growers and producers continued despite the risk that thin public support might evaporate and that the ethanol fuel market might stagnate. But the risk was modest, since the marginal investment in adding an annex distillery was not huge and because the grower-producer could fall back on sugar production if necessary. Indeed, this investment in annex distilleries was attractive to growers and producers in that it diversified their investment and reduced the overall risk to existing plantation and sugar mill investments. With sugar prices low in the 1976–1978 period, it is not surprising that sugar growers and producers were happy to divert some of their cane to government-purchased alcohol.

By late 1978 it was becoming obvious that the period of easy and haphazard growth was at a critical juncture. The oversupply of excess sugar capacity was more than used up and the 20 percent ethanol/gasoline limit was being approached. If the program was to be anything more than a subsidy for the sugar industry, then some aggressive action and leadership were necessary. The sugar industry bailout scheme would need to be transformed into an energy scheme in which alcohol fuel pumps would have to be put in place, alcohol vehicles built, and autonomous distilleries become the major source of fuel production. This step would represent a break in the historical link between sugar and alcohol production.

ALCOHOL BOOM AND BUST

The energy issue gained renewed attention in December 1978 when Iran, Brazil's second-largest petroleum supplier, was disrupted by revolution, causing upward pressure on world oil prices. However, despite this heightened attention to energy, ethanol fuel activities continued to be plagued by bureaucratic conflicts—centered mostly on the efforts of Petrobras to wrest control of the program away from various other organizations. Pending a resolution of the leadership and direction of Proalcool and the level and type of support to be given it, all distillery proposals were frozen within the IAA during the first five months of 1979 and few others were submitted.[16]

The paralysis was broken on June 6, 1979, the day OPEC announced a 37 percent price hike. This price hike on top of Brazil's ballooning foreign debt created a national crisis—which the new administration of President Figueiredo immediately used as justification to assert itself. That same day, the government raised the annual production target of alcohol from the already achieved 790 million gallons to 2.8 billion gallons by 1985. It also set a goal of $5 billion (U.S. $) to be invested during the next six years in fuel production and distribution facilities and created a higher-level body (replacing the ineffectual National Alcohol Council) to implement the program. Figueiredo portrayed these moves as elements of a "war economy." These initiatives were widely supported as a strategy for reducing dependence on increasingly expensive foreign oil, enhancing national security, protecting the future of the domestic auto industry (which was viewed as the leading sector of the "economic miracle" of the 1960s and 1970s), providing new investment opportunities for the private sector, and appealing to nationalism by providing a uniquely Brazilian response to the energy crisis. A strong consensus formed regarding these initiatives; Petrobras was forced to back off. Most important, the government initiatives encouraged the auto industry to make the crucial decision in 1979 to begin manufacturing alcohol vehicles.

ROLE OF THE AUTO INDUSTRY

Brazil has a large automobile industry consisting primarily of subsidiaries of U.S. and European corporations. During the early stages of the alcohol fuel program (1975–1978), the automobile industry played a quietly supportive but noncommittal role. For many years Brazilian cars had been designed to burn gasoline with up to about 20 percent ethanol, so the initial target of blending up to 20 percent ethanol required no response or action from the automakers. Although privately they remained skeptical of the government's resolve,[17] automakers investigated the possibility of producing vehicles that operated on straight alcohol. Some parent companies, especially Ford and Volkswagen, already had initiated research in their home countries and had transferred that initial development work to their Brazilian subsidiaries. By August 1976 General Motors, Ford, and Volkswagen were able to announce that no major technological barriers precluded the production of alcohol cars.[18] However, the industry was unwilling to invest in retooling factories without an assurance of continued and reliable fuel supply.

By 1979 the auto industry had become more interested in producing alcohol cars. The early alcohol production targets had already been attained and the auto industry was convinced of the government's commitment to alcohol fuel. Moreover, the previously rapid growth in auto sales had slowed (declining for the first time ever in 1977) and the future of gasoline prices and gasoline availability was uncertain. The prospect of government incentives to purchase alcohol cars and thereby stimulate car sales was attractive. In addition, some of these companies, especially Volkswagen and Ford, were eager to develop ethanol vehicle technology with the hope of becoming exporters, an aspiration that meshed with Brazil's growth-oriented development objectives.

Although, beginning in August 1978, the automakers had made various pronouncements of plans to begin selling alcohol cars (presumably to promote the faltering Proalcool program without risk to themselves), it was not until mid-1979 that they became firmly committed. In exchange for incentives to consumers to purchase alcohol cars and guaranteed low ethanol price ceilings, the auto companies began to manufacture alcohol cars within several months.

In January 1980, 980 alcohol cars were sold, representing 1.2 percent of new cars sold that month.[19] Given the short lead time, automakers obviously had made only the minimal changes necessary to existing gasoline vehicles—incorporating only those changes necessary for the car to run satisfactorily, but not efficiently, on straight alcohol. Those initial changes were the following: replacement of materials that were incompatible with ethanol, increases in engine compression ratios to take advantage of the higher octane rating of ethanol (from about 7:1 to about 11:1), adjustment of carburetors, and placement of one-to-two quart gasoline tanks under the hood to assist in cold starts. Gradually automobile technology was improved to increase fuel efficiency, ease startability problems in cold weather (i.e., down to around 0° C), and enhance performance.

MARKETING STRATEGIES AND CONSUMER RESPONSES

Initial demand for alcohol cars was higher than expected. Sales of alcohol cars increased from 1 percent of total car sales in January 1980 to 73 percent by December of that year (fig. 5). The stock of alcohol cars was further expanded by retrofitting existing gasoline cars; when performed by government-certified mechanics the conversion cost was about $250, but by early 1980 these mechanics were so overwhelmed with orders that many motorists turned to unauthorized mechanics who charged as little as $60.[20] Many of the early retrofitted vehicles were taxicabs that, because of their high usage, had the most to gain from subsidized fuel prices.

Consumers were attracted to alcohol cars for several reasons. First, the government made alcohol car purchases attractive relative to gasoline cars by reducing the registration tax on them and providing easier credit in the form of longer payment periods and smaller down payments. Second, it was announced that ethanol prices would be capped at 65 percent that of gasolone (assuring the consumer that ethanol would be a less expensive fuel than gasoline, since ethanol has two-thirds the energy density of gasoline). Third, there was a fear that gasoline prices were likely to continue increasing in the future and that gasoline would become less available. The Iran/Iraq war, which began in September 1980, reinforced this perception of uncertain availability,

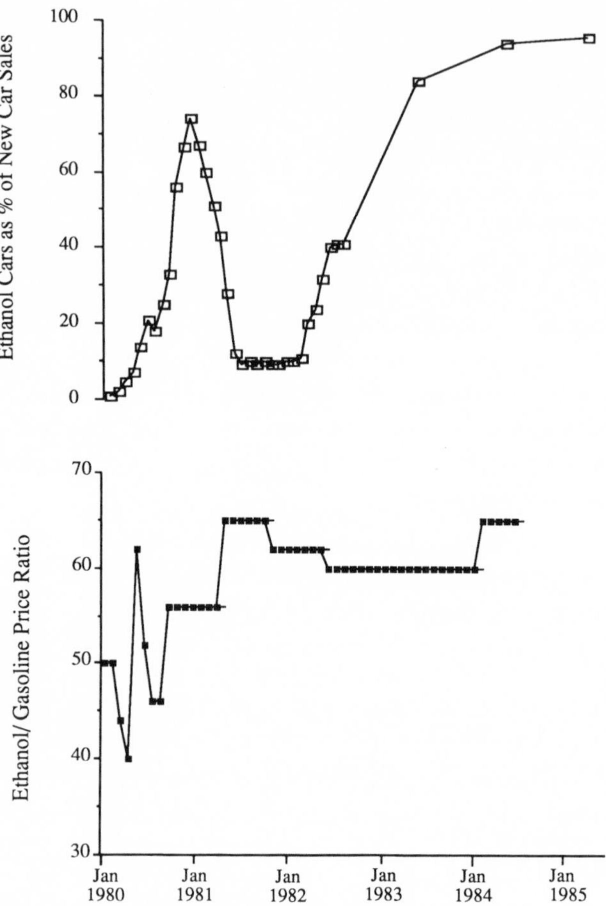

Figure 5. Market Penetration of Ethanol Vehicles and Ratio of Anhydrous Ethanol and Gasoline Prices

since Iraq had been supplying about 40 percent of Brazil's oil imports. Fourth, Brazilians felt pride in their locally developed vehicles and fuel.

As might be expected with a new product, problems arose—most troublesome being the tendency not to start. Other reported problems were corrosion of engine parts, buildup of deposits in the carburetor, and high fuel consumption. However, most of the problems, especially those that involved starting and corrosion, were associated with converted cars, not with the new cars designed for alcohol. Many conversions were done by unauthorized amateurish mechanics who had little experience or expertise with alcohol fuel. Little or no distinction was made by the media or consumers between retrofitted cars and cars designed for alcohol. One observer suggests that a primary source of these bad news stories was taxicab drivers (many of whom, as mentioned, had converted their cars), who passed on disaster stories of (converted) alcohol cars to their customers.[21]

At the same time that these stories of technical problems were circulating, several other concerns were coming to the forefront. It appeared that alcohol supplies would be inadequate to supply the growing number of alcohol cars (because alcohol car sales were much higher than expected), that large profits could be gained by exporting at least limited quantities of ethanol, and that generous credit to new distilleries was creating inflationary pressures on the economy. There was also a growing awareness that ethanol was replacing gasoline, not diesel fuel. As a result, it was necessary to continue importing petroleum in order to produce diesel fuel—with the unwanted effect of also producing unneeded gasoline. The gasoline was being replaced by ethanol and was unmarketable abroad because of its low octane except at discount prices. In 1980, 67 million gallons of gasoline were exported at a discount of about eight cents per gallon. (By 1983 exported gasoline surpluses amounted to over 500 million gallons.) The perception was also growing that the production of alcohol was more expensive than the importing and refining of petroleum.[22]

For the above reasons and also in response to rising world sugar prices, the government began to increase ethanol prices in late 1980, raising them from levels as low as 40 percent that of gasoline toward the 65 percent limit. Consumer demand for alcohol cars began to slacken immediately. As reports of technical problems circulated and as government leaders began making public comments regarding short-term ethanol supply shortages and the questionable economic viability of ethanol fuel,[23] demand for alcohol cars evaporated. A decision in June 1981 to suspend credit subsidies to distilleries confirmed the worst fears of lack of a government commitment. In the seven months between December 1980 and July 1981, alcohol car sales dropped from 73 percent to only 9 percent of new car sales. The public was far more responsive to government signals than had been expected.

RECOVERY AND STABILITY

Throughout 1981 and 1982 the government's involvement with ethanol was reassessed, the result being a renewed and more unified commitment, as will be explained later. To restore consumer confidence, the government increased economic incentives, using tactics such as reducing alcohol prices and taxes on alcohol cars.[24] Promises were made that ethanol prices would be kept at or below 59 percent of those of gasoline for at least two years. Previous incentives were retained, such as favorable credit conditions, lower annual vehicle taxes, and ethanol availability on Saturday.

Taxicab drivers in Brazil, who generally own their vehicles, were given special additional benefits, including elimination of taxes on the purchase of alcohol-fueled taxis; the price of an ethanol taxi was about 50 to 60 percent that of the price of an equivalent gasoline car. (These buyers were required to keep their taxis for three years to avoid resale of the vehicles to the general public, but these restrictions were not enforced. A large but unknown number of people became instant "part-time" taxicab drivers, thereby increasing the impact of the tax breaks for taxicabs.)

The cost of owning and operating ethanol cars once again became attractive. One analysis found that the cost of owning and operating a car in early 1980 was much lower for an alcohol car than for a comparable gasoline car, but that those costs rose substantially in 1981 and then, in response to new government initiatives, fell back in mid-1982 to a level roughly equal to that of operating a gasoline car.[25] While some observers suggest that this analysis overstates the cost of alcohol cars, it nonetheless demonstrates the swings in costs that took place in the early 1980s. The same study found that alcohol cars used as taxicabs were far more attractive than gasoline cars, primarily because of the lower excise taxes placed on alcohol-fueled taxicabs, and secondarily because taxis accumulate more mileage than the average car and thus derive greater benefit from lower fuel prices. Also, according to that analysis, alcohol-fueled taxis cost about one-third less than comparable gasoline-fueled taxis to own and operate during late 1982.

In summary, the initial offering of alcohol cars received an enthusiastic response from consumers. Some buyers were undoubtedly motivated by patriotic reasons (patronizing home-grown fuel and locally developed vehicle technologies), and some presumably purchased the vehicles because they like to be trendsetters. But most people undoubtedly bought the cars because they appeared to be a better investment—from an economic and fuel-availability perspective. The fragility of these motivations became very apparent when sales crashed in 1981.

Since 1982 alcohol fuel production and alcohol car sales have in-

TABLE 9 ETHANOL PRODUCTION IN BRAZIL, 1930 TO 1987

Crop Year[a]	Total (10^6 gallons)	Anhydrous Share (%)[c]	Hydrous (neat) Share (%)[d]
1930/31	9	0	100
1940/41	34	53	47
1950/51	37	20	80
1960/61	121	38	62
1970/71	169	40	60
1975/76	147	42	58
1976/77	176	45	55
1977/78	389	80	20
1978/79	651	84	16
1979/80	898	80	20
1980/81	980	57	43
1981/82	1121	34	66
1982/83	1541	61	39
1983/84	2081	31	69
1984/85[b]	2394	25	75
1985/86[b]	2960	-	-
1986/87[b]	2900	-	-

SOURCE: Compiled by Kahane, "Economic Aspects of Brazilian Alcohol Program," based on data from the Sugar and Alcohol Institute of Brazil.

[a]June-May.

[b]Estimated from unofficial Petrobras statistics.

[c]Anhydrous ethanol contains less than about 1% water (i.e., it is 198 proof) and is used for blending with gasoline.

[d]Hydrous ethanol contains about 4% to 5% water and is used straight (i.e., without blending with gasoline).

creased at a strong and steady pace. Alcohol production rose from 900 million gallons during the 1979–1980 sugar cane season to over 2.4 billion in the 1984–1985 season (see table 9). Total domestic sugar cane production meanwhile increased from 8.3 million metric tons in 1976 to 20.0 in 1983.[26] By the end of 1983, alcohol car sales accounted for 88 percent of new car sales and the market-share percentage was continuing to increase.[27] Ethanol was providing about one-half as much energy as gasoline and that share was increasing rapidly (see table 10). At the end of the 1984–1985 sugar cane harvesting season, 400 fuel distilleries were in operation.[28]

As stated previously, revival of the alcohol program was accomplished by restoring consumer confidence and by providing attractive credit terms to producers. The two initiatives go hand in hand, since each is dependent on the other. Consumer confidence was restored in part by the cap on ethanol prices and reduced vehicle taxes, but also by

TABLE 10 FUELS USED FOR ROAD TRANSPORTATION IN BRAZIL BY PERCENTAGE, 1979–1985

	Diesel Fuel	Gasoline	Anhydrous Ethanol	Hydrous Ethanol
1979	45%	47%	8%	negl.
1980	49	41	.8	2
1981	50	41	4	5
1982	51	37	7	5
1983	51	32	7	10
1984	51	28	7	14
1985	50	26	4	20

SOURCE: Ministerio das Minas e Energia, National Energy Balance (Brasilia: 1986).

evidence of abundant supplies of ethanol that had accumulated when the alcohol car market collapsed in 1981 and 1982. About 935 million gallons were in storage at the end of 1982.[29] They were also assured by new inducements that had been offered in 1982 and 1983 to build additional distilleries. Moreover, it was obvious that low world sugar prices (fig. 4) removed any temptation to abandon alcohol and to divert sugar cane to the sugar export market. Finally, with Brazil facing a severe foreign debt problem in the 1980s, the prospects for imported oil were uncertain (even though world petroleum prices were shrinking). Indeed, fears of gasoline rationing grew in 1983 on the basis of reports that Petrobras was having difficulty financing its oil purchases. Thus by 1983 consumers were confident that ethanol fuel would continue to be available into the future, and at a reasonable price.

Alcohol production received favorable treatment in 1982 and 1983 even though world oil prices were dropping and credit was tightening in Brazil in response to the foreign debt crisis and to conditions attached to emergency financial assistance received from the International Monetary Fund.[30] Alcohol distilleries were exempted from a tax on agricultural production in July 1982, and from reductions in credit subsidies imposed across the economy in 1983. Investment subsidies, less substantial than during earlier periods and unavailable for many other activities, were available for the alcohol sector. About 17 percent of the 1983 monetary budget was said to be allocated to the alcohol sector.[31]

As Brazil's financial situation worsened, support for alcohol fuel increased. Even Petrobras began praising Proalcool; this turnaround included an offer by Petrobras to provide capital and to purchase surplus stocks of alcohol from producers. Strong and widespread support of alcohol fuels was motivated by the perception that increased alcohol production would improve the balance of trade by reducing petroleum imports.[32] While different studies came to different conclusions on this point,[33] what is important to note is the widespread conviction that Proalcool was generating balance of payment benefits for the country.

EPILOGUE

Ethanol production approached 3.2 billion gallons in 1986 and over 90 percent of new cars were powered by ethanol. About 2.5 million of the 12 million vehicles on the road used alcohol.[34] As oil prices tumbled in late 1985 and early 1986, ethanol production became clearly and definitively uneconomic. The government response was to ban construction of new distilleries. In a decision announced February 28, 1986, proposals to increase the price of ethanol from 65 percent to 75 percent that of gasoline were rejected, perhaps temporarily, along with proposals to reduce other subsidies (the price had been raised to the 65 percent level in 1984). While Brazil can be expected to slow its transition to ethanol fuels as long as petroleum prices remain low, it is unlikely that Brazil will reduce ethanol production (and consumption). Brazil has invested billions of dollars in ethanol fuel and created a large and powerful constituency—which includes not only sugar cane growers and distillers but also the estimated 1.7 million new jobs created by the domestic ethanol fuels industry.[35] A powerful inertia has developed that restrains Brazil from turning back to gasoline.

CONCLUSIONS

The Brazilian case demonstrates the perseverance and diligence with which government must act to assist a transportation energy transition. In Brazil the president called for a war economy to rejuvenate the ethanol initiatives at about the same time President Carter was proposing the expenditure of tens of billions of dollars to create a synthetic fuel industry. The Brazilian initiatives were sustained because of a continuing crisis atmosphere—first the Iran/Iraq war, which threatened its petroleum imports, and then the continuing foreign debt problems. In contrast, Carter's 1979 synthetic fuel program gradually atrophied as the crisis faded, as government assumed a more passive role, and as the national consensus for action disintegrated. Government plays a critical role in moving a transition beyond the initial blending stage. Without the stabilizing influence of a coherent public sector, uncertainty in the marketplace will be so great as to discourage investments in nonpetroleum fuels until well after the market sends price signals that a particular production investment is economically attractive.

When and how government should act to initiate and accelerate the transition is problematic. It is the subject of this book. The inherent instability of the world petroleum market and the large role petroleum plays in modern industrialized economies makes the petroleum situation unique; in practice, no one has the knowledge or prescience to identify the optimal timing for a transition. Indeed, Brazil was acting in an ad hoc (though aggressive) manner to resolve difficult economic, political, and energy problems. It was not actively pursuing social or sociopolitical goals. It was not following a strategic plan. Nonetheless, the country shifted in a remarkably short period of time to a transportation fuel dissimilar to petroleum—something no other country has accomplished. In hindsight Brazil erred on the side of starting too soon, but in doing so has provided the rest of the world with important insights regarding the transition to nonpetroleum transportation fuels.

APPENDIX: EVALUATION OF PROALCOOL

The extent to which Brazil's biomass fuel path will prove superior to a path of continued dependence on petroleum is unclear. An examination of this question in terms of economic efficiency, distributional effects, and environmental externalities produces mixed findings.

First, consider the nonmarket effects on air, land and water quality. The effects of monoculture sugar cultivation on long-term soil quality are speculative.[36] The effects on air quality are also uncertain. On the one hand, emissions from process plants are significant, although the effect on air quality is generally small, since the plants are usually located in rural areas. On the positive side, alcohol vehicles have less hazardous emissions than gasoline and diesel vehicles and therefore improve urban air quality, although the effect has not been quantified.[37] Water quality has reportedly been seriously degraded by effluents from distilleries; for each liter of ethanol produced, another twelve to thirteen liters of distillery waste is generated. These wastes have an extraordinarily high biochemical oxygen demand ($BOD_5 = 30{,}000–40{,}000$ mg/l); the wastes from the production of 10 billion liters is equivalent to the sewage generated by about 200 million people. In the early years of the alcohol program, wastes were discharged into rivers and lakes, causing severe degradation, but regulations prohibiting that practice are now enforced;[38] most wastes are now spread on land (reducing the need for fertilizers), though some illegal direct discharges to waterways apparently continued, mostly in the Northeast.[39] In summary, ethanol fuel has had a mixed effect on environmental quality. Greater attention to regulatory and enforcement activities could greatly improve the environmental performance of ethanol fuel, but those improvements would come at some cost.

Distributional effects of the ethanol program tend to be somewhat more negative. Concerns related to the distribution of benefits and costs are important. Because of the already existing large income disparities in Brazil, and because the alcohol fuel program aims to serve the more affluent car-owning population, the distribution of costs and benefits deserves attention. A review of the fairly sparse literature on distributional effects[40] suggests that increased alcohol production has the following distributional effects: (1) an increased concentration of land ownership, (2) an increase in low-skilled employment (compared to the average effect of new industrial investments), and (3) a diversion of some land from food to energy production and a small diminishment of food availability. The food versus fuel issue is unsettling, but may not be important in Brazil because of the vast amounts of prime agricultural land that is still unused.

The critique of income distribution effects of cane-based alcohol activities is inspired by the comparison with cassava-based ethanol production. Those who are concerned with equity effects tend to be strong advocates of cassava alcohol. These advocates point out that cassava, in contrast with sugar cane, grows well under a variety of soil and climatic conditions, is grown primarily by resource-poor farmers, and has no sharply defined harvest seasons, thereby allowing year-round utilization of equipment and year-round employment. Sugar cane production has a five-to-six-month harvest season in Brazil and about two-thirds of the total workforce is seasonal labor.[41] Cassava is also easily dried (for storage and shipment to distilleries) with rudimentary technology. Thus, according to these advocates, the greatest economic opportunities with cassava farming are provided not to a few large plantations, as with sugar cane, but to a large and dispersed group of relatively poor

TABLE 11 COST COMPARISON OF ETHANOL AND GASOLINE, BRAZIL, 1982 U.S.$

Ethanol	Value
Sugar Cane cost	$10-$12/ ton
Ethanol yield	17.2 gallons/ton
Distillation cost	$0.34-$0.42/gallon
Ethanol production cost	$1.00-$1.12/gallon
Replacement ratio for ethanol vehicle	1.2 gallons of ethanol per gallon of gasoline
Ethanol cost	$50-$56/barrel gasoline replaced
Gasoline	
Imported petroleum	$31/barrel (including $2 for transport)
Import surcharge	$6/barrel
Refining gasoline	$10/barrel
Total gasoline cost	$47/barrel

SOURCE: Based on World Bank analysis (see text).

farmers and other rural residents. Initial cassava-based activities have not been successful, however. Small-scale activities have been hampered by the logistical and contractual difficulties of assuring a steady supply from numerous small farms, and by the low yields (and correspondingly high production costs) associated with small farms—problems that result from poor organization and poor knowledge of cassava crop production.[42] Likewise, large scale monoculture plantations of cassava in Brazil (and elsewhere) have also been failures, largely because of persistent problems with diseases that rarely affect smaller plots.[43]

A definitive economic analysis is confounded by changing exchange rates and internal currency adjustments, complex subsidy arrangements, uncertain opportunity costs for land and capital, widely varying costs from one region to another, and foreign debt considerations. As suggested in note 22, various cost estimates have been made with varying assumptions. A representative analysis is presented in table 11; it is in rough agreement with an analysis done by the World Bank.[44] The analysis in table 11 eliminates all subsidies but includes an import surcharge of 20 percent on petroleum that the government applied to all companies using foreign currency to purchase foreign goods.

Table 11 shows that when oil was priced at $29 per barrel, ethanol fuel production was somewhat more expensive than petroleum imports (especially in the Northeast, where production costs were reportedly

$10 per oil-equivalent barrel higher than elsewhere). Clearly, when oil prices dropped to $15–$20 per barrel in 1986–1987, ethanol production became uneconomic. At those low prices, ethanol investments were not economically viable, even excluding sunk capital costs (which account for about 25 percent of total production costs).

It is still not possible to determine whether the ethanol initiative was a mistake. The ethanol program has provided indirect benefits to the economy of increased employment and industrial investment and has helped reduce the cost and risk of depending upon foreign oil. However, in the mid-1980s, even with those benefits, the ethanol program must be considered an economic failure. But who can say if this evaluation will still be valid in the 1990s?

FIVE

New Fuels in New Zealand, South Africa, and Canada

This chapter has three parts. First, it reviews the use of compressed natural gas and synthetic gasoline in New Zealand, second, the production of liquid fuels from coal in South Africa and, finally, the experience with oil sands and compressed natural gas in Canada.

NEW ZEALAND: SYNTHETIC GASOLINE AND CNG

New Zealand and Brazil have been the most adventurous and aggressive countries in the world in replacing petroleum transportation fuels. As of 1986, about 10 percent of "gasoline" demand in New Zealand was met by compressed natural gas, about 3 percent by liquefied petroleum gases, and about 35 percent by synthetic gasoline. New Zealand, more than any other country, has promoted a multiplicity of alternative fuels. New Zealanders were motivated to do so by their abundance of natural gas and dearth of petroleum. While they acted hastily and without a clear choice of energy path, their experience nonetheless provides unusually rich insights into the risks and barriers confronting new fuels.

New Zealand is one of the most isolated countries in the world. It comprises two main islands stretched 800 miles from tip to tip about 1,200 miles southeast of Australia. Only 3.3 million people (1986) live on the sparsely populated islands. The population is well educated and has a per capita income of $9,000 per year. The economy, in transition from a dependence on agriculture to a more diversified economy, has been stagnant since 1973, when it suffered the double blow of high oil prices and loss of favored trade status with Great Britain (a result of that country's entry into the European Economic Community).

New Zealand moved faster than any other country in the world in introducing alternative fuels. There are two principal explanations for its quick response: a long history of intervention by the centralized government in economic activities and a strong independent "do-it-yourself" tradition. Thus, when high oil prices in 1973 and again in 1979 threatened the security and economy of the nation, there was strong support for a vigorous response.

NATURAL GAS—THE KEY RESOURCE

The country is blessed with abundant energy resources—except for petroleum. It has abundant hydropower, biomass, geothermal energy, coal, and natural gas (see table 12). Various investigations in the 1970s indicated that the most attractive of these resources for transportation energy production was natural gas.[1] Maui, the first and only large natural gas field discovered in New Zealand, was discovered in 1969 and developed jointly in the mid-1970s by the government-owned oil com-

TABLE 12 ENERGY RESOURCES IN NEW ZEALAND

Resources	Quantity (10^{12} Btu): Proven Reserves	Quantity (10^{12} Btu): Indicated and Inferred Reserves[a]
Coal	6650	83,200
Natural Gas		
Maui	5300	--
Other Fields	460	--
Natural Gas Liquids	870	--
Oil Shale	---	200-4000
Petroleum	50	--
Biomass[b]		
Forests	60-80/yr	200-250/yr
Agriculture	4-9/yr	6-15/yr
Waste	minor	minor

SOURCE: International Energy Agency, Synthetic Fuels (Paris: IEA, 1985), p.12.

[a]See chapter 7 for definition of terms.

[b]Quantity available for transportation energy production.

pany (51 percent) and a private consortium (49 percent). The government signed a "take-or-pay" contract with the gas field owners. Whatever the motivation may have been for signing such a contract, the result was that new markets had to be created quickly. The largest single investment was a synthetic fuel plant that converted the natural gas into methanol and then into gasoline. Other smaller projects were built including a plant that produces chemical methanol, mostly for export, and another that produces ammonia-urea fertilizer. The natural gas was also used for electricity generation and in compressed form as a transportation fuel.

ENERGY POLICY

The stated goal of the government has been to gain energy self-sufficiency by developing indigenous energy resources and substituting them for imported oil.[2] A principal motivation of this goal was economic development: to reduce the cost of energy to the economy and to create the potential for export products. The following statement by the governing political party in 1981 summarizes the government's policy at that time:

> The oil shocks of the 1970s brought home to us how vulnerable we are. In 1973, oil cost us $93 million, this year a massive $1,500 [million], over 30 percent of our import bill. While we remain dependent on imported oil, the oil producing countries effectively undermine our level of economic independence. The National Party believes this is quite unacceptable to the people of New Zealand. Our aim is to ensure an orderly transition from an economy dependent on import oil to an economy relying on our own diversified energy resources. The long term goal is for New Zealand to be totally self-sufficient in energy by the turn of the century.[3]

This goal of being self-sufficient by 2000 proved to be disastrous. It led to preference being given to large projects that would have immediate impacts. It encouraged synthetic gasoline and discouraged CNG and methanol.

The New Zealand government, in addition to participating financially in most energy activities, also directly controlled most energy prices until 1987. The government regulated prices for gasoline (including synthetic gasoline), diesel fuel, natural gas, and other fuels by determining an allowable rate of return on investments. Since January 1987 there has been a move toward a totally deregulated energy system.

SYNTHETIC GASOLINE: A BLUNDER?

In 1979 the government decided to invest in what for New Zealand was a huge project—a $1.475 billion plant to convert natural gas into gasoline. Construction began in 1982 and the plant began operating in 1985. The plant was built on time and under budget. It is a joint venture between the New Zealand government (75 percent) and Mobil Oil (25 percent). The plant produces about 14,000 barrels per day of gasoline.

The plant is the first (and perhaps last) of its kind. The basic process consists of, first, converting natural gas into methanol using standard technology (described in chap. 6). The methanol is then passed through a special catalyst that converts the methanol into hydrocarbons that resemble gasoline. This second stage of the process, the conversion of methanol into gasoline, had only been tested in a small four b/d pilot plant, but there were few technical uncertainties since the full-size plant replicated the depth of catalysts used in the pilot plant.[4] It was expected that the project would be a huge financial success—and it would have been if the price of oil had continued to increase in the 1980s. Cost estimates at the time indicated that if crude oil prices averaged $28 per barrel in 1980 dollars over the life of the project, then the project would be commercially successful. In 1987 world oil prices were about $18 per barrel.[5] The government was losing hundreds of millions of dollars a year.

What is equally interesting is their decision to build that particular type of plant. In effect, New Zealand decided to produce a high-octane fuel, methanol, and then to degrade it by producing at additional cost a more hazardous, lower octane, and higher polluting fuel. The government decided that the disadvantages of producing a more costly, lower octane, and dirtier fuel was preferable to having to modify the nation's vehicle fleet and fuel distribution infrastructure to handle methanol. In some sense the decision to produce synthetic gasoline instead of methanol was a lack of political will and imagination and a single-minded devotion to the goal of rapid replacement of petroleum.

New Zealand was limited in its options, however. The country does not have its own auto industry, as does Brazil, or widespread expertise in automotive engineering. Its domestic automotive market is tiny—about 90,000 new automobile sales per year—and thus it had limited leverage in convincing multinational automobile manufacturing companies to produce methanol cars.

This dilemma of choosing between, on the one hand, lower quality and more expensive fuels that are compatible with existing vehicle technology and fuel distribution infrastructure and, on the other hand, higher quality and less expensive fuels that are not compatible with existing vehicles and infrastructure is one of the central questions addressed in the remainder of this book. It is a question of how and where technologies are developed and adapted.[6]

CNG

The decision in New Zealand to promote the use of compressed natural gas (CNG) indicates that in this case the country indeed was willing to undertake the task of building a new fuel distribution infrastructure and modifying motor vehicles. The reason New Zealand was willing to promote CNG but not methanol is that the cost of retrofitting a vehicle for CNG is somewhat less than it is for methanol, there was considerably more experience worldwide with CNG conversions (mostly in Italy) than with methanol conversions, and a gas pipeline network was already in place. In terms of infrastructure, the only cost was for establishing a network of fuel outlets.

So even though there was no auto manufacturer ready to design and build production-line CNG vehicles for New Zealand, the government was convinced that after-market retrofits were an acceptable and feasible option. CNG has gained a larger market share in New Zealand—10 percent in 1986—than in any other country. It is used only on the North Island where the Maui gas field is located. The natural gas pipeline distribution system has not been extended to the less populated South Island.[7]

The government began aggressive promotion of CNG in 1979.[8] Its first actions were to work with industry to set up standards for vehicle conversions and refueling stations, to establish training courses and certification procedures for mechanics, to convert post office and other government vehicles, and to provide grants to CNG refueling stations to cover 25 percent of the cost of the equipment. After the number of CNG conversions began to falter in mid-1980, the government announced a package of incentives to stimulate additional conversions. These incentives included a grant of $135 to motorists for each conversion (all values are in U.S. dollars at the prevailing exchange rate), reduction of road user taxes for CNG, and extension of the 25 percent grants to cover other refueling station expenses. At the time a goal was set of 150,000 conversions by 1985. The government also committed itself to a publicity and information dissemination campaign.

Later, additional incentives were provided. The government offered loans of $350 to $3,000 to vehicle owners for each conversion at a subsidized interest rate of 10 percent with no down-payment requirements. The government also committed itself to keeping the price of CNG at half or less that of gasoline into the foreseeable future. One other important incentive was an offer from the gas utilities of $200 worth of vouchers for the purchase of CNG fuel to each person who converts his or her vehicle to CNG.

The cost of a conversion was about $1,500 N.Z., which was equivalent to $1,000 U.S. in 1981 and $750 U.S. after a 1984 devaluation of the currency. This initial cost could be quickly repaid in fuel savings. For most motorists, the repayment period was one to two years. Given the economic advantages of CNG, it is instructive that conversions did not occur at a faster rate.

The number of vehicles converted to CNG each year increased steadily as shown below:[9]

1979	1,600
1980	4,700
1981	10,500
1982	15,700
1983	19,000
1984	25,000
1985	25,000
1986	25,000

By the end of 1986, over 120,000 vehicles had been converted, representing about 10 percent of light-duty gasoline vehicles, and about 300 refueling stations were supplying CNG.

Motorists in New Zealand adopted CNG more slowly than Brazilian consumers adopted ethanol. The explanation has to do with less fuel availability, higher initial cost outlays by consumers, less commitment by the government, and less similarity of the fuel to gasoline—almost all Brazilian fuel stations provided ethanol while only a small proportion of stations carried CNG in New Zealand; CNG conversions required a large initial investment while ethanol cars cost about the same as gasoline cars; and ethanol is more similar to gasoline than CNG and therefore more acceptable to consumers. These differences and their importance in the transition process will be explored in depth later in the book.

CONCLUSIONS

Just as in Brazil, it was the government of New Zealand that led the transition to nonpetroleum fuels. In the case of the gas-to-gasoline plant the result was a financial (but not technical) disaster. In the case of the CNG initiative, the government's financial risk was much smaller, and less petroleum was replaced, but the government's monetary outlays were also much smaller since the CNG was being sold at market prices. The government was providing only modest subsidies to owners of CNG vehicles and CNG refueling stations. From a technology change perspective, the CNG experience suggests that it is possible to introduce fuels that are very different from petroleum on a small-scale incremental basis—but only if the consumer sees substantial economic benefits. In the New Zealand case, greater emphasis was given to synthetic gasoline than CNG (or methanol) because of contractual obligations to purchase large amounts of natural gas and the goal of achieving self-sufficiency. The conversion of vehicles to CNG and methanol is a slower and more complex process than simply substituting synthetic gasoline into the fuel stream. In hindsight, New Zealand probably emphasized the wrong fuel in its haste to develop its gas resources and attain self-sufficiency.

SOUTH AFRICA: LIQUIDS FROM COAL

Since the 1950s the Republic of South Africa has been producing gasoline- and diesel-like fuels from coal. Because the fuels are essentially identical to petroleum fuels, no technical changes in fuel distribution infrastructure or motor vehicle technology are necessary. What is remarkable about the South African program is the large scale of the investments and the significant improvements that have been made in improving the efficiency of a basically inefficient conversion technology. This review of South Africa's coal liquids activities is brief because the South African government releases very little information regarding domestic energy activities.

South Africa has huge supplies of coal but practically no petroleum. The country depends upon coal for over 80 percent of its energy needs and imported petroleum for most of the remainder.[10]

Soon after World War II the South African government began to create an oil-from-coals industry. In 1950 it formed the Sasol company using government funds. The company subsequently sold stock on the open market; 70 percent is now owned by the private sector and 30 percent is owned by the central government.

The first small plant built by the company was opened in 1955. It uses the Fischer-Tropsch process developed decades earlier in Germany. The plant produces a wide range of products for use as fuels, chemicals, and fertilizers. At full production, it yields about 3,800 oil-equivalent barrels per day. This plant had many technical and economic problems. But as the threat and then the reality of an international petroleum embargo emerged (as a protest against South Africa's apartheid policies), the government remained committed to the project, even though it was suffering large financial losses.

When oil prices increased in 1973–1974, the government decided to build a much larger second plant, followed in 1979 by an identical third plant. The two plants proved to be very expensive, costing about $6 billion to build.[11] Sasol II reached full production in 1982 and Sasol III in 1985. The two plants together produce about 70,000 oil-equivalent barrels of fuel per day.

South Africa is not willing to make any information available on production costs, which is unfortunate because the Sasol technology has been in operation far longer than any other modern technology that produces liquid or gaseous fuels from coal. Engineering experts familiar with the Sasol plants note that with constant experimentation and engineering modifications the South Africans have significantly improved the efficiency of the plants. Unsubstantiated claims have been made that the two more modern plants produced gasoline in the early 1980s for a little over $1 per gallon (devaluation of the South African currency in the mid-1980s more than doubled the cost as measured in U.S. dollars).

Whether or not the costs were reduced to that level, the important observation is that over time the South Africans apparently were able to make significant improvements in the efficiency and design of the plants. As shown in chapter 8, these benefits of "learning by doing" tend to be substantial for new conversion processes, but as suggested in South Africa, it takes decades of continuing engineering modifications for those learning benefits to be fully realized. There is no substitute for actual working experience with full-scale production plants.

The long lead time for developing an efficient engineering design is critical because it reduces the ability to respond quickly to an energy crisis. Even though no other country is likely ever to use the Fischer-

Tropsch process for converting coal into liquid fuel, this lesson of slow incremental design improvements should not be ignored.

CANADA

Because of its cold climate, immense distances, and high income, Canada is the largest per capita user of oil in the world. About 1.5 million barrels of oil are consumed each day. Fortunately for Canada, it is richly endowed with abundant supplies of energy resources. It has large amounts of natural gas, petroleum, heavy oil, oil sands, coal, biomass (especially forests), and hydropower. Petroleum reserves and production, although they are substantial, have been diminishing since 1970, except for a brief recovery in the 1980s. A study by the government of Alberta, where 85 percent of Canada's oil is produced, predicts that conventional reserves in that province will drop from 3.16 billion barrels in 1985 to 1.04 billion in 2007.[12] Canada was a net oil importer until 1983, when oil exports from western Canada exceeded imports to eastern Canada. The resurgence in the 1980s was due in part to increased production of fuels from oil sands and heavy oil.

Dating back to its 1980 National Energy Program, the Canadian government has devoted considerable effort to developing coherent energy strategies. Its national energy policy in the 1980s has been based on three fundamental principles:[13] security of supply and ultimate independence from the world oil market, greater participation by Canadians in energy industries, and equitable distribution of energy benefits and costs. The federal and provincial governments have participated actively in energy activities, more so than in the U. S., as equity holders and regulators, and through incentives and subsidies.

This high level of intervention lessened somewhat in the mid-1980s when oil prices decreased and a conservative government came to power. Even so, provincial governments have continued to aggressively promote the development of local resources, oil sands (and heavy oil) in the case of Alberta, and compressed natural gas in the cases of Quebec and British Columbia. Coal has been virtually ignored as a source of transportation energy at all levels, and interest in biomass has receded as oil prices dropped. Alcohols from natural gas, coal, and biomass have received relatively little attention in Canada, except in blends with gasoline, partly because of the inherent difficulty of starting alcohol engines in very cold weather.

COMPRESSED NATURAL GAS

Canada's huge reserves of natural gas dwarf domestic demand. At 1986 production levels, natural gas reserves would last thirty years. As a result, the country has sought to introduce natural gas into the transportation market so that the displaced petroleum could be used to reduce imports or expand exports. Initially, as part of the 1980 energy plan, liquefied petroleum gases (LPG) were given greater priority than compressed natural gas (CNG) as a transportation fuel. That was because LPG is more similar to gasoline than CNG and therefore faced fewer technical obstacles. Indeed, propane has become an important fuel in Canada; about 125,000 vehicles have been converted to LPG (70,000 received $300 grants) and about 5,000 refueling stations (public and private) have been established. But, as elaborated upon in later chapters, LPG (which comes from the light components of petroleum and heavy components of natural gas) has limited potential because it constitutes only a small percentage of oil and natural gas production.

A large number of incentives were instituted to stimulate the use of CNG.[14] The federal government offered a grant that covered 50 percent of the capital costs of a refueling station up to a maximum of $50,000 and $500 for each vehicle converted to (or built for) CNG. The Quebec, Ontario, and British Columbia provincial governments exempted natural gas vehicles from road taxes, and Ontario eliminated the sales tax (up to $100) on new CNG vehicles and CNG vehicles converted within one month of purchase. Also, natural gas companies in those same three provinces offered grants and/or low interest or interest-free loans for vehicle conversions, and various attractive financing arrangements to refueling stations.

As of early 1987 about 13,000 natural gas vehicles were operating in Canada and CNG fuel was available at 148 retail refueling stations.[15] Ninety-five conversion centers were registered with the government. About one-half of the vehicles and stations were in British Columbia (mostly in the Vancouver area). In 1985 about one-fifth of the vehicles were owned by individuals, and the remainder mostly by various business fleets.[16] About half the vehicles were light-duty trucks; 7 percent were school buses and most of the rest were autos.

The price of natural gas was controlled at 85 percent of that of crude oil from 1975 to 1981 (on an energy basis) and at 65 percent from 1981 to 1985. Since then the price has been gradually decontrolled, with prices staying well below the 65 percent level.

As will be shown in later chapters, CNG is highly attractive from a societal perspective—it displaces a more valuable fuel, it is clean burning, and it is safe. Recognizing these virtues, the activist Canadian governments have sought to promote this fuel. But consumers and vehicle fleet operators have been slow to convert their vehicles, despite major financial benefits. The reasons for this slow market penetration will be explored in chapter 14.

OIL SANDS

Over 40 percent of all oil sands in the world are located in Canada, almost all of it in the Province of Alberta, mostly in the Athabasca deposit. The Athabasca deposit covers 16,000 square miles and contains about 1.1 trillion barrels. Of this, about 65 billion barrels are estimated to be recoverable with existing technology at oil prices of up to $28 per barrel.[17] About 200,000 barrels per day of liquids were being produced from oil sands in the mid-1980s, representing about 10 percent of domestic oil consumption.

Oil sands are a mixture of sand grains, water, and bitumen. The bitumen is a member of the petroleum family and can be processed into petroleum products. Recovery of the bitumen can be a difficult and expensive process, and for that reason oil sands are categorized as an alternative energy resource. It is easiest and least expensive to extract oil sands that are close to the surface (within about 150 to 250 feet); in this case the ground cover above the bitumen is scraped away and large surface mining equipment scoops out the oil sands mixture and places it on a conveyor system that delivers the mixture to adjacent upgrading facilities. There the bitumen is separated from the sand and other impurities and upgraded into a refinable petroleum-like liquid. About half of the 65 billion barrels of recoverable bitumen is extractable with surface mining techniques. As of 1985, 183,000 barrels per day of oil were being produced from two surface mining operations in Athabasca.[18]

One of the operations, Suncor, is 75 percent owned by the Sun Oil Company of the U.S. and 25 percent owned by the Ontario provincial government. It began operations in 1967 and now produces about 57,000 barrels per day. The other facility, Syncrude, is owned by a joint venture of eight Canadian and U.S. companies and the Canadian and Alberta governments. It produces almost 130,000 barrels per day and has been operating since 1978. Several other projects have been proposed and reached advanced stages of design but have been suspended or postponed because of falling oil prices. A 130,000-barrel-per-day multibillion dollar project was abandoned in 1982 by Shell, a 140,000 b/d project was canceled the previous year, and several smaller projects were suspended in 1986.

Deeper deposits of oil sands are and can be extracted using various in-situ processes. These processes are the subject of considerable research and development. Small quantities of bitumen are now extracted using a simple in-situ process in which the bitumen is softened with injections of steam and then pumped to the surface of the oil sands deposits using conventional wells. The largest oil sands project using conventional oil wells began operation in 1985 with production of 19,000 barrels per day, with plans to expand to 75,000 barrels.

The cost of producing bitumen at existing oil sands facilities and upgrading it to crude-oil quality has been steadily reduced over time. Initially, both plants had numerous technical problems that kept production levels down and operating costs high. Industry experts estimate that production costs in the mid-1980s were about $11 to $15 per barrel.[19] The product's value is a few dollars per barrel less than light crude oil. Thus at oil prices above $20 per barrel or so, many of the existing operations are still profitable. Higher prices will probably be needed to elicit further investments, however.

Most of the bitumen produced in Canada is mixed with natural gas liquids and shipped to petroleum refiners in pipelines. Only refineries equipped to process heavy petroleum can handle this bitumen; many of these refineries are located in the U.S.

The future of oil sands is mostly dependent on world oil prices, but it also depends on the willingness of the Alberta government to support the industry. Alberta owns the mineral rights to almost all the oil sands and collects royalties on all production. In the past it also has controlled the price of the oil sands. If world oil prices stay low, the Alberta government has the option of reducing or eliminating royalty payments, dipping into its large cash reserves to help out, and regulating the price of oil at an artificially high level.

REFLECTIONS AND QUESTIONS

Although, as argued earlier, the transportation sector is resistant to new fuels, we have seen in the last two chapters that there are some instances in which nonpetroleum transportation fuels indeed have made significant inroads. Over one-third of all gasoline in New Zealand is manufactured from natural gas, a significant proportion of South Africa's diesel and gasoline fuel is made from coal, and about 10 percent of Canada's oil consumption is provided by oil sands. Even more impressive, in terms of barriers that were overcome, are Brazil's success in replacing over half of all its gasoline with ethanol and New Zealand's success in introducing CNG. So examples do exist of nonpetroleum fuels penetrating transportation markets, but they are the exceptions rather than the rule. Even so, these experiences raise some important and troubling questions that must be addressed in the very near future.

1. Which fuels are most attractive—economically, environmentally, politically—and most acceptable to the consumer?

What are the national security effects of replacing petroleum imports with natural gas or methanol imports? Are environmental reasons sufficient to justify an accelerated introduction of clean fuels? How does one determine the economic attractiveness of alternative fuels when there are so many indirect effects and when the future price of petroleum is so uncertain? Even if we as analysts or policymakers determine that a particular choice is attractive and urgent, we still do not know how consumers will value the different attributes associated with different fuels and engines, and under what conditions they would be willing to purchase them. We need a better understanding of consumer utility in order to choose, for example, between synthetic gasoline and methanol. What makes this choice process so difficult is that the preferred choice will not be uniform—it will differ from one region and one social group to another.

2. What is the appropriate role of government?

What role should or could government play in introducing new vehicles and fuels? Who should receive R&D funding and for what? What government rules and regulations inhibit the introduction of new fuels and engines (e.g., emissions testing, fuel quality, safety, antitrust rules against interindustry collaboration)? What actions could government take to accelerate the introduction? What would be the costs, political and social as well as economic, of taking those actions? How does the role of government in the fuel transition process vary from one country to another?

3. What planning is necessary—when and how should the transition begin?

Under what conditions are individuals likely to purchase a nonpetroleum vehicle? Under what conditions would a fleet owner purchase nonpetroleum vehicles? Which types of fleets are more likely to purchase nonpetroleum vehicles? What are the perceived disbenefits to the vehicle owner and operator of a limited network of retail fuel outlets? Under what fuel availability conditions would motorists be willing to purchase a nonpetroleum vehicle? What would be the most cost-effective strategy for building a network of retail fuel outlets? Which strategy is more effective and less costly: to subsidize or mandate the introduction of dedicated (i.e., fuel-specific) nonpetroleum vehicles, or to subsidize or mandate the establishment of a network of fuel outlets? What are the most effective policies and actions for reducing consumer and industry uncertainty? What are the costs of carrying out those policies and actions? What can be done to encourage those countries that have abundant natural gas, coal, hydropower, and biomass to accelerate their transition? What would be the benefits to other countries in terms of reduced world oil demand and lower petroleum prices? What technology transfer and technology development problems must be overcome in order to introduce alcohol or natural gas vehicles in developing countries? What are the economic costs to the nation of initiating a transition too late, or too soon? What effect would the development of a viable nonpetroleum alternative have on world petroleum prices? What government actions would be most effective, inexpensive, and acceptable for initiating a transition?

These questions will be explored throughout the book. In some cases definitive answers will not be provided—either because there is too much uncertainty or because further research is needed. But many of the questions can and will be answered.

PART III

Feedstocks and Fuel Production

The list of options for replacing petroleum in the transportation sector is long. These options may be categorized as follows:

Electricity (battery vehicles and electrified roadways)
Fuel cells
Petroleum-like liquids from coal
Petroleum-like liquids from oil shale and oil sands
Methanol and natural gas fuels from natural gas
Methanol and gaseous fuels from coal
Methanol and ethanol from biomass
Hydrogen.

As will be expanded upon in chapter 12, electric-powered vehicles are not likely to play a major role in future transportation systems. Battery vehicles may become popular for certain specialized uses; but their short driving range and high cost will continue to make them uncompetitive in the marketplace and unacceptable to most drivers. Electrified roadways, in which vehicles gain all or some of their energy from circuits within the pavement, would require tremendous capital outlays, create major disruptions, and exacerbate the peaking problem of electricity suppliers.

An alternative technology for using electricity in vehicles with a much higher ratio of power output to weight is fuel cells. It is a technology that was improved for use in the U.S. space program. It converts hydrogen, methanol, or some other fuel into electricity. It produces negligible emissions and is more efficient than the internal combustion engine, but it is also bulkier and many times more expensive. Fuel cells will not be price competitive with other alternative fuels for highway transportation in the foreseeable future. In any case, since the most likely fuels for fuel cells are methanol and hydrogen, which are addressed in these next few chapters, further discussion of fuel cell technology is postponed to the chapter on motor vehicle technology.

All other options for supplying transportation energy are linked to the internal combustion engine. They are the most viable transportation energy options at least for the next 75 years and probably much longer. They are the subject of the next four chapters.

Transportation fuels may be made from a large number of plentiful materials. Possible feedstocks include grain, sugar crops, wood, grass crops, crop residues, municipal solid waste (garbage), coal, oil shale, oil sands, natural gas, and practically any other organic material, including petroleum. Many of these materials are renewable; others, such as oil shale and coal, exist in such large quantities that they could supply our energy needs for many hundreds of years. But the extracting, harvesting, and processing of these materials carries a cost—economic, social, and environmental. No one feedstock or production technology is inherently superior in all situations.

A review of past choices is not necessarily a good guide for the future. Past choices were often based on distorted and now out-of-date political conditions, on inaccurate perceptions of the state of production and end-use technologies, and arbitrary assessments of future risk. Making informed energy choices is a difficult challenge because the best choice for one region is not necessarily the best choice for another. Specific circumstances in a particular region will determine which option is most attractive. Considerable empirical knowledge, much of it location-specific, is necessary to make the best choice.

The next four chapters are devoted to generating that base of knowledge. An analysis of the fundamental features of production options using renewable feedstocks is given in chapter 6; those of mineral-based options are addressed in chapter 7 (see fig. 6). In chapter 8 the single most important consideration, economic cost, is specified for each set of options, and in chapter 9 other important production considerations are addressed. Environmental impacts are considered in chapters 15 and 16.

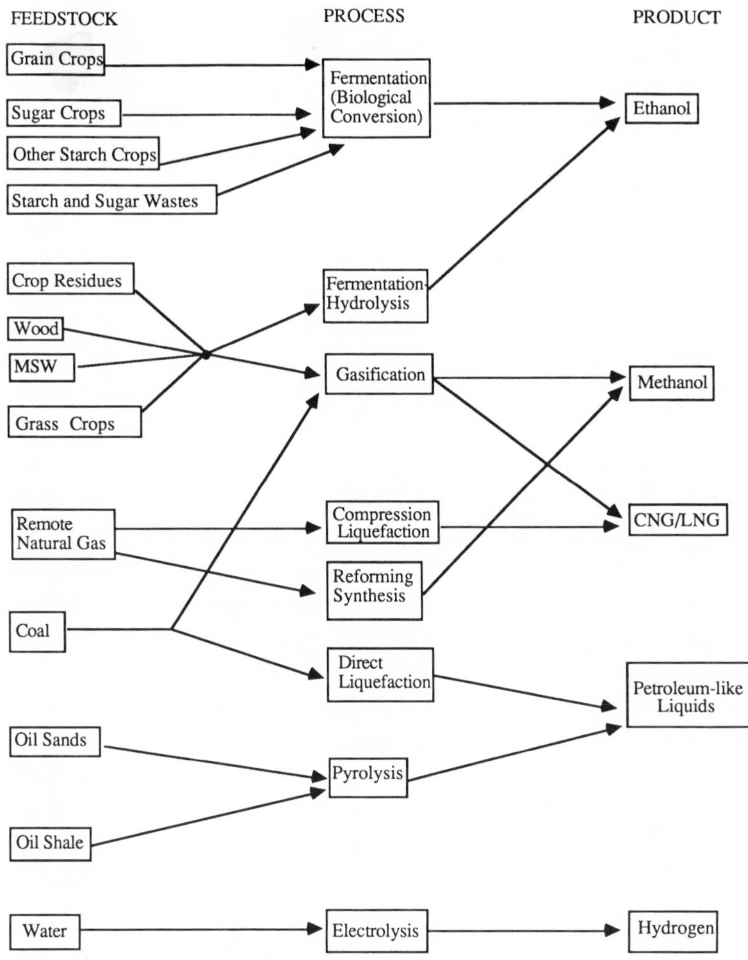

Figure 6. Feedstock-Production Systems

On the basis of the material presented in the next four chapters, one begins to gain an appreciation of the difficulty of comparing different options. For instance, the decision to pursue small-scale decentralized biomass fuel activities is fundamentally different from the decision to pursue large-scale centralized oil shale and coal projects.

Before proceeding, I caution once again that this book is not a taxonomic treatment of all fuel options. Rather, it addresses those energy options with the greatest potential for replacing petroleum. One option that will not be addressed further is the extraction of oil from plants for use as a substitute fuel in diesel engines. A fairly large number of plants produce oil, including peanuts, cottonseed, soybean, palms, and sunflowers. Plant oils ignite relatively easily when compressed and therefore are better suited to diesel engines than are alcohols. Over 250 million gallons (16,000 b/d) of plant oils are produced annually in the U.S., none of it for fuel. The high value of these oils for cooking—over $2 per gallon in the U.S. for the least expensive plants oils—renders them economically unattractive for fuel production. Some exceptional situations and market niches may exist where using plant oils as fuels is attractive, such as in remote areas in developing countries where there is no local cooking-oil market and where certain plants with high oil yields such as African Palms grow particularly well. But since plant oils do not have the potential to be a major source of transportation energy, they are not considered further.[1]

Other options not considered further are liquefied petroleum gases (LPG) and methane from sewage and manure. LPG is composed primarily of propane (C_3H_8) and butane (C_4H_{10}), the lighter hydrocarbon compounds in petroleum and the heavier compounds in natural gas. These hydrocarbons are extracted from petroleum at the oil refinery and from natural gas before it is put in pipelines. LPG is widely used in a variety of stationary and vehicular applications, but its production is permanently linked to local production of petroleum and natural gas and is limited to a small percentage (on an energy basis) of oil and gas production. Manures and sewage may be used to generate methane gas (CH_4), the principal constituent of natural gas, and then compressed for use as a transportation fuel, but the total production potential is negligible compared to other options addressed in the next two chapters.[2]

In summary, the next few chapters investigate in detail the most important transportation fuel options.

SIX

Biomass and Hydrogen Fuels

Two types of renewable fuels may be attractive for motor vehicles: fuels made from biomass and hydrogen made from water with solar energy. Hydrogen can be produced from a large number of materials, not just water. Hydrogen has several disadvantages, principally high production cost and low energy density, but it has the advantages, when produced from water using photovoltaics for the energy required in the manufacturing process, of being produced from renewable feedstocks using a very clean production process and of being a very clean-burning fuel. Hydrogen will not be an attractive option at least until well into the twenty-first century, however, although it may emerge as the dominant fuel thereafter.

The bulk of this chapter is devoted to renewable fuels that already are commercially used and that have much more potential in the near- and medium-term future—fuels made from biomass. Hydrogen is addressed at the end of the chapter.

"Biomass" refers to materials of biological origin, including plants associated with agriculture, forestry, and aquaculture, as well as plant material that has been processed, such as paper and food-processing wastes. In many countries the potential exists to produce very large quantities of fuel from biomass. As shown in figure 7, countries that might find biomass energy attractive are those with surplus agriculture production and insufficient domestic production of transportation energy. Countries located in the upper left quadrant of figure 7, including the United States and Brazil, meet the initial criteria. Other countries that continue to provide subsidies to their agricultural sectors, even as

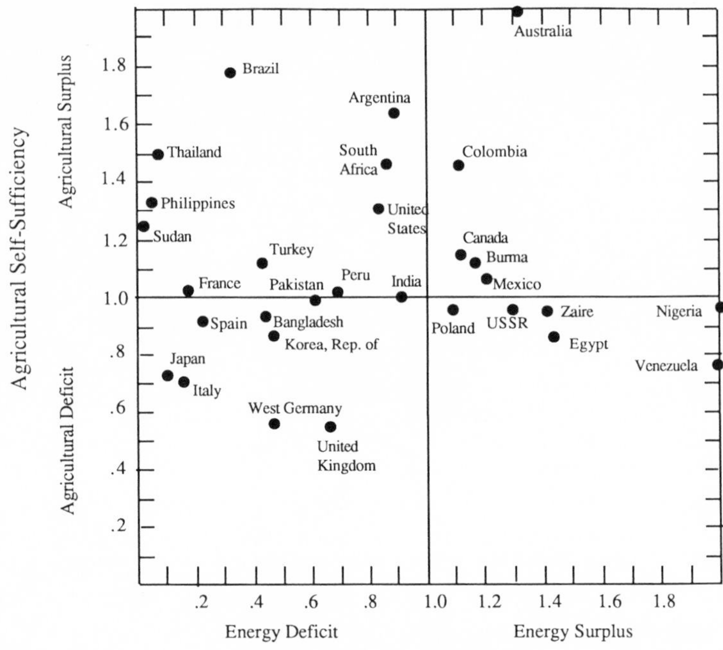

Figure 7. Energy and Agricultural Self-Sufficiency Ratios for Selected Countries, 1976

Source: World Bank, *Alcohol Production from Biomass in the Developing Countries* (Washington, D.C.: 1980), p. 51.

agricultural productivity continues to increase and populations stabilize, can be expected to move into this upper left quadrant.

An analysis of land use in the United States provides a more vivid indication of the potential to grow biomass for energy. As shown in table 13, 90 percent of the land in the forty-eight contiguous states is devoted to agriculture and forests. Much of the land, especially the forest, pasture, and range land, is poorly managed and/or lightly utilized.

Clearly, then, opportunities exist to produce large amounts of biomass fuel for transportation—but those opportunities have a cost, economic as well as environmental. The central question is: Do the benefits of renewable biomass fuels outweigh the costs? In most but not all situations the answer appears to be no. This chapter investigates biomass feedstock opportunities and provides a technical overview and assessment of production technologies for converting biomass into

TABLE 13 LAND USE IN THE CONTIGUOUS FORTY-EIGHT STATES OF THE U.S.

Land Use	Thousand Square Miles	Percentage
Commercial forest	765	26
Noncommercial forest	360	13
Harvested cropland	470-550	16-19
Idle cropland	50-240	2-8
Pasture and rangeland	780-940	26-32
Other (urban, wetlands, preserves, etc.)	310	10
Totals	2970	100%

SOURCE: U.S. Congress, Office of Technology Assessment, Energy from Biological Processes, Vol. II (Washington, D.C.: GPO, 1980); John Ferchak and E. Kendall Pye, "Utilization of Ethanol Fuel as a Gasoline Replacement," in U.S. Congress, Joint Econ. Comm., Subcomm. on Energy, Farm and Forest Produced Alcohol, 96th Cong., 2d sess. (Washington, D.C.: 1980), p. 38.

transportation energy. It provides the background for answering the question: "When and where do the benefits outweigh the costs?" Later chapters will elaborate upon and compare the economic, environmental, and various societal implications of biomass fuels with other options, providing a more complete evaluation of biomass fuel options.

The first part of this chapter is an analysis of the simplest and presently the most common means of producing a nonpetroleum transportation fuel—the fermentation and distillation of starch and sugar materials to produce ethanol. In most situations it is also one of the most expensive means of producing a nonpetroleum transportation fuel. The second part of the chapter investigates a biomass option that should prove much more attractive—the conversion of cellulosic materials into alcohol (either methanol via thermochemical processes or ethanol via hydrolysis).

BIOLOGICAL CONVERSION

The technologies for fermenting and distilling starch and sugar materials have been known for thousands of years. Current ethanol fuel production in Brazil, the U.S., and elsewhere relies exclusively on this process. Fermentation can be represented as follows:

$$C_6H_{12}O_6 \rightarrow 2C_2H_5OH + 2CO_2 + \text{heat}$$
Sugar Ethanol

Sugar is fermented into a "beer" that contains dilute ethyl alcohol (ethanol) (C_2H_5OH) in water. The ethanol is removed from the beer by vaporizing and condensing it, a process referred to as distillation. The distilled ethanol is about 190 proof (95 percent alcohol, 5 percent water) and an excellent fuel for engines adapted to it. If the alcohol is to be blended with gasoline rather than used straight, it must be dehydrated to be made at least 199 proof (or it will not dissolve in the gasoline). The additional cost for dehydration is about five to ten cents per gallon.

Starch is a polymer of sugar ($C_6H_{10}O_5$)n and can be fermented and distilled only if it is first pretreated to transform it into simple sugar. Pretreatment of starch is simple and inexpensive and can be accomplished by boiling (a hydrolysis process also accomplished by, for instance, chewing the starches in the presence of enzymes secreted in our mouths).

Another method to produce ethanol is to pretreat celluosic materials and then ferment and distill them. Since cellulose is a more stable polymer than starch, it is more difficult to hydrolyze. Although the reaction is the same as that for starch, cellulose requires extensive cooking with strong acids or special enzymes. Unlike the pretreatment of starch, cellulose pretreatment is complex, costly, and has not been commercially proven in the U.S.[1] Acid hydrolysis plants for converting wood into ethanol were used throughout the world for many years (and still are in the Soviet Union), but the conversion efficiencies are very low and the production costs are therefore presumably very high. Hydrolysis of cellulose is appealing because it expands the bioconversion feedstock base beyond starches and sugars.

Hydrolysis processes will likely prove superior to the biological conversion of starch and sugar crops in the long term, but recent research developments suggest that a superior method for converting cellulose into liquid fuel is via thermochemical processes (even though for political reasons the U.S. Department of Energy continues to put much more of its research dollars into biological conversion).[2] The preference for thermochemical conversion or for hydrolysis of cellulose feedstocks is not crucial to this book, however, because the distribution, marketing, and end use of fuels and the strategies and opportunities for pursuing a transition to alternative fuels are not very different whether one converts cellulose into ethanol via hydrolysis or methanol via thermochemical processes. A full analysis of cellulose feedstocks is deferred until later in the chapter when thermochemical processes and methanol production are addressed.

Some of the prime starch-bearing materials for producing ethanol are grains such as corn, wheat, and grain sorghum, and tuberous plants such as potatoes and cassava (manioc); those bearing sugar include sugar beets, sugar cane, and molasses. As discussed in an earlier chapter, sugar cane is the predominant feedstock for ethanol production in Brazil. Sugar cane is attractive in Brazil because of low production costs due to low labor costs, efficient management, and high yields (as well as other political and macroeconomic reasons). In the U.S. corn is by far the most common feedstock for producing alcohol; almost all biomass ethanol fuel produced in the U.S. uses corn as a feedstock. Corn's attractiveness is due to its relatively high yield per acre, high starch content, slow perishability (important because of the need to store seasonal harvests in order to operate ethanol production facilities beyond the harvesting season), and high value of coproducts.

Coproducts are those nutrients and other constituents that are separated from the original feedstock during pretreatment and fermentation, and that have value for other uses. The protein in corn is the most valuable coproduct that emerges from the processing of any feedstock. The coproducts of other grains and tuberous plants are much less valuable; those from sugar crops are almost valueless, although the roughage from sugarcane (bagasse) can be burned as the fuel source in the distillation process.

The high value and volume of coproducts resulting from the fermentation of corn is particularly critical. Using a "dry milling" process, about seven pounds of high protein material (dried distiller's grain) are produced along with each gallon of ethanol; using a wet milling process, about 5.8 pounds of gluten feed (21 percent protein), 1.2 pounds of gluten meal (60 percent protein), and 0.7 pounds of corn oil are produced. The high-protein coproducts are used for animal feed. Their value per gallon of ethanol produced was about \$0.30 to \$0.50 in the early and mid-1980s.

In the U. S. other attractive starch and sugar feedstock sources besides corn are food-processing wastes from cheese, fruits, and sweet corn. But these sources are limited by availability, high transport costs, and rapid perishability. Cheese whey, for instance, is only about 6.5 percent solids and is therefore costly to ship. Food wastes are economically practical only for small alcohol plants located within about fifty miles of the food processing plant.[3]

Two crops that are still being experimented with and show promise as high yield, inexpensive alcohol feedstocks are the Jerusalem artichoke, a tuberous plant that grows readily in a wide range of soils, and sweet sorghum, a high-yield sugar crop that has been planted for many decades but has never enjoyed commercial success. Although they have received considerable interest, there has been very little commercial activity. The lack of investment is probably due to uncertainty and risk: uncertainty about continued government subsidies and the risk of no major secondary market, especially for Jerusalem artichokes.

Various estimates have been made of the potential for producing starch and sugar feedstocks in the United States. For instance, the National Alcohol Fuel Commission, created by the U.S. Congress (PL 95-599) to investigate and report on a broad range of issues related to alcohol fuels, estimated that starch and sugar crops could be expanded to produce about 7.5 billion gallons of ethanol annually (489,000 b/d),[4] which represents less than 10 percent of gasoline energy consumption. The only resources that were considered were those presently unused and those that were determined to be economically obtained. But as will be shown, all crops in the U.S. will continue to be uneconomical for fuel production into the foreseeable future.

A general consensus has developed that corn-to-ethanol production of less than one to two billion gallons per year would not significantly affect grain prices or the supply of corn to its principal markets, but that two to four billion annual gallons, while physically attainable, would create disruptions.[5] Four billion gallons require about 1.6 billion bushels of corn, approximately 20 percent of domestic production. At a level of four billion gallons, prices and supplies of grains would be strongly impacted, and environmental degradation increased as a result of intensive use of irrigation and fertilizers on previously unused land. It is estimated that at this level, real prices of corn would increase 15 to 40 percent.[6] Estimates of changes in corn prices are based on shifts in the use of farmland, land productivity, domestic utilization of crops, and the mix of agricultural exports. They are based on estimated price elasticities between substitutable farm products and are the result of long-run market analyses. However, considering the difficulty of predicting relationships between so many volatile variables, the estimated price effects are probably good guesses at best.

The major cost in ethanol production is usually not from fermentation and distillation but from cultivating and collecting the feedstock. The cost breakdown for what would be a medium-sized corn-to-ethanol plant with a capacity of 3,600 barrels per day (55 million gallons per year), assuming corn at a moderate price of $2.50 per bushel, is roughly as follows:[7]

Corn	61%
Operating costs	20%
Capital Costs	19%
Total	100%
Coproduct credit	(−23%)

The coproduct item is revenue derived from the sale of high-protein material used for cattle feed. While this coproduct had a high value in the early 1980s, if corn fermentation activities continue to increase, the supply of this high-protein coproduct will also increase, thereby saturating the feed market and lowering the price of the coproducts. Local markets will become saturated quickly because the material has a high water content, making transport very expensive. Forty miles is usually considered the economic limit for this wet material. The market area is expanded by drying the material so as to reduce its weight and therefore its shipping cost, but drying is expensive and market potential is limited. The most attractive market for the dried product has been in Europe (about 2.7 million tons were exported to Europe in 1982), where prices for grain and other animal feed components are maintained at an artificially high price.[8] If the supply of these high-protein coproducts increases significantly or if the European Common Market imposes restrictions or tariffs on them (as they have threatened), then the prices could drop significantly, reducing even further the economic viability of corn ethanol production.

The reason for the large feedstock cost component (61 percent in the case cited above) is the high opportunity cost of using corn (and other sugar and starch crops). It is expensive to grow those crops, and they also have high value as food for humans and animals. Table 14 presents costs and yields of prospective ethanol feedstocks in the United States. The total cost of producing ethanol from corn in large plants, after crediting coproduct revenues, was about $1.40 to $1.50 per gallon in the early to mid 1980s.[9] This is equivalent to over $2 per gallon of gasoline, since ethanol has only two-thirds the energy content of gasoline (each gallon of ethanol contains 75,670 Btu versus 115,400 Btu per gallon of gasoline). Because the feedstock component represents such a

TABLE 14 AVERAGE COST AND PRODUCTIVITY OF SUGAR AND STARCH FEEDSTOCKS, UNITED STATES, 1979–1981

Feedstock	Feedstock Cost ($ per gallon ethanol)	Yield (gallons ethanol per acre)
Corn	1.10	250
Wheat	1.40	82
Grain sorghum	1.00	149
Potatoes	5.00	872
Sugar cane	1.60	600
Sweet sorghum[a]	1.00	380
Cheese whey	0.25[b]	--
Citrus wastes	0.85[b]	--

SOURCE: Adapted from U.S. National Alcohol Fuel Commission, Fuel Alcohol (Washington, D.C.: GPO, 1981); and U.S. Congress, Office of Technology Assessment, Energy From Biological Processes, vol. 1 (Washington, D.C.: GPO, 1980).

[a]Not a major commercial crop.

[b]Assumes on-site conversion to alcohol (no transport costs).

large portion of the cost, even major improvements in production technologies will not have a large effect on total cost; as a result, the fermentation of starch and sugar crops, including corn, is generally not an attractive option for producing an alternative fuel in the U.S.

Some attractive but small market niche opportunities might exist, however. These opportunities occur when feedstocks can be procured at little or no cost. Food-processing wastes (which have a negative opportunity cost) are one source. Feedstock prices depressed far below production cost are another. Depressed prices might be found, for example, in remote areas where the transport of farm products is difficult and costly, when oversupply occurs, or when products are damaged or do not otherwise meet quality standards. A specific example of a market niche opportunity is in California, where up to 30 percent of the state's orange crop is reportedly discarded in some years because it does not meet minimum size requirements.[10] Another opportunity to gain inexpensive feedstocks for fermentation is the approximately 5 percent of grain produced each year that is distressed and not sold.[11] Other opportunities exist because governments in the U.S. and Europe have long-standing policies to support farm prices by buying surpluses, imposing rigid quality control (which results in discarded produce, as in the California oranges case), and even paying farmers not to grow crops. Thus opportunities do exist to procure low-cost feedstocks, but taken together their total production potential is modest at best.

THERMOCHEMICAL CONVERSION OF LIGNOCELLULOSE

Another category of biomass suitable for energy production is lignocellulose. Included in this group are wood, crop residues, grass (forage) crops, and municipal solid waste (MSW). Since these materials are available in far greater quantities and often at far less cost than sugar and starch feedstocks, they represent a much greater opportunity for replacing petroleum.

Lignocellulosic materials contain three major constituents: cellulose, hemicellulose, and lignin. As described earlier, the cellulose and hemicellulose may be hydrolyzed into sugar, but lignin is relatively inert and unresponsive to hydrolysis treatment. Hydrolysis processes are attractive because they expand the potential feedstock base beyond starch and sugar materials to include lignocellulose—but thermochemical processes are potentially even more attractive because they convert not only cellulose and hemicellulose, but also the lignin (which constitutes about 30 to 40 percent of the mass of wood).

Various themochemical processes are available for producing liquid and gaseous fuels from lignocellulose (hereafter referred to as cellulose for the sake of brevity). Numerous direct liquefaction processes, including pyrolysis, may be used to produce a large range of petroleum-like oils, but these liquids tend to be dense, corrosive, and best suited as boiler fuels. These low-quality oils could be upgraded for use as transportation fuels, but at considerable cost. A more attractive option is indirect liquefaction in which the biomass is gasified (using oxygen) to produce methanol. First, the biomass is decomposed into a synthesis gas comprising carbon monoxide, hydrogen, carbon dioxide, and small amounts of other gases. Next, the mixture of gases is cleaned by removing undesired impurities such as tars, oils, and sulfur, the ratio of carbon monoxide and hydrogen is adjusted, and then the gas mixture is pressurized and catalytically reacted to form methanol.

This combination of gasification and chemical synthesis technologies for converting biomass into methanol is the same generic process used to convert coal into methanol (and is related to the indirect coal liquefaction processes used in South Africa). Advanced gasification technologies have been developed for coal, but considerable engineering is needed to adapt them to biomass gasification. Chemical synthesis technologies, however, are well known. They have been widely used for several decades in the petrochemical industry for converting natural gas into methanol and can be applied with little or no modification to biomass and coal synthesis gases.

CELLULOSIC FEEDSTOCKS

Wood, municipal solid waste (MSW), crop residues, and grass (forage) crops are all widely available and, in contrast to starch and sugar, relatively inexpensive. In some cases there are no competing uses; the opportunity cost is zero. A large part of their cost is the cost of harvesting and then transporting them to the fuel processing plant. The low value-to-weight ratio of all these cellulosic materials, including conventional timber (about zero to three cents per pound) makes the concept of national supply functions unworkable both in theory and practice. Because there are no national markets for MSW, wood residues, grass crops, or crop residues, cellulosic biomass costs are more appropriately analyzed on a site-specific basis. Collection and transport costs dominate the feedstocks' cost to the extent that rarely would it be feasible to haul these materials (except wood as a high-value finished product) even as far as 100 miles; in contrast, the density and high value of corn kernels makes it feasible to transport them hundreds and even thousands of miles.

In the following pages each potentially important feedstock is briefly identified. In order to impart some sense of magnitude to the following discussion of feedstock quantities, it may be conservatively assumed that 2.4 barrels (100 gallons) of methanol is produced from each dry ton of feedstock material. Note that U.S. gasoline consumption is over 6.5 million b/d (100 billion gallons per year). Estimates of biomass fuel production potential are summarized in a table at the end of this chapter.

CROP RESIDUES

Crop residues are the plant material left in the field after a crop is harvested. Every year nearly 400 million tons of these crop residues are produced in the U.S. in the form of stalks, stems, leaves, and husks.[12] A certain amount of these crop residues must be left on the ground to protect the soil against wind and water erosion (but contrary to popular opinion, it has relatively little value as a fertilizer). The remainder is excess and has little or no environmental or economic value.[13] A group of Purdue University researchers estimated the volume of residues generated that could be considered "excess," taking into consideration soil type, topography, wind characteristics, rainfall, and soil nutrients.[14] They found that from a technical and environmental perspective about 20 percent of crop residues generated in the U.S. could be removed for energy production. More recent estimates are that up to 60 percent or more could be safely diverted to energy production without affecting long-term soil productivity.[15] The Purdue study estimated that corn and small grains comprised 18 of those 20 percent of usable residues, and that most of the excess residues were located in the Cornbelt area of Illinois, Indiana, Minnesota, and Iowa.[16] Those 20 percent of residues would produce about 520,000 barrels per day (b/d) of methanol if they were all used. The cost of these feedstocks to fuel producers is determined mostly by the cost of harvesting and hauling the residues to the alcohol plant, since excess crop residues have little or no economic value even for direct combustion. The feedstock cost, then, is highly site-specific, depending primarily on crop type and residue yield per acre.

A major factor discouraging the harvesting of crop residues is timing. In the Cornbelt and elsewhere in northern areas, the harvest period is short. Corn and other crops are harvested as late in the season as possible, before the onset of freezing weather. Crop residues would have to be harvested during this same period. The concentration of these two major activities—crop harvesting and residue harvesting, as well as autumn plowing in many cases—during a brief period of time in autumn severely strains a farmer's labor and equipment resources. Considering this timing conflict, the economics would have to be highly favorable for a farmer (or outside agent) to engage in residue collection, especially in the northern and midwestern areas, where the great majority of usable residues are located.

GRASS AND LEGUME CROPS

These are perennial crops that are grown solely for the feeding of animals. These forage crops present a major opportunity to expand the biomass resource base. They include such grasses and legume plants as orchard grass, clover, and alfalfa (many of these grasses are commonly known as hay).

In general, hay and pasture lands in the eastern U.S. (east of the Mississippi River) are producing at only a fraction of their potential.[17] At present farmers manage the land only to the extent of growing sufficient animal feed for local consumption.[18] Little or no attempt is made to gain high yields. The prospect of alcohol fuel production would be an incentive to intensify crop production on pasture and hay land in the eastern part of the U.S. where irrigation would not be necessary.

Forage crops provide two advantages. One is that they have excellent erosion-control properties and can be grown productively on rolling lands that are unsuitable for tillage of row crops. All of the harvestable material can be removed without creating an erosion hazard because the herbage sod remains in the field after harvest, thereby protecting the soil from wind and water erosion. A second advantage is that the cutting and hauling of these forage crops need not conflict with crop harvests.

The cost of grass crops for energy production is estimated to be greater than the cost for crop residues, principally because of the need for increased fertilization and more frequent cuttings.[19] The costs for harvesting and hauling of crop residues and forage crops would be similar.

MUNICIPAL SOLID WASTE (MSW)

Currently almost all MSW (i.e., garbage) is either incinerated or buried in landfill areas. MSW consists of about 50 percent organic material, 25 percent inorganic material, and 25 percent moisture.[20] The inorganic materials include glass and metal, which may be recycled. Organic materials include mostly paper and some plastics; they have no use except as a feedstock for energy production.

Per capita generation of organic solid waste is about 1,900 pounds per year in the U.S.; that quantity of material could be converted to about 1.2 barrels of methanol. Given that the smallest commercially feasible plant for methanol production is about 300 to 600 b/d, a city of at least about 100,000 people would be needed to support one methanol plant. If one assumed that all MSW in cities over 200,000 population were diverted to methanol production, then total methanol production from MSW would be about 250,000 barrels per day.

MSW is a potentially attractive energy feedstock because it otherwise has negative value (due to the cost of disposal) and because it is readily available in centralized locations. MSW may be directly burned to produce steam (for heat or to produce electricity), thermochemically transformed into methanol, or biologically digested to form gaseous fuels (methane). The major problem for all these options, especially methanol production, is the huge variability of the materials in MSW. It may never be possible to develop gasification technology that can handle that variability. For that reason MSW is not a likely candidate for liquid fuel production.

PEAT

The largest quantities of peat are found in the U.S.S.R. and Brazil, with smaller but still very substantial quantities in Scandinavia, Ireland, Poland, West Germany, Canada, and the U.S. Estimates of economically recoverable peat range from about 15 to 25 million tons,[21] which is the energy equivalent of about 80 to 150 billion barrels of petroleum. These estimates are highly unreliable, however, because peat has not been widely used as a fuel and therefore has not been inventoried.

Estimates of total reserves in the U.S. range from 13 to 120 billion tons, but it is unknown what proportion is economically recoverable.[22] About half the U.S. reserves are in Alaska and about one fourth are in Minnesota. The remainder are spread through forty other states. The U.S. consumed about 1.3 million tons in 1985; worldwide production was between 30 and 45 million tons.[23] Most peat is used, after drying, as a fuel in electricity-generating power plants. Ireland, Finland, and the Soviet Union are the leading producers and users of peat in the world.

Peat has little or no economic value except for energy production; in fact, it has a negative value in some locations because it covers fertile soil that might otherwise be prime agricultural land. A major peat-to-methanol plant was proposed for construction in North Carolina and came close to gaining funding in the early 1980s from the U.S. Synthetic Fuel Corporation.

Peat will not be a dominant energy feedstock because only a few locations exist where sufficient quantities of peat are available to supply a methanol plant over its twenty- to thirty-year life. Even where enough peat is available, its removal might disrupt other activities already established on that land. The proposed peat methanol plant in North Carolina, for instance, would scrape peat from twenty-three

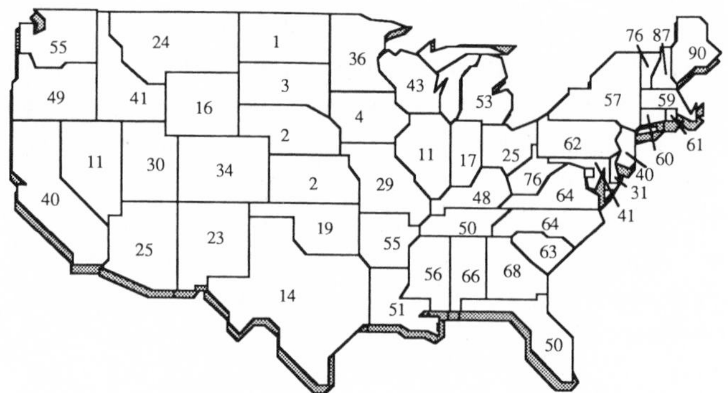

Figure 8. Forest Land as a Percentage of Total Land Area

Source: Forest Service, U.S. Department of Agriculture, *An Assessment of the Forest and Range Land Situation in the U.S.* (Washington, D.C.: USDA, 1980)

square miles of land but would produce only 4,400 b/d.[24] It is difficult to imagine many areas in the eastern and Great Lakes regions of the United States—where much of the peat is located—that have undeveloped land areas of over twenty-three square miles where peat removal would not be disruptive. The environmental impacts of such obtrusive actions would be highly visible and, as in the North Carolina case, would likely attract opposition from environmental groups.

WOOD

Wood is the largest potential source of biomass for energy production, in both the short and the long term. Almost two-fifths of the land area in the U.S. is covered by forest, most of which is available for commercial production (see fig. 8). In all, about one billion dry tons (DT) of woody biomass is added to the commercial forests each year through natural growth. Of that, only 200 million DT of wood are cut for the paper and wood products industry. And of that 200 million DT, only 110 million is ever used. A material flow diagram is presented in figure 9.

In the near term, the following wood resources are most attractive for transportation fuel production:

Logging residues

Concentrated residues at mill sites

Surplus growth.

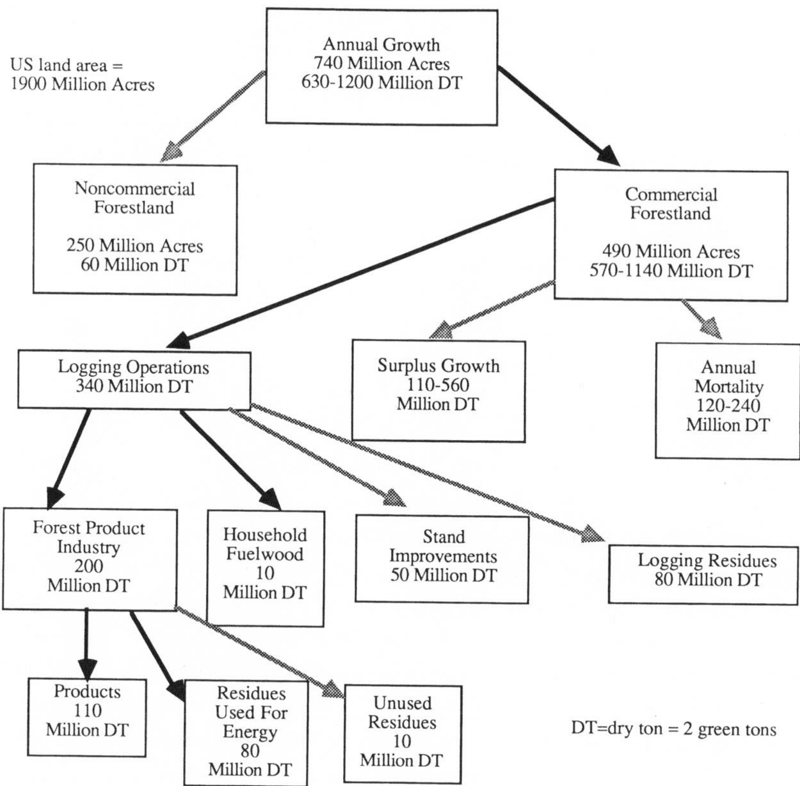

Figure 9. Material Flows of Forest Biomas in U.S.

Source: Derived from U.S. Congress, Office of Technology Assessment, *Energy from Biological Processes*, Vol. II (Washington, D.C.: 1980).

Wood resources are most attractive when they have few competing uses and are therefore of low value, and when they are fairly accessible. Logging residues include branches, small trees, rough wood, and other materials abandoned at a logging site. They can be readily collected because they are clustered.

Mill residues are the most accessible, but they are already used on-site for direct energy production (as sawdust and chips in boilers) or are recycled to make particle board and other wood products. The forest products industry is about 50 percent self-sufficient in energy mostly because it uses mill residues.[25] Large amounts of mill residues are therefore unlikely to be diverted to alcohol or any other energy use outside the forest products industry.

Surplus growth is the third major source of wood. It is attractive for alcohol production because it is not used for lumber and pulp and can be harvested in clusters.

Other near-term sources of wood are residues from stand improvements (e.g., thinning), mortality, and noncommercial forestland. Wood from stand improvements and mortality is less attractive because it is widely dispersed and therefore expensive to collect. Wood is classified as noncommercial if its production yield is low (mostly in the western U.S. between the Mississippi River and the western coastal states), if it is inaccessible (as in Alaska), or if it is withheld from timber production for other reasons (e.g., parks and designated wilderness areas). Generally, noncommercial forestland will not be diverted to energy production.

Expanded wood production, beyond what is currently available, is possible with more intensive forest management practices. The simplest and easiest improvement to implement is more efficient and diligent collection of fallen and decaying branches and trees. The next level of sophistication would be to stock land with faster-growing species and to thin trees more regularly. As an example, companies in the southeastern United States are overstocking forests prior to harvesting; they plant 1,000 stems per acre rather than 600. This allows for an early "energy thinning" and a later "pulpwood thinning."[26] The land can then produce wood for both energy and traditional products. A still more intensive management practice would be the use of fertilizers and genetic hybrids to increase growth rates. And finally, the most intensive approach would be silvicultural farms in which fast-growing hardwood species are closely planted and harvested on a short rotation basis (e.g., every three to ten years).[27]

Typical wood harvests in conventional forests now yield about one dry ton per acre per year (see fig. 9). Although the potential of intensive forest management has not been fully demonstrated, small-scale experiments in many parts of the U.S., Canada, Sweden, and New Zealand indicate that intensive management will produce sustained yields many times higher than those achieved with conventional forestry practices—up to around ten dry tons (DT) per acre-year, and perhaps more than twenty DT with further genetic improvements and more advanced crop and land management practices.[28] Forestry experts tend to believe that yields of up to about five DT per acre are the upper limit for large commercial plots during this century. Higher yields are attainable but changes are implemented slowly. Paper companies already have instituted more intensive management practices to increase wood-pulp yields, and Christmas-tree growers have also learned to increase their monoculture yields, but wide-scale implementation of rapid rotation harvesting of hybrid species is not expected at least until the late 1990s.

A large study of wood-energy farms prepared for the U.S. Department of Energy estimated in 1977 that by 2005 about 260 million DT per year could be produced from silvicultural farms (yielding about 1.7 million barrels of methanol daily), and that by 2025, utilizing still more advanced management practices, 488 million DT per year could be produced—based on an assumption that 10 percent of the land presently used for permanent pasture, forest range, land rotated between hay and pasture, hay land, and open land formerly cropped is all diverted to silvicultural farms.[29] This estimate is undoubtedly overstated, but nevertheless suggests the great potential that exists to produce alcohol from wood.

Actual production of fuel from biomass would, of course, represent an equilibrium between what buyers are willing to pay (which in turn depends on petroleum prices until petroleum becomes a minor fuel) and the price at which feedstock owners are willing to sell the material (as represented by a feedstock supply curve). Some estimates will be presented of what those equilibrium quantities might be, but first the structure of feedstock supply curves is addressed.

Aggregate supply curves, as noted earlier, cannot be constructed for cellulosic biomass because of the dominance of site-specific factors. Supply curves must be empirically specified for each region and/or location. The general shape and structure of those cost curves are influenced by three principal factors: competing uses for feedstock materials and land, seasonality of the material, and feedstock collection costs.

Competing uses for woody biomass were included in figure 9.

Seasonality considerations, the second factor, are crucial because many plants are available only during short harvest seasons—crop residues are an extreme example that was described earlier. Transportation fuel, however, must be available year round. Thus either the biomass feedstock or the fuel must be stored. Ideally the feedstock is stored so that the process plant can be fully used throughout the year. Most cellulosic feedstocks are not perishable and can be stored, unlike most starch and sugar feedstocks. (Since sugar cane is harvested only five to six months per year and is highly perishable, process plants in Brazil close for extended periods during the off-season.)

Collection costs are the third principal factor in determining feed-

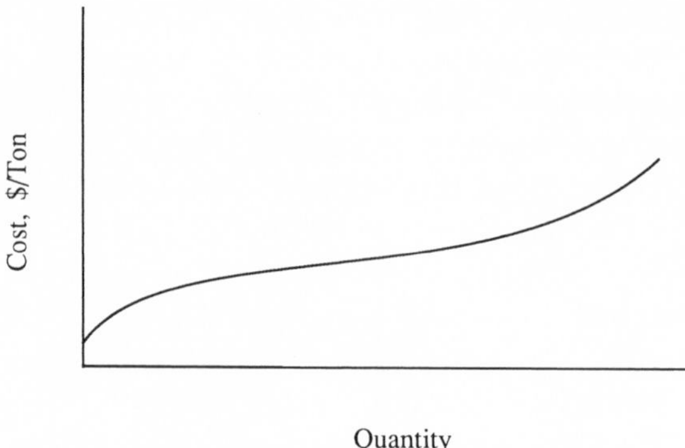

Figure 10. Biomass Supply Curve for a Centrally Located Process Plant

stock supply costs. They are the dominant cost component for residues, wastes, and other cellulosic matter that tends to be dispersed or has a low yield per unit area of land. All solid feedstocks have a low weight-to-volume ratio relative to the liquid or gas into which they are converted, and are bulky and costly to transport. As a result, feedstock transport costs are high and create diseconomies of scale for the producer; the effect is to reduce the economically optimal scale of biomass process plants and to induce fuel producers to locate process plants in or near dense concentrations of feedstock materials.

The general structure of a site-specific supply curve for biomass materials is presented in figure 10. The shape of this curve is based on two phenomena: distance-related transport costs, and diminished feedstock availability as distance from process plant increases.[30]

At sites where large quantities of feedstock materials are readily available, initial supplies of feedstock would be inexpensive. As larger quantities are sought, the cost per marginal unit of feedstock would increase, but at a decreasing rate, because harvest area increases exponentially while radial distance increases linearly ($A = \pi r^2$). Beyond a certain volume, however, another phenomenon intervenes. As larger volumes are sought, one must proceed ever farther from the centrally located plant site. Less feedstock becomes available and the cost curve twists sharply upward. As one moves away from this site, a smaller proportion of land tends to be devoted to the particular feedstock crop or material, roads tend to be more circuitous, and land ownership tends to be fragmented. More remote feedstock owners are frequently less willing to establish a contractual relationship in which they commit their biomass to the process plant, since other opportunities intervene when the distances are greater. The result is that as one moves away from the initial site, less feedstock becomes available, and thus the feedstock cost curve twists sharply upward as greater volume is sought.

Thus the original plant site is initially selected because feedstock owners in that area indicate a willingness to sell the material on their land (whether it be residues, surplus, or the full yield). But as the distance from the plant site increases, less land is available to supply feedstock, and, accordingly, the marginal cost of retrieving each unit of feedstock increases and the cost curve begins turning sharply upward. The effect is to create large diseconomies of scale for biomass plants.

THERMAL PROCESS PLANTS

The process of converting raw materials into methanol may be characterized as follows:

$$\text{Feedstock} \xrightarrow{\text{gasified}} H_2 + CO + \text{gases} \xrightarrow{\text{catalyst}}$$

$$CH_3OH + \text{hydrocarbon gases}$$

In principle, the operating parameters in the gasification and methanol synthesis processes may be adjusted to accommodate a wide range of feedstock materials (as well as to alter the mix of hydrocarbon gases and alcohol in the output stream). In practice, however, engineering considerations will likely dictate that each process plant be closely matched to one feedstock. The opportunity may exist to build considerable feedstock flexibility into the technology, but once the physical plant is built it would generally be excessively expensive to modify it for a significantly different feedstock.

Somewhat more flexibility in operating parameters is possible on the output end. One well-known example of process fiexibility on the output end is the alternating production of ammonia and methanol. In the past many natural gas conversion plants would produce both ammonia and methanol on a seasonal basis by changing catalysts and some of the process parameters.[31] During World War II, twelve small methanol/ammonia plants were distributed around the country by the government.[32] A methanol/ammonia plant is particularly attractive in rural areas where ammonia is used as fertilizer.

Those World War II plants were expensive and inefficient, however. At present no commercial plant exists for producing methanol from biomass using modern thermochemical conversion processes. Research is currently being directed at developing commercial technologies by raising conversion efficiencies and reducing capital and operating costs, but much of this research is proprietary and not disseminated.[33] Although considerable engineering is still needed to adapt the basic gasification and synthesis technologies to the particular characteristics of different biomass feedstocks, with an intensive R&D effort, commercial biomass-to-methanol (and possibly cellulose-to-ethanol) process technologies could probably be developed in a few years.

The manufacture of methanol (using any feedstock) has steep economies of scale. These economies of scale are substantial because of the efficiency improvements gained by using large compressors and other large industrial-size equipment from the chemical process industry. The minimum economic size of methanol plants into the foreseeable future is about 300 b/d; a smaller plant would require an extremely expensive type of compressor that is not currently available. Despite these large economies of scale, the most efficient biomass methanol plant would be relatively small by petroleum and chemical industry standards—principally because of the large diseconomies of scale in biomass collection. Diseconomies of scale in feedstock collection offset scale economies in processing.

The case of a wood-to-methanol plant in a heavily wooded area is represented in figure 11.[34] The wood cost curve in that graph was empirically derived for a typical site in the area of Mississippi and Louisiana. The processing (i.e., manufacturing) costs are based on the conventional chemical engineering wisdom of the late 1970s. The net effect of decreasing processing costs and increasing feedstock costs for larger plant capacities, as shown in figure 11, is an intermediate size range in which biomass fuel plants are most efficient. That is, the total cost of producing biomass fuel is lowest for intermediate-size plants.

In most cases, the most efficient plant size for a biomass-to-methanol system would be smaller than is suggested by figure 11. One reason is that the feedstock supply curve shown there represents an especially favorable situation in which forest land is ubiquitous, wood growth is rapid, large tracts of land are available, and the forest products industry is well established. In less favorable circumstances the feedstock supply curve would be steeper, causing even more severe diseconomies of scale in feedstock supply. Methanol plants that use crop residues would tend to be even smaller than wood-based plants because of the lower density at which residues are generated.

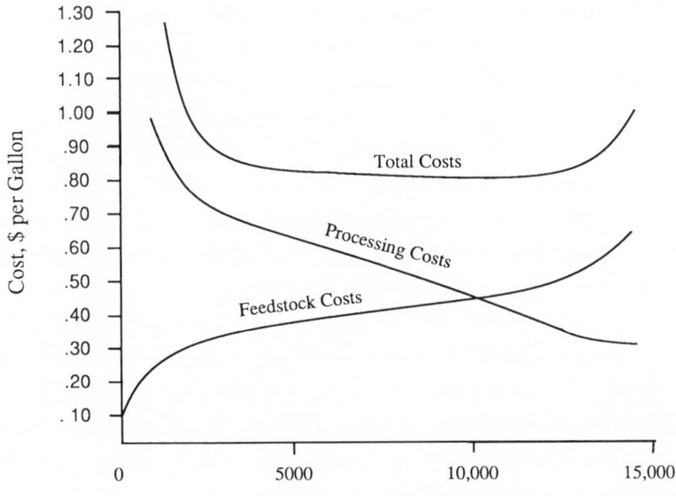

Figure 11. Cost of Producing Methanol from Wood (1986 $)

Sources: Processing Costs adapted from SRI, *Alcohol Fuels Production Technologies, and Economics*. (Menlo Park, California.: 1978), pp. X-58; and Mitre Corp., *Silvicultural Biomass Farms*, Vol. V (Springfield, Va.: NTIS, 1977, MTR-7374); Feedstock costs adapted from Mitre Corp., "Potential Availability of Wood as a Feedstock for Methanol Production," in U.S. DOE, *The Report of the Alcohol Fuels Policy Review: Raw Material Availability Reports* (Springfield, Va.: NTIS, 1979, DOE/ET-0114/1), pp. 35–36.

A second reason that biomass methanol plants are likely to be smaller than the 3,000 to 12,000 b/d range shown in figure 11 is the probably exaggerated steepness of the processing cost curve. That cost function for processing was developed in the 1970s by engineers who were borrowing from experiences with natural gas–based methanol production and relying on optimistic technological forecasts of coal-to-methanol plants. In reality it took chemical engineers several decades to increase commercial natural gas plants from 500 to 15,000 b/d of methanol output. The engineering problems are more difficult with biomass than with natural gas (because it is a solid) and much more difficult than with coal (because it has many impurities), and therefore economies of scale for biomass plants will be modest, at least initially.

A third reason biomass plants are likely to be smaller is the possibility of mass producing (or, more accurately, serially manufacturing) process plants. The sizing analysis in figure 11 assumed that the plants were individually designed and constructed on site. An alternative is to design smaller plants that can be constructed as modules in a factory, with simple assembly on skids at the site. Serial production generates large cost savings in labor and construction. The disadvantage of small serially produced plants is higher operating costs (excluding feedstock) per unit of output than for larger plants. These small plants may prove to be the preferred size because they can be sited to take advantage of situations in which only small quantities of feedstock are available—for instance, residues from a mill or factory, or where only small areas of land could be assembled to supply feedstock. International Harvester, a farm-equipment manufacturer, announced plans in the early 1980s to begin marketing small mass-produced plants around 1985.[35] Its marketing plan called for manufacturing fifty units per year by about 1988, each producing 300 to 600 b/d of methanol at a cost of $10 million each with a per gallon production cost of $0.60 per gallon (1981 $, using wood that costs $35 per ton). While these cost estimates were not documented and cannot be verified, there is every reason to believe that in most situations the cost of producing methanol in small process plants manufactured in a factory would be cost competitive with (if not superior to) much larger plants. In the case of International Harvester, the combination of decreasing gasoline (and methanol) prices and the near financial collapse of the huge company halted plans for developing and commercializing the wood-to-methanol plants.[36]

In summary, for the three reasons stated, biomass methanol plants would tend to be smaller than suggested in figure 11. They would be one-of-a-kind plants in the range of 3,000 to 10,000 b/d or small serially manufactured plants of about 300 to 600 b/d.

The potential quantity of methanol that could be produced from biomass cannot be accurately estimated. Better knowledge of the future, a national biomass inventory, and precisely defined criteria for determining what is economically recoverable is needed. We are not prescient, a detailed biomass inventory has never been assembled, and no study has formulated precise criteria. Acknowledging these uncertainties and information gaps, but recognizing the need to give some sense of magnitude to the biomass-to-methanol production systems, I have assembled in table 15 the best available estimates. These projections of methanol production potential represent the conventional wisdom of around 1979 and 1980, a period of optimism regarding alternative fuels. They are really little more than educated guesses, reflecting some

TABLE 15 POTENTIAL METHANOL PRODUCTION FROM BIOMASS[a]

Feedstock	Potential Annual Production (thousands of barrels per day)[b]
Crop residues	500
Forage (grass) crops	650-1300
Wood resources	250-2000[c]
Municipal solid waste	50-250
Peat	0-100[d]
Total	1450-3550

SOURCES: U.S. Congress, Office of Technology Assessment, Energy from Biological Sources (Washington, D.C.: GPO, 1980); SRI, Alcohol Fuels "Production Technologies and Economics," by J. Jones et al., Menlo Park, Calif.: November 1978; Aerospace Corporation, "Alcohol Fuels from Biomass, Appendix A," in U.S. NAFC, Fuel Alcohol, Appendix (Washington, D.C.: 1980), pp. 1-253; Mitre Corporation, Silvicultural Biomass Farms (Springfield, Va.: NTIS, 1977); Mitre Corporation, "Potential Availability of Wood as a Feedstock for Methanol Production," in U.S. DOE, The Report of the Alcohol Fuels Policy Review: Raw Material Availability Reports (Springfield, Va.: NTIS, 1979, DOE/ET-0114/1); SRI, "Availability and Cost of Agricultural and Municipal Residue for Use as Alcohol Fuel Feedstocks," in U.S. DOE, The Report of the Alcohol Fuels Policy Review: Raw Material Availability Reports, pp. 47-50.

[a]These estimates were mostly developed in the late 1970s and are highly optimistic. Although they were intended to be estimates of production potential for the period 1990 to 2000, a more realistic view is to treat the estimates as ultimate production potential to be realized sometime in the next century.

[b]Projections adjusted by conservatively estimating that 100 gallons are produced from each dry ton of biomass. If the feedstock is not used as the energy source to power the process plant, then the yield would increase by about twenty gallons per ton.

[c]Lower quantity refers to wood residues; larger quantity to silvicultural energy farms.

[d]No published projections available; this is author's best guess.

hazy criteria related to physical availability, competition for feedstock materials by other uses, and economic cost.

The estimates for MSW in particular are unrealistic and would probably never be attained because of previously mentioned difficulties in developing process technologies to handle the disparate ingredients of MSW. The other projections of potential production are highly overstated for the near- and medium-term future although they may prove realistic for a more distant time period, perhaps after 2030 or so.

CONCLUSION

In conclusion, biomass fuels have some attractive features: they are renewable, in many regions can be produced in large quantities, and can be manufactured in relatively small production plants (unlike the mineral fuels addressed in the next chapter). In a few cases, such as in Brazil, biomass fuels are a viable option, but in most situations, beyond some fringe market opportunities, the promise may never be realized. Later chapters will compare the economic, environmental, and sociopolitical implications of alternative energy choices and address when and where biomass fuels are likely to be attractive and should or should not be pursued.

HYDROGEN: A LONG-TERM OPTION

One last important renewable energy option is hydrogen. It is a long-term option in two senses: it will not be economically attractive in the short and medium term, and when (and if) it does become economically attractive, it could be a permanent long-term solution.

Elemental hydrogen occurs freely on Earth in only negligible quantities but is chemically active and is found in many compounds in combination with other elements. Hydrogen fuel would be produced by extracting hydrogen from compounds. Candidate feedstocks must be abundant, accessible, and contain dense quantities of hydrogen. The feedstocks meeting these criteria are coal, natural gas, petroleum, and water. Of these water is the most promising.

HYDROGEN FROM HYDROCARBONS, COAL AND BIOMASS

Hydrogen is produced today from oil, coal, biomass, and natural gas as part of various industrial processes (e.g., chemical and steel production). One means of production is to "crack" methane directly into carbon and hydrogen by heating it in the presence of a suitable catalyst, but a more common means of production is "steam reforming," in which methane and water vapor are combined to produce hydrogen and carbon monoxide ($CH_4 + H_2O \rightarrow 3H_2 + CO$). Steam reforming of natural gas is also the first (and most expensive) step in converting methane into methanol.

This process route for manufacturing hydrogen fuel is not attractive, because if natural gas were to undergo steam reforming, then it would make more sense to take the process one additional step and synthesize the hydrogen and CO products into methanol. While the cost of producing hydrogen along this process route would be somewhat lower than the cost of producing methanol, methanol is a superior choice for motor vehicles because it is much easier and less expensive to store on board the vehicle (see chap. 12).

Hydrogen may also be manufactured from petroleum using a partial oxidation reaction that produces CO, CO_2, H_2, and CH_4—or from coal and biomass using a gasification process that produces H_2 and CO. With the exception of biomass conversion, all these processes for producing hydrogen are well established and could be adjusted readily to produce hydrogen for transportation energy purposes. While it would cost roughly as much to produce hydrogen as to produce methane or methanol, the cost of transporting the hydrogen and storing it on board the vehicle would be much higher.

HYDROGEN FROM WATER

Hydrogen may also be extracted from water—either thermally, using high temperatures, or by electrolysis. Thermal processes are not promising, however, because it would be very difficult and expensive to build containment vessels that could withstand the very high temperatures (2500° K) used in the process to decompose water into hydrogen.[37]

At lower (but still high) temperatures, water can be decomposed in a series of chemical reactions in a closed cycle in which H_2 and O_2 are produced and the chemical intermediaries are regenerated. Several different thermochemical processes are under investigation; the more promising ones utilize iodine, bromine, or sulfur.[38] The reactions require large amounts of heat (which require an external source of energy), and the chemicals involved are corrosive and/or toxic. An inexpensive heat source could be nuclear energy.[39] Research has not advanced far enough to provide reliable output and cost estimates for this process. While there is no reason to abandon research in this area, many scientists believe that electrolytic water-splitting techniques are far more promising.

In electrolysis an electric current is applied to water that has been made more conductive by the addition of an electrolyte. The electric current splits the water ($H_2O \rightarrow 2H^+ + O^{-2}$) and the hydrogen gas collects at the negative electrode (cathode). Electrolysis processes are clean and well known but generally much more expensive than processes in which hydrogen is extracted from coal and hydrocarbons.[40] Although some researchers expect significant efficiency improvements to be forthcoming, the cost of electrolytic hydrogen production ultimately depends on the cost of electricity. It is pointless to use coal, oil, natural gas, or biomass to make electricity to split water since it is cheaper and more efficient to convert those materials directly into hydrogen. It is therefore crucial that alternative sources of energy be developed to produce electricity—ideally, sources that are benign. Most promising are solar-electricity options.

Current research has focused on photolytic production of hydrogen and the development of solar photovoltaic systems to supply energy for electrolysis. Photolytic production of hydrogen using organic or biological catalysts that simulate the function of chlorophyll is the most elegant solution to the hydrogen production problem.[41] These photolytic systems are still very inefficient, however, utilizing less than 1 percent of the incoming energy.

Photovoltaic technology provides another potential energy source for electrolysis but, again, major cost reductions (efficiency improvements) are needed in order to reduce costs. Significant improvements are likely, but at present, photovoltaic electrolysis processes are several times more expensive than coal gasification processes that produce hydrogen.[42] If hydrogen fuel is to become viable, the most likely route will be photovoltaic electrolysis.

This emphasis on clean, benign energy sources for hydrogen production is critical, because as will be shown later, the private market cost of producing hydrogen fuel is considerably greater than the cost of producing other liquid and gaseous fuels from coal or even biomass. Hydrogen fuel is a viable choice only if society decides to value and reward energy systems that do not degrade the environment by generating large amounts of toxic wastes, carbon dioxide, and other pollutants. That possibility exists but is highly uncertain at this point. Thus hydrogen fuel production cannot be considered a near- or medium-term option; it will not be viable for many decades.

SEVEN

Mineral-Based Fuels

Biomass is attractive because it is renewable and therefore has potential to be a permanent source of energy. But a large number of *nonrenewable* feedstocks are available in such huge quantities that they could satisfy world transportation energy demand for many centuries. The most important of these abundant nonrenewable materials are heavy oil, oil sands, oil shale, coal, and natural gas. For ease of discussion and to distinguish them from biomass, this group of candidate feedstocks will be referred to hereafter as minerals. Note that fuels derived from "mineral" resources are commonly referred to as synthetic fuels in popular usage.[1] "Synthetic" is an inadequate descriptor and is not used here because all processed fuels are synthetic to some extent, and because the term "synthetic" fails to distinguish between identical fuels made from different types of feedstocks (e.g., it fails to distinguish between methanol made from coal and methanol made from biomass).

What most distinguishes mineral-based feedstock-production systems from biomass-based systems is the dense physical concentration of the feedstocks. As a result of this concentration, the marginal cost of extracting or collecting larger quantities from any location is small—that is, there are few or no diseconomies of scale in the mining and extraction of mineral resources. The absence of these diseconomies encourages the construction of very large process plants. The large scale—and therefore the large investment in single plants—has far-reaching consequences: it affects what type of investors are likely to participate, how and where the fuel would be marketed and distributed, and the location and type of environmental stresses that would result.

This chapter describes and analyzes the salient characteristics of each important mineral feedstock and process technology. It will be suggested that in the foreseeable future the most attractive mineral-based options for producing liquid transportation fuels in the U.S. are the conversion of remote natural gas and coal into methanol. Oil shale in some cases might also be attractive.

MINERAL FEEDSTOCKS

Some mineral feedstocks, such as oil shale, oil sands, and heavy oil, are available in huge quantities because they have remained virtually unexploited. Other potential feedstocks, such as coal and natural gas, are already widely used as energy sources, but only a small fraction of their total supply has been exploited.

Estimating the quantity of mineral resources that would be available for the production of liquid and gaseous fuels is a difficult task. Because of misunderstandings of terminology and misleading comparisons, there is considerable confusion over the potential availability of different materials. A widely used conceptual framework for measuring resource quantities is presented in figure 12. This McKelvey diagram, named after the former director of the U.S. Geological Survey who first used the diagram, establishes a set of measurement categories. In that diagram "total resources" are the quantity in place. "Reserves" are those resources that could be economically recovered; they must have geologic and other characteristics similar to other resources currently being produced. "Proved" reserves are a subset of total reserves.

Estimates of world energy resources are presented for selected countries in table 16. Countries included in table 16 are those with particularly large reserves of one or more nonpetroleum resources. Most OPEC countries, except Venezuela, have negligible quantities of resources other than petroleum and natural gas. All estimates are defined as *proved* resources except for oil sands and oil shale, which are expressed as *identified* resources. As indicated by the McKelvey diagram, identified resources is a large category that includes all proved, indicated, and inferred resources, including those that are both economic and uneconomic. Thus the oil shale and oil sands estimates are not equivalent to the other estimates in the table. The larger "identified" category is used for oil shale and oil sands because insufficient informa-

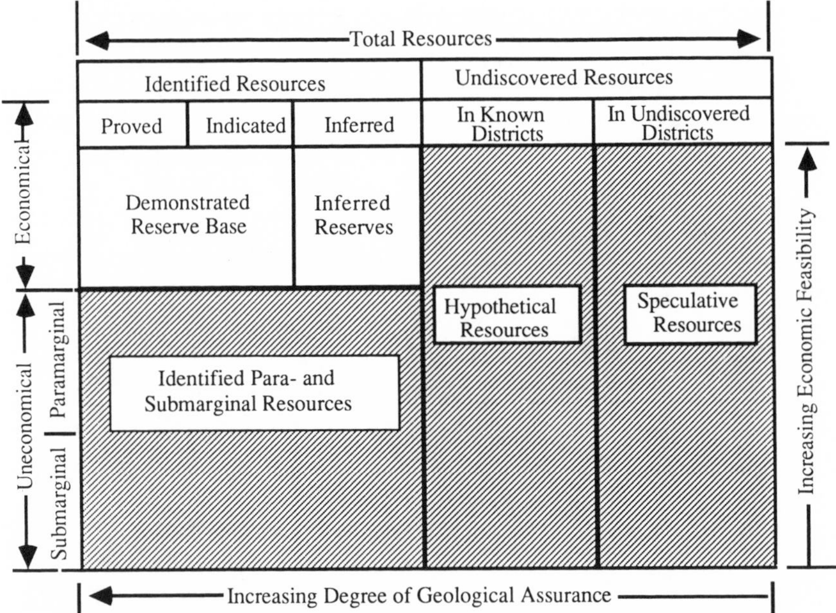

Figure 12. McKelvey Diagram

Source: V. McKelvey, "Mineral Potential of the United States," in E.N. Cameron, ed., *The Mineral Position of the United States* (Madison, Wisconsin: University of Wisconsin Press, 1973).

tion is available to determine what portion should be classified as proved (and economically recoverable). As an illustration of the relationship between these different categories, note that in the U.S. identified coal resources are about fifteen times greater than proven resources.

Estimates of energy resources vary considerably and therefore the values presented in table 16 should not be considered definitive. The reasons for imprecision in these estimates are differing assumptions regarding which mining or extraction methods are used, proportion of material lost or left behind during extraction, laws and rules affecting which resources can be extracted and how, environmental considerations, etc. The point of these numerical estimates is to indicate the huge amounts of resources that are still available for energy production. A determination of the precise quantity is less important, however, than a determination whether these resources should be exploited and, if so, when and where.

TABLE 16 ESTIMATED WORLD'S ENERGY RESOURCES IN BILLIONS OF BARRELS OF OIL EQUIVALENT FOR SELECTED COUNTRIES WITH LARGE NONPETROLEUM ENERGY RESERVES

	Petro-leum (proven)	Coal (proven)[a]	Oil Shale (identi-fied)[b]	Oil Sands (identi-fied)[b]	Natural Gas (proven)	Heavy Oil (proven)
U.S.A.	27	667	2100	small	34[c] (902)[d]	110
Canada	7	NA	small	945	16	200
U.S.S.R.	63	1000	small	605	249	NA
China	19	370	small	--	5	NA
Venezuela	26	--	--	693	9	700
Western Europe	24	147	small	--	36	--
World Totals	698	2560	3150	2270	585[c]	Uncertain

SOURCES: Sam Schurr, Joel Darmstadter, Harry Perry, William Ramsay, and Milton Russell, Energy In America's Future (Baltimore, Md.: Johns Hopkins University Press, 1979); Ronald Probstein and R. Edwin Hicks, Synthetic Fuels (New York: McGraw-Hill, 1982); G. O. Barney, The Global 2000 Report to the President (New York: Penguin Books, 1982); Oil and Gas Journal, December 31, 1984 (for oil and gas estimates).

[a]Assumes 10,000 Btu per pound.

[b]Only a small percentage of these resources would be economically recoverable due to geological constraints and low concentration of some of the reserves.

[c]Conventional sources only.

[d]Nonconventional sources (e.g., geopressured gas, gas from tight-sand formations and coal seams).

NATURAL GAS

Natural gas is a naturally forming gas composed of a mixture of hydrocarbon molecules (mostly methane, CH_4). It is particularly interesting because of the diversity of geologic and geographic settings in which it is found, and our ignorance of how much is available, especially outside the U.S. Historically, most natural gas has been found mixed with crude oil, in which case it is referred to as associated gas. Some was found in self-contained underground fields, but this is the exception rather than the rule. Most natural gas that has been exploited is associated gas that was found as an incidental (and often bothersome) byproduct of petroleum production. In the U.S. (which has the most highly developed system of natural gas production and consumption in the world), the economic utilization of natural gas did not become prominent until after World War II, when pipelines were built to transport it from oil and gas fields to end-use markets. Before then, and even afterward in some areas, the gas was burned ("flared") so that the associated crude oil could be extracted.

In many parts of the world, gas is still flared. The World Bank estimated in 1982 that about 43 percent of all gas produced in developing countries—a quantity equivalent to about 1.2 million barrels per day—was either flared, or, to a lesser extent, reinjected into oil fields.[2] The U.S. Energy Information Administration (EIA) estimates that 1.8 million b/d of oil-equivalent gas was flared or vented worldwide in 1983 and that an additional 2.5 million b/d of oil equivalent (2.48 billion cubic feet per day) was reinjected, mostly because local markets were not available.[3]

Estimates of natural gas reserves are much less certain than estimates for petroleum and coal for the reason that relatively little effort has been directed at identifying and exploiting natural gas. There appear to be vast reserves of natural gas in remote areas of the globe and in unconventional and deep formations. Estimates of conventional natural gas even in the relatively well-explored U.S. are fairly uncertain—high estimates are almost three times greater than low estimates.[4] Estimates of natural gas reserves in less-explored regions and estimates for unconventional natural gas sources such as in tight sands, coal deposits, organic-rich shales, and dissolved in subsurface waters are even less certain. The quantity of unconventional natural gas resources in the U.S. is huge, many times greater than petroleum resources; what is not known is how expensive it would be to extract these resources.[5] These unconventional sources of natural gas are spread over much of the U.S.

Another source of uncertainty in estimates of natural gas reserves is a revolutionary theory that the primary source of natural gas (and petroleum) is not decayed biological matter but rather gases released from the core of the Earth.[6] According to this contested and unverified theory, there may be a virtually unlimited supply of hydrocarbons, mostly in the form of gas. Of course, even if this theory were proven true, the cost of extracting this deeply buried gas would be high.

In any case, there is no doubt that much more natural gas will be found. From 1975 to 1985 proven reserves of natural gas doubled, while those of petroleum stayed constant. Worldwide, natural gas with a total energy content approaching that of oil has been found in about fifty countries. Natural gas has also been found in about thirty countries in which no petroleum has been discovered. World reserves of natural gas will probably surpass those of petroleum by 1990 and will continue to expand into the foreseeable future.[7]

Currently, most proven reserves of natural gas are in countries where it was found in association with oil. In 1984 proven reserves were distributed as follows:[8]

Soviet Union	39.9%
Iran	12.6%
United States	6.3%
Qatar	3.8%
Algeria	3.5%
United Arab Emirates	3.5%
Canada	2.9%
Mexico	2.4%
Saudi Arabia	2.3%
Norway	2.3%
Other	20.5%
	100%

As exploration efforts to find natural gas are initiated, the percentage of proven natural gas reserves found in oil producing countries will diminish.

Natural gas resources have not been exploited for several reasons. In the U.S. they were not exploited in recent years because of price regulations that kept prices low and discouraged commercial development. The broader and more fundamental reason for the lack of development of natural gas resources (indeed, the reason they have been wastefully flared) is the difficulty of transportation. The two practical means of transporting natural gas are by pipeline or at very low temperatures in containers. Pipeline transport is much less expensive and much safer. But pipeline is not practical if there is no nearby market for the gas, if the gas field is located offshore, or if the gas is found in quantities too small to justify the construction of a pipeline. For instance, large gas reserves exist in Algeria, Oman, Malaysia, Nigeria, and many other countries where domestic markets for the gas are small. Since natural gas has only about 0.2 percent as much energy per unit volume as petroleum, the gas much be compressed to a fraction of its original volume to be transported economically in a container. This is done by cooling the gas to near absolute zero and transporting it in specially built cryogenic ship tankers as LNG.

The remaining option for commercial development of natural gas is to convert it into methanol and transport it by conventional tanker ship to far-off markets. While this methanol option tends to be more expensive than the LNG option, it is safer and is amenable to smaller production plants. Saudi Arabia, Indonesia, Libya, and eight other countries decided in the late 1970s and early 1980s to build gas-to-methanol plants to take advantage of their large domestic reserves of associated and nonassociated gas. The result was an oversupply of methanol for the chemical market, with resulting downward pressure on methanol prices.

The conversion of natural gas into methanol represents a major opportunity to expand liquid fuel production. Natural gas, found in small remote gas fields and offshore locations (hereafter referred to as remote natural gas [RNG]) is an especially attractive feedstock for methanol production.

OIL SANDS, OIL SHALE, AND HEAVY OIL

Abundant reserves of oil sands, oil shale, and heavy oil are also scattered around the world. These resources are more plentiful than petroleum.

The U.S. possesses two-thirds of the world's oil shale resources; smaller oil shale deposits are located in Brazil, the Soviet Union, Zaire, Canada, and elsewhere. The greatest concentration of high-grade oil shale in the world is in Colorado. Oil shale may be thought of as an incompletely developed crude oil. It was formed over long periods of time by the accumulation of plankton plants and animals and their mixture with sediment on the floors of lakes. Oil shale contains kerogen, a solid organic material almost totally insoluble in all common organic solvents, and bitumen, a highly viscous crude hydrocarbon. The major part of synthetic shale oil is derived from the kerogen. (Oil shale refers to the raw feedstock; shale oil refers to the liquid product.)

While oil shale may be the most abundant form of hydrocarbons on earth, most of the deposits are not considered viable because they are too deeply buried, occur in very thin layers, or their hydrocarbon content is too low. Oil shale is defined as having a yield greater than 11 gallons (42 liters) per ton, but may have as much as ten times that yield.

The richer oil shale deposits in Colorado have an organic content of about 13 to 21 percent and yield about 25 to 40 gallons (100 to 160 liters) per ton.[9] By comparison, the much higher organic content of coal typically ranges from 75 percent to over 90 percent by mass. Consequently, the amount of raw oil shale to be processed will range from 3.5 to 7 times the amount of coal for an equivalent hydrocarbon output and thermal conversion efficiency. For this reason very large quantities of processed shale must be disposed of as a solid waste.

Oil sands, sometimes called tar sands, are normally a mixture of sand grains, water, and bitumen. Bitumen is a member of the petroleum family; it dissolves in organic solvents and decreases in viscosity when heated. Oil sands are somewhat less abundant worldwide than oil shale. About 40 percent of the reserves are located in Canada, 30 percent in Venezuela, and 25 percent or so in the U.S.S.R. The U.S. has negligible quantities, most of which are located in Utah.

Another minor set of feedstocks for the production of alternative fuels are low-value petroleum hydrocarbons. Two minor sources are excess residual oil and refinery residuals. Refinery residuals are solid or dense liquids that result from the severe processing of conventional crude oil at oil refineries. They have low value and often cannot be easily marketed. In Europe refinery residuals have been converted to methanol since about 1950.

Another possible source of petroleum hydrocarbons that might be used is heavier grades of crude oil, which contain large amounts of sulfur and heavy metals. Heavy oil may be subjected to intensive processing to produce petroleum products or may be gasified and synthesized to produce methanol. Heavy oil is not suited to conventional oil refineries (although some refineries in the U.S. have been modified at great expense to handle these heavy crudes). Most existing oil refineries, especially in the U.S., have been designed to handle the light, low-sulfur crudes that can be converted relatively easily to high-value products such as gasoline. Heavy oil is an appealing option because it is available worldwide in quantities as great or greater than conventional light crudes. Major reserves are found in Venezuela, Mexico, and Canada; much smaller quantities are located in California and elsewhere in the U.S. Heavy oils are difficult to extract and refine and, indeed, are largely unexploited for that reason. These low-value petroleum hydrocarbons—excess residual oil, refinery residuals, and heavy oil—have been introduced here to illustrate the range of opportunities for producing alternative fuels. They will play an important but modest role in providing transportation fuels. Because they are so similar to conventional petroleum and because they will not be a dominant energy source in the future, they will be addressed only in passing in the remainder of the book.

COAL

Coal is a complex, highly variable mineral that represents about 80 percent of the world's proven energy resources. Its energy value derives from the high concentration of carbon, but it also contains nearly every known element, as well as many complex chemical compounds. The United States, U.S.S.R., China, and a number of other countries are endowed with huge reserves of coal, only a small fraction of which have been exploited. The U.S. has proven reserves of over 200 billion tons; at 1985 production rates of about 900 million tons per year, these coal reserves would last almost three hundred years. The location of major U.S. coal deposits (and other mineral resources) is shown in figure 13. The cost and quality of coal varies greatly from one coal field to another. These differences, some of which are presented in table 17, have a major influence on the siting of energy facilities.

In the past most coal was supplied by underground mines in the Appalachian Mountain region, but newer coal production has and will come from surface mines in the Great Plains and Rocky Mountain

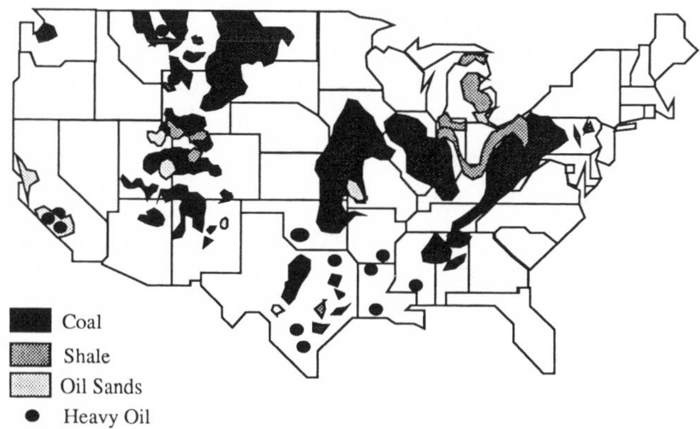

Figure 13. U.S. Mineral Fuels Resources

Source: U.S. Synthetic Fuels Corp., *Comprehensive Strategy Report, Appendices* (Washington, D.C.: 1985).

TABLE 17 CHARACTERISTICS OF COAL IN MAJOR COAL-PRODUCING REGIONS

Region	Energy Density (Btu/lb)	Estimated Cost (1978 $ per Million Btu, FOB Mine*)	Sulfur Content (%)
Appalachia	12,000	$1.42	1-4
Midwest	11,000	$0.98	3-5
West	9,000	$0.78	0-1

SOURCES: U.S. Congress, Congressional Research Service, *National Energy Transportation*, Vol. I (Washington, D.C.: GPO, 1977); and U.S. Department of Energy, *Annual Report to Congress*, Vol. III (Springfield, Va: NTIS, 1982, DOE/EIA-01783180/3). See U.S. EIA, *Annual Energy Review*, for annual updates of average coal prices.

*These costs increased an average of about 16% in real terms between 1978 and 1984; region-specific changes in prices are not readily available.

areas. Coal production west of the Mississippi, almost all of it from surface mines, is projected (as of 1985) to reach about 490 million tons by 1995, representing 40 percent of total domestic production.[10] (A liquid fuel industry producing 1 million b/d would require about 150 to 200 million tons of coal per year.)

New coal mines are mostly being located in the western part of the country because the sulfur content of the coal is low and coal seams are located near the surface, making it easier and less expensive to mine. Sulfur content is important, because when coal is directly combusted (for instance, in powerplants for electricity generation) it releases sulfur oxides that react chemically in the atmosphere to form acid rain. Restrictive air quality regulations have made midwestern coal unattractive because of its high sulfur content, and therefore coal mining in that area is declining.

Methanol (and, more generally, coal gasification) may be the savior of the high-sulfur coal fields. In coal gasification and methanol production, sulfur is removed easily and inexpensively as a routine part of the process, whereas for all other energy options, special add-on equipment is needed to remove it from the pollutant stream or fuel product. Thus methanol and substitute natural gas may be attractive production options for midwestern coals.[11]

Western coal suffers from the disadvantages of low energy density (i.e., Btu per pound) and remoteness from major coal markets in the Midwest and Pacific Coast regions. These disadvantages are countered by its very low extraction costs, which are due to its thick seams that are located near the surface and that are amenable to the use of surface mining techniques. For instance, the cost of transporting coal 1,200 miles from Montana to Chicago by rail is about $10 to $15 per ton. The mining cost is $10 to $15 per ton, resulting in a total cost of about $20 to $30 per ton. Midwestern coal also sells for $20 to $30 per ton but its transport costs are much less. On-site utilization of western coal to produce liquid fuels would be attractive because the cost of transporting liquids long distances by pipeline is much lower than the cost of transporting solid coal by railroad or slurry pipeline.

The lowest-quality coal in the U.S.—that with energy densities less than 8,300 Btu/lb.—is located mostly in Alaska, the western states of Montana, North Dakota, Wyoming, Utah, New Mexico, and Colorado, and in the Gulf Coast states of Texas, Mississippi, Louisiana, and Arkansas. These "low-rank" coals represent a major but largely untapped energy resource.[12] Coal gasification technology, such as that used in methanol production, can be designed to operate efficiently on any type of coal, and therefore can take advantage of the low cost of low-rank coals. The Great Plains SNG Plant (see chap. 3), the first commercial-scale "synthetic" fuel plant in the U.S., and the DOW coal gasification plant both are designed to use low-rank coal.

PROCESS TECHNOLOGIES

The principal process routes for converting mineral resources into transportation fuels are (1) direct hydrogenation, (2) pyrolysis, (3) gasification, and (4) chemical synthesis of gases into liquids (see fig. 14). These are thermochemical processes in that externally applied heat is used to alter the chemistry of the materials.

These technologies are at different levels of development. As elaborated upon later, the most mature technologies are those that convert methane and synthesis gases into methanol. They have been widely applied and their cost and performance characteristics are well known. At the other extreme are processes for direct liquefaction of coal; they are still in the pilot testing stages, and their costs and performance are far less certain.

The production of liquids (or gases) from oil shale, oil sands, and coal is not necessarily in competition with other uses for those re-

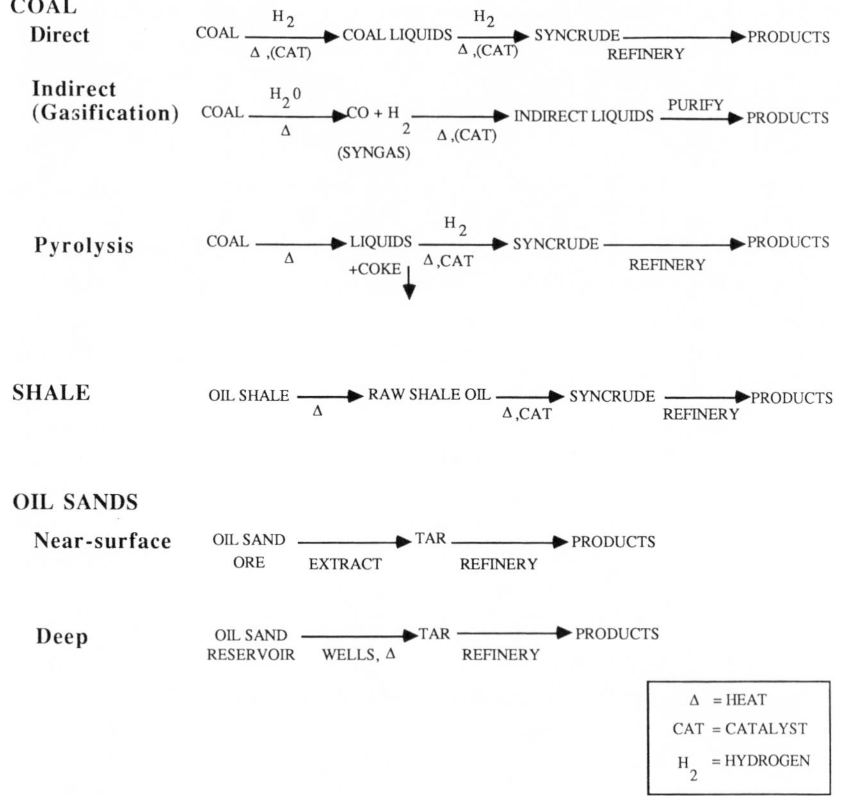

Figure 14. Processing Routes

sources. Oil shale and oil sands have no economically attractive alternative use. They can be directly combusted, but because of their low energy density and remote location it would be too expensive to transport them in raw form to locations where they would be burned. Thus the development of oil shale and oil sands resources is almost completely dependent on their use as liquids.

Coal, on the other hand, because of its relatively high energy density, is an attractive fuel for direct combustion. Indeed, the predominant use of coal in the 1980s is as a raw fuel for electrical generation and industrial plants. Coal reserves are so vast, however, that competition between direct combustion and liquid fuel production is not likely to be an issue in the U.S. for some time, either nationally or regionally.

The manufacture of transportation fuels from these three materials is essentially a process of extraction and hydrogenation of organic matter.

TABLE 18 RATIO OF CARBON TO HYDROGEN AND PERCENTAGE OF INERT SOLIDS FOR SELECTED MATERIALS

Fuel	C/H Mass Ratio	Inert Solids (%)
Bituminous coal	~15	6-12
Oil sands (organic part)[a]	~ 8	~83
Oil shale (organic part)[a]	~ 8	80-93
Crude oil	~ 7	<1
Wood[b]	~ 7	<1
Gasoline	6	0
Methanol	3	0
Methane (CH_4)	3	0

SOURCE: Ronald Probstein and R. Edwin Hicks, Synthetic Fuels (New York: McGraw-Hill, 1982).

[a]High grade.

[b]Inserted here for comparative purposes.

The raw feedstocks contain large amounts of inert material and have a low hydrogen content that must be raised to levels similar to those for gasoline and natural gas. The mass ratio of carbon to hydrogen for a variety of materials is shown in table 18. Generally, the more hydrogen that must be added—or, alternatively, the more carbon that must be removed—the lower the overall conversion efficiency in the production of the fuel. Converting oil shale and oil sands into liquid fuel is technically much simpler than it is for coal because the organic part of those materials has a carbon/hydrogen mass ratio of about eight compared to fifteen or so for coal.

The complicating feature of oil shale and oil sands is the relatively small proportion of organic material; before liquid processing can even begin, the 15 percent or less of the feedstock which is organic must be extracted from the inert nonorganic material.

Coal has much more organic content than oil shale (about 90 percent), and therefore the upgrading of the carbon-hydrogen ratio of coal through direct hydrogenation is potentially efficient; however, the need to supply a source of hydrogen can be costly. In indirect liquefaction, the source of hydrogen is steam, which is much cheaper to provide than the almost pure hydrogen needed for direct hydrogenation.

OIL SHALE CONVERSION

Oil shale is converted into liquid fuels by pyrolysis using retorts, which are chambers in which the shale is decomposed by heat into liquids. Retorts are located either above ground, below ground (in-situ), or in a hybrid configuration. Above-ground retorts may use oil shale mined by either underground or surface mining methods. In-situ methods essentially heat the oil shale in place underground. Modified in-situ processes are a hybrid method in which part of the oil shale is first removed (and retorted above ground) so that explosives can be used to shatter the unmined oil shale above the mined-out area. The shattered oil shale falls into the mined-out section, where it is ignited and retorted in place. The in-situ methods create less severe environmental problems (if they can be prevented from contaminating underground aquifers) and use less water than the above-ground retorts.

Of these methods, above-ground retorting is in the most advanced stage of development. The Union Oil Co. and the abandoned Colony and Paraho projects are all based on above-ground retorting. In the 1970s a great deal of research on in-situ methods was carried out, but the technology is considered to be at least five years behind above-ground retort methods.[13] The pyrolysis (and liquid recovery) of oil sands is similar to that of oil shale, though mostly above-ground retorts are used because the oil sands are generally unconsolidated and structurally weaker (see chap. 5).

Generally speaking, the pyrolysis techniques for extracting liquids from oil shale and oil sands are simpler than the processes for converting coal into liquids. Also, the liquids produced from oil shale and oil sands are higher in hydrogen content than coal-based liquids and are therefore better suited as transportation fuels—but oil shale and oil sand liquids are also high in nitrogen, sulfur and, in some cases, arsenic, and their production generates much larger quantities of solid waste.[14]

COAL CONVERSION

The most attractive techniques for converting coal into transportation fuels are direct and indirect hydrogenation. Pyrolysis processes are technically feasible, but they tend to produce heavy liquids from coal and are therefore not efficient for producing transportation fuels.

DIRECT HYDROGENATION

Modern direct hydrogenation processes are descended from the work of Bergius earlier in the century (see chap. 3). There are two basic direct hydrogenation processes, hydroliquefaction and solvent extraction; they are commonly referred to as direct liquefaction processes. In hydroliquefaction, the coal is mixed with a slurry of coal and coal oils and, together with hydrogen, fed to a high-pressure catalytic reactor where the hydrogenation of coal takes place. The most highly developed hydroliquefaction process is the H-Coal process. A small pilot plant was built and operated for a few years in the early 1980s.

In solvent extraction (also known as solvent refining), coal and hydrogen are dissolved at high pressure in a recycled coal-derived solvent; the solvent transfers the hydrogen to the coal. The gas and liquids are then separated, the coal liquids are cleaned of impurities, and then upgraded by conventional refinery procedures to high-quality liquid fuels. The two most advanced process technologies of this type are the Solvent Refined Coal (SRC) and Exxon Donor Solvent (EDS) processes. Two SRC pilot plants began operation in 1974, one sponsored by a Gulf Oil subsidiary, the other by several electric utility companies, and an EDS pilot plant was started up in 1979 by an Exxon affiliate.[15] Larger-scale direct liquefaction plants have not been built since World War II.

INDIRECT HYDROGENATION

The second process route, indirect hydrogenation, is more advanced commercially. Indirect processes have two principal stages: gasification of the solid coal, and then synthesis of the manufactured gases into liquids. In practice, indirect liquefaction requires a large number of steps (see fig. 15). Indirect liquefaction of coal is conceptually identical to indirect liquefaction of biomass. As with biomass, first the carbon bonds are broken during gasification, and then they are reconstructed to form a liquid. The gasification technologies are less developed and more difficult to engineer than the chemical synthesis technologies.

A number of distinct technologies for gasifying coal are available. They use a range of pressures and reaction temperatures (from about 700° C to 1925° C) to produce different mixes of gases. The production of low-quality gas, often called producer gas, is relatively easy; the gasifiers are technologically simpler, generally operate with lower

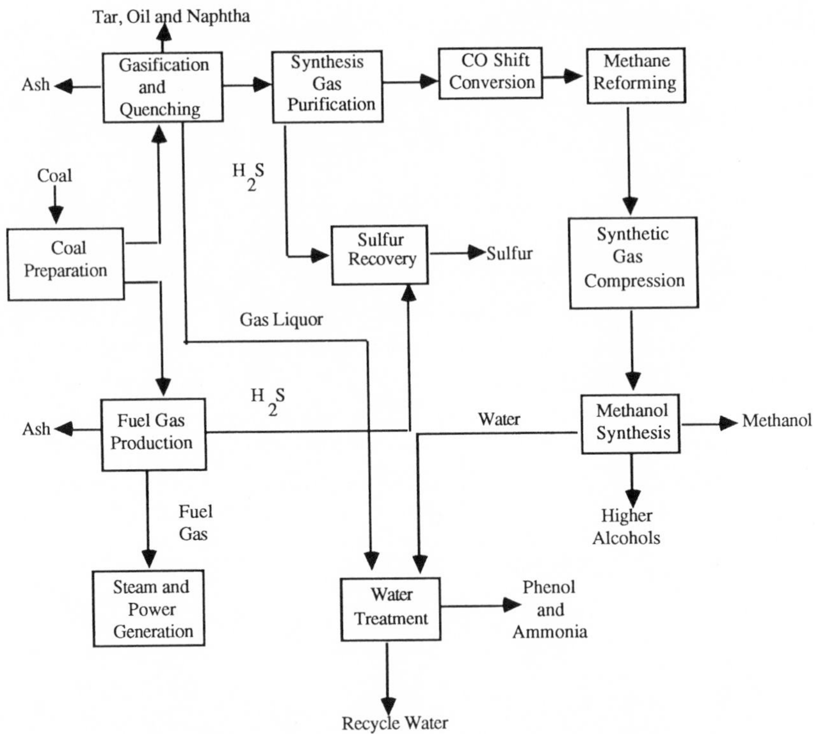

Figure 15. Flow Diagram for Methanol From Coal Process Plant
Source: SRI, *Synthetic Liquid Fuels*, Vol. II (Washington, D.C.: GPO, 1976), p. 116.

temperatures and pressures, and use air instead of oxygen in the reaction process. Producer gas has a low energy density (150–300 Btu per cubic foot) and is too impure to be feasibly synthesized into liquids or upgraded into pipeline-quality gas. The production of synthesis gases with higher energy densities and no impurities is far more challenging. The chemistry of coal (and biomass) gasification is extremely complex and the underlying theory is not well developed.[16]

As indicated in chapter 3, coal gasification is an old, well-established technology. By 1850 approximately fifty-six commercial coal gasification plants were producing coal gas in the U.S. for lighting and heating in urban areas.[17]

Considerable improvements have been made in gasifier technology since those times, however. The gasifiers of the late 1800s and early 1900s operated at atmospheric pressure with steam and air and produced a low-Btu gas (about 150 Btu per cubic foot). Large-scale commercially proven coal gasification processes in operation in the early 1980s (Lurgi, Koppers-Totzek, and Winkler) were developed in the 1930s and 1940s in Germany and, to a lesser extent, from the late 1940s to the mid-1950s in the U.S. More than one hundred of these large-scale gasifiers are now in operation: seventy-two Lurgi gasifiers at the South African Sasol Plants (which produce gas that is converted into over 70,000 b/d of gasoline and diesel fuel); fourteen Lurgi gasifiers in the Beulah, North Dakota, Great Plains project; and fourteen Koppers-Totzek gasifiers at chemical plants located in India, South Africa, and Zambia.[18] Other more primitive versions of the Lurgi, Winkler, and Koppers-Totzek gasifiers were used through the 1970s (and reportedly into the 1980s in some cases) in Turkey, Yugoslavia, East Germany, the U.S.S.R., and Bulgaria to produce various fuel and chemical products.[19]

The South African coal plants and the Great Plains SNG plant in the U.S. use Lurgi gasifiers to produce large amounts of methane (CH_4), which is excellent for compressed natural gas fuels but not well-suited to methanol production. These older Lurgi gasifiers are less efficient and "dirtier" than newer gasifiers being developed in the 1980s (producing more tars, oils and trace pollutants; see chap. 15). Their prime virtue has been many years of reliable operation at the Sasol plants. More advanced coal gasifier technologies have been under development in the 1970s and 1980s (e.g., Texaco, Lurgi slagger, high-temperature Winkler, Shell-Koppers, KILN GAS, and Westinghouse); the advanced Texaco gasifiers are being used in the successful Tennessee Eastman and southern California Coolwater plants that began operations in 1984, and a variation of that technology is being used in the Dow plant in Louisiana.[20]

Indirect liquefaction processes take the carbon monoxide and hydrogen molecules produced during the gasification stage and chemically synthesize them into liquids. There are three distinct choices for synthesis and liquefaction: Fischer-Tropsch hydrocarbon synthesis, Mobil methanol-to-gasoline conversion, and methanol synthesis. The Fischer-Tropsch process is well established, having been used in South Africa for many years (in conjunction with the Lurgi gasifiers), but is relatively inefficient. It produces a large slate of hydrocarbon products that includes high-quality products such as gasoline, but also a large proportion of low-quality products. It is not considered a serious option for future plants.

The second indirect liquefaction option, the Mobil process, uses a catalyst to convert methanol into a mixture of hydrocarbons that are similar in composition, octane number, boiling range, and other specifications to high-quality gasoline. A plant employing the Mobil process began operations in 1985 in New Zealand.

The most straightforward and well-developed of the major process technologies is methanol synthesis, in which the feed gas may be either natural gas or synthesis gas from gasification of coal (or biomass). If natural gas (which is mostly methane, CH_4) is the feed, it must be combined with steam and "reformed" into a synthesis gas of CO, H_2, and CO_2, the same synthesis gas that is produced through gasification. Sulfur and other impurities are removed either before or after reforming. The ratio of H_2 and CO is adjusted, the gas is compressed and heated, and then the gas is synthesized with a catalyst into methanol.

Methanol synthesis was first introduced around 1923 and has been widely used in the petrochemical industry since World War II. The process has been subject to a series of improvements. Currently available designs are now nearing a level of thermal efficiency close to the theoretical maximum.[21] The major single improvement was the development by Imperial Chemical Industries (ICI) during the 1960s of a new process and catalyst suitable for much lower pressures (50 to 100 atmospheres, compared with previously used pressures of 275 to 375 atmospheres) and slightly lower temperatures (240–260° C versus 300–400°C). This new process lowered the cost of producing methanol from natural gas by 15 to 20 percent and has contributed to a major reduction in the price ratio between methanol and gasoline since the early 1970s.

Possibly the most attractive near-term option for producing nonpetroleum transportation fuels is barge-mounted plants that produce methanol from natural gas.[22] The concept of barge-mounted process plants is not new—small oil refineries, wood pulp plants, and liquefied petroleum gas plants have all been mounted on barges in the past[23]—although barge-mounted methanol plants have never actually been built. Barge-mounted methanol plants are virtually identical to land-based plants. They are attractive because they provide access to offshore natural gas fields that otherwise could not be exploited. The plants also can be moved from one field to another, thereby making it feasible to recover gas even from very small gas fields. Since these offshore gas fields have no alternative use, the opportunity cost of the gas feedstock is very low (wells must be drilled, connecting pipelines built, gas "cleaned," etc., but these costs are minor. No one knows how much "remote" offshore natural gas exists, since little exploration has been carried out, but geologists suspect the quantities are huge.

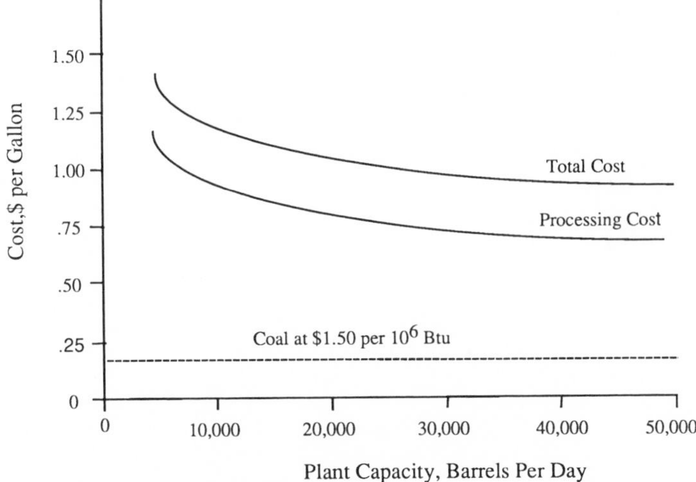

Figure 16. Cost of Producing Methanol from Coal (1986 $)
Source: JET Propulsion Laboratory, *California Methanol Assessment* (Pasadena, Calif.: 1983), chapter 4.

PLANT SIZE

Process plants using coal, oil shale, oil sands, and hydrocarbons as feedstocks would be much larger than biomass plants because of the concentrated availability of the mineral feedstocks. The per unit cost of extracting coal, natural gas, oil shale, and oil sands at a particular site does not increase with quantity. If large quantities of gas, for instance, are not available, perhaps because only small pockets are available, then a process plant probably would not be built at that site. The effect of a flat feedstock curve (i.e., no diseconomies of scale in feedstock collection) and a decreasing processing cost curve, as shown for a coal-to-methanol plant in figure 16 and for a natural gas-to-methanol plant in figure 17, is to encourage the construction of very large process plants for mineral fuels. Economies of scale in processing for coal-to-methanol plants increase up to about 35,000 b/d. Up to that size there is a single production train in key subsystems. Above that size, barring dramatic and unexpected breakthroughs in the engineering of process technologies, there is and will be a paralleling in production subsystems. Other economies of scale, especially in transport and distribution, provide mild incentives to increase capacity above 35,000 b/d and to cluster plants together.

Natural gas-to-methanol plants have a substantially simpler con-

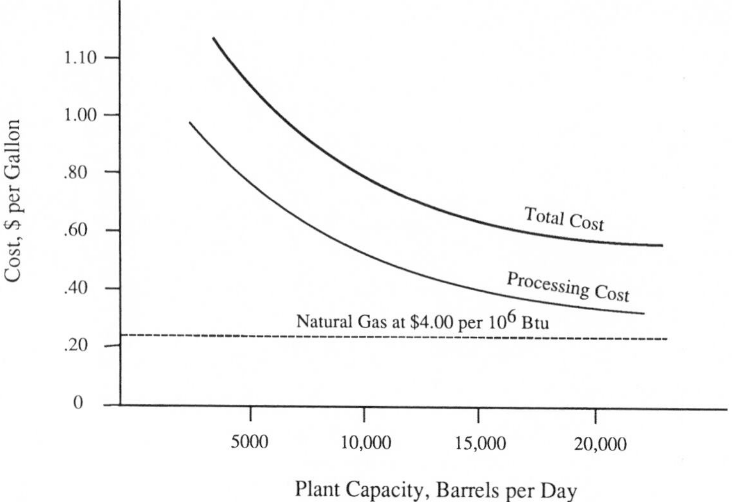

Figure 17. Cost of Producing Methanol from Natural Gas (1986 $)

Source: Adapted from Davey McKee data in World Bank, *Emerging Energy and Chemical Uses of Methanol* (Washington, D.C.: 1982); and Jet Propulsion Laboratory, *California Methanol Assessment* (Pasadena, Calif.: 1983), chapter 4.

figuration and simpler process technologies than the coal-to-methanol plants discussed above; a gas-to-methanol plant using single production train commercial-scale technology components has a capacity of about 10,000 to 20,000 b/d. Offshore floating methanol plants would be comparable in size to land-based plants.[24] Coal hydrogenation, oil shale, and oil sand plants would be roughly the same scale as the coal-to-methanol plants, though the simpler pyrolysis technology of oil shale and oil sand plants make somewhat smaller plants economically efficient. The Union Oil above-ground oil shale process plant had an initial production capacity of 10,000 b/d, but a full commercial-size plant would probably be considerably larger. In the early and mid-1970s many companies planned to build mineral process plants in the much larger 50,000 to 200,000 b/d range; these designs were apparently based on the expectation that coal and oil shale liquids would be diverted through existing oil refineries for upgrading, in essence filling the capacity for the petroleum it was displacing.[25] That perception no longer appears likely to be a dominant factor, and is unlikely to be so at least during the early years of the industry. Methanol would not go through oil refineries; oil shale liquids could, but some refinery modifications would be needed; and coal liquids also could, but even more extreme modifications would be needed.

METHANOL VERSUS DIRECT LIQUEFACTION

An important historic note that should be inserted here is the changing perception of the attractiveness of direct coal liquefaction processes relative to coal-to-methanol processes. Until well into the 1980s, public and private sector R&D was heavily weighted toward direct liquefaction of coal; R&D for methanol production received much less attention and funding.[26] This preference was the result of three perceptions: that direct liquefaction was more efficient, that methanol production from natural gas was a mature technology, and that incompatibility was an overwhelming problem for methanol fuels.

In theory, direct liquefaction techniques are significantly more efficient at converting coal into liquid fuel than the indirect processes; less hydrogen is required, and the coal undergoes less structural change. It was commonly perceived during the 1970s that the thermal efficiencies for producing fuels using indirect processes would be about 40 to 50 percent, while the direct processes would have efficiencies of 65 to 75 percent. This perception that indirect liquefaction was an inferior option was reinforced by evidence from the commercial-scale Fischer-Tropsch plant in South Africa, which had conversion efficiencies of less than 40 percent and produced a large slate of products, many of low quality and low economic value. Even the gasoline was of a low quality (low octane rating). Clearly the Fischer-Tropsch option was unattractive.

The other indirect liquefaction option, methanol synthesis, was received with skepticism because of the problem of incompatibility. Although the methanol route was superior to Fischer-Tropsch processes in terms of thermal efficiency, there was a concern that the costs (and institutional problems) of making motor vehicles, pipelines, tank cars and trucks, storage tanks, fuel pumps, and various other elements of the fuel distribution infrastructure compatible with methanol were overwhelming. Investors and producers were also concerned that a market would not appear for methanol. A January 1976 report by ERDA,[27] the predecessor agency of DOE, rated synthetic gasoline a far more promising alternative than methanol (produced from either coal or biomass), arguing that

> the only private institutions likely to undertake synfuel ventures are the oil companies, because they have the most compelling incentives—an existing business that requires a continuing supply. . . . This dominating interest by oil companies will inevitably shape the choices of synfuels to be produced.

The report continued, stating that the adoption of methanol by the oil companies would be extremely unlikely, since "methanol would not fit readily into the existing marketing system." Conservative and cautious perceptions and conservative institutional and technological forces merged to restrain the development and introduction of nonpetroleum-like alternative fuels.

The two justifications for suppressing the methanol option appear much less convincing in hindsight. The first reason for favoring direct liquefaction processes—higher theoretical conversion efficiencies—may not have been well founded. The higher theoretical efficiencies of the direct hydrogenation plants have in practice been offset by the costs of providing a free (external) hydrogen supply (which is not necessary with gasification and synthesis), unfavorable reaction kinetics, and problems in controlling sulfur and nitrogen impurities.[28]

Even according to an efficiency criterion, methanol is comparable to direct liquefaction if one expands the boundaries of the analysis to a broader systems perspective. As shown in table 19, direct liquefaction processes do indeed have an advantage in theoretical conversion efficiency—but this ignores the quality of the fuel output. Directly liquefied (hydrogenated) coal produces a synthetic crude oil that is best suited to the production of medium- and high-density oils. To upgrade the syncrude to light gasoline-grade fuel requires additional energy and cost. Methanol, however, does not need to be upgraded.

Moreover, even after upgrading the coal liquid product from direct liquefaction, it is still a lower-octane fuel than methanol. In addition, an engine designed for alcohol has a thermal efficiency of about 10 to 20 percent greater than that of a gasoline engine (see chap. 12). As a result of methanol's superior octane and because of the trend toward using petroleum mostly for transportation, the thermal efficiency argument for direct liquefaction is tenuous.

The second major concern, incompatibility, remains credible in the post-1970s era; but as will be argued in this book, incompatibility is not sufficient reason for ignoring methanol (or any fuel dissimilar to petroleum products). Incompatibility is an important consideration, but it should not dominate energy choices.

TABLE 19 THERMAL EFFICIENCIES OF SELECTED CONVERSION PROCESSES

Feedstock	Fuel	Process	Thermal Efficiency(%)
Petroleum	Gasoline/diesel	Conventional refining	88-92
Natural gas	LNG	Liquefaction	90
Natural gas	Methanol	Reforming/synthesis	65-70
Coal	Synthetic crude	Direct liquefaction	58-70
Coal	Gasoline/diesel	Direct liquefaction	40-60
Coal	Medium/high-Btu gas	Gasification	65-80
Coal	Gasoline/diesel	Fischer-Tropsch	35-45
Coal	Gasoline	Mobil	35-55
Coal	Methanol	Gasification/synthesis	40-60
Coal	Electricity	Steam turbine	30-40
Oil shale	Synthetic crude	MIS/Surface retort	54-74[a]
Oil shale	Gasoline/diesel	MIS/Surface retort	45-65[a]
Oil sands	Syncrude	Pyrolysis	52-64[b]
Oil sands	Gasoline/diesel	Pyrolysis	43-57[b]
Biomass[c]	Medium/high-Btu gas	Gasification	50-80
Biomass[c]	Ethanol	Fermentation[d]	40-50
Biomass[c]	Methanol	Gasification/synthesis	40-55

SOURCE: Adapted from Probstein and Hicks, Synthetic Fuels (New York: McGraw-Hill, 1982), p. 450; and National Research Council, Refining Synthetic Liqiuds from Coal and Shale (Washington, D.C.: National Academy Press, 1980), p. 13.

[a]High value for surface retorts; low value for modified in situ (MIS).

[b]High value for Canada; low value for Utah.

[c]Excluding energy for farming.

[d]Starch or sugar feedstock.

LNG VERSUS METHANOL

The near- and medium-term preference for methanol hinges on the development of remote natural gas. Remote gas may be reformed and synthesized into methanol or cooled into liquefied natural gas (LNG). As explored in the next chapter, the cost of converting natural gas into

LNG is significantly lower than the cost of converting it into methanol. However, LNG is more dangerous and difficult to handle, requires larger production facilities, and is more expensive to transport.

As indicated earlier in the chapter, the technology for producing methanol is well proven. The technology of liquefaction is also well established, but LNG plants are less suited to offshore use (in barges) and small gas fields because of safety considerations and economy-of-scale requirements.

LNG also requires specially designed cryogenic facilities at the origin, specially designed cryogenic ships to transport the liquefied gas, and regasification facilities at the tanker destination. In contrast, methanol can be shipped in conventional tankers, in some cases in the same tankers used for petroleum transport, and does not require additional processing facilities. These differences between LNG and methanol production will be assessed in following chapters.

At this point, it is noted that the RNG-to-methanol option is suited to small and offshore gas fields, and allows the use of smaller plants and smaller commitments of capital. LNG is better suited to large gas fields and to situations in which a long-term contract can be signed (of at least ten years or so). LNG is less expensive, as will be shown, but because of the large investments in production facilities at the production and tanker termination site, and in the special tankers, the LNG option is less resilient and flexible.

A TRANSPORTATION PERSPECTIVE OF NEAR- AND MEDIUM-TERM MINERAL FUEL OPTIONS

A transportation perspective of mineral fuels is very important because some fuel products are better suited to transportation applications than others, and because as shown in chapter 1, transportation accounts for a large and growing proportion of petroleum consumption. From a transportation and environmental perspective, the least attractive process is direct coal liquefaction. Direct liquefaction liquids have high levels of sulfur and nitrogen and need to undergo considerable refining to be suitable as transportation fuels. Refining these liquids may be possible in conventional oil refineries, but only after additional processing units have been installed to remove the nitrogen and sulfur from the fuel. Moreover, major upgrading of the fuel will be necessary to make it suitable as a transportation fuel. For instance, only about 39 percent of the H-coal liquid output is light fractions, and of that only about half would be readily suited to upgrading to gasoline-grade fuel.[29] Liquids from direct liquefaction are best suited to replacing heavier boiler fuels, not gasoline-like fuels. The cost of upgrading coal liquids to gasoline-grade quality has been estimated at about $1.30 to $3.80 per million Btu ($0.15 to $0.45 per gallon).[30]

Indirect liquefaction processes, on the other hand, especially those that produce methanol and are based on advanced gasifiers, have the following advantages: sulfur is inexpensively removed even from high-sulfur coal as a routine part of the gasification process, lower-quality (rank) coals may be used as feedstock, higher-octane fuel is produced, and the methanol needs no additional refining. The main disadvantage of methanol production is that the use of methanol requires modification of end-use and distribution technologies.

Oil shale production is more attractive for transportation use than direct liquefaction, but still has significant drawbacks. Like directly liquefied coal liquids, shale oil liquids must be upgraded for use as transportation fuel. Shale oil has high concentrations of nitrogen and sulfur, similar to petroleum, and contains various oil-soluble trace elements, of which arsenic is the most troublesome. These contaminants create refining problems and diminish the environmental attractiveness of the fuels. Nonetheless, shale oil liquids are somewhat more attractive than directly hydrogenated coal liquids, since a larger proportion of the liquids can be readily upgraded to diesel and gasoline fuels. A typical split of shale liquid products is expected to be 10 percent naphtha (which is suited to gasoline production), 50 percent to 60 percent middle distillates (suited to diesel and jet fuel), and the remainder, low-quality "bottoms."[32]

In conclusion, of all the mineral-based feedstock-production options considered in this chapter, the most attractive options are those that use remote natural gas. Remote natural gas is attractive from an energy-systems perspective because it has no other use except as methanol or liquefied natural gas, and from a technical perspective because the technologies for converting gas into methanol and LNG are well known and relatively inexpensive. The next chapter will show that natural gas fuels and methanol made from natural gas are less expensive than other liquid fuel options, and later chapters will show that natural gas fuels and methanol are environmentally attractive.

EIGHT

Production Costs

(WITH MARK FARMAN)

Precise cost estimates have been generated elsewhere for each feedstock-production option addressed in the two previous chapters. In many cases the cost figures are in close agreement. But they are not necessarily accurate, and for any particular project they may be irrelevant. The first part of this chapter reviews conventional wisdom to present cost estimates that are based on consistent and reasonable assumptions. The second part of the chapter reconstructs step by step the cost estimate made for one particular project in order to illustrate the sensitivity of these estimates to various assumptions.

An objective of this chapter is to demonstrate that production cost estimates of experimental production processes cannot be accurately or precisely specified, and that evaluations of specific projects and long-term cost assessments of generic options should be treated as little more than crude estimates. Cost estimates reflect the biases and views of the future held by the organization that conducts the analysis; even when these biases and forecasts represent a broader collective wisdom, they should be regarded with a healthy dose of skepticism. There have been enough surprises in the 1970s and 1980s to cast doubt on any specification of future technological, pricing, and financing conditions.

CONVENTIONAL WISDOM

Endless problems arise in comparing cost estimates for different energy options. Every cost study uses different parameters and assumptions.

For each plant, one must consider the specific configuration of the process technologies, operating parameters, feedstock characteristics, feedstock costs (especially tricky for cellulosic materials), plant capacity, product quality (i.e., slate of products), coproducts, interest rates, debt-equity ratio, and a host of other factors. Moreover, for most of these process technologies there is little or no commerical experience to draw upon in carrying out these cost analyses. Indeed, some process technologies are still not commercially available.

Table 20 is a summary of a range of production costs published or made public by various organizations. The cost estimates have been adjusted from their primary source to standardize the assumptions as much as possible. The costs are plant gate costs; they do not include transportation. Costs do not consider the quality or end-use efficiency of fuels. The estimates may be treated more or less as the costs for first- and second-generation process technologies (except for oil refining, bioconversion, and gas-to-methanol processes, which are mature technologies). As indicated later in the chapter, costs for third-generation and later technology can be expected to be much lower than for second-generation technology. The costs presented in table 20 should be treated as suggestive, not definitive.

Of all the cost estimates in table 20, probably the least reliable are those for hydrogen and for gasoline derived directly from coal. The water electrolysis and direct liquefaction technologies are the least well developed of all those presented and the costs are therefore subject to considerable uncertainty. The estimates for methanol from coal may be somewhat high, based on the promising initial experience with the first-generation Great Plains project (see chap. 3).[1] Cost estimates of methanol and CNG from remote gas are uncertain because of uncertain feedstock costs. LNG/CNG costs do not include regasification at destination (about $0.75 per million Btu).

It can be surmised from table 20 that the least expensive options are CNG/LNG and methanol from remote natural gas (RNG), followed by liquids from oil shale. However, these energy options are likely to be constrained somewhat: RNG because of uncertain transportation fuel markets for methanol and natural gas, and oil shale because of environmental considerations (see chap. 16) and unresolved technical problems. The least attractive options (based on these cost estimates and earlier analysis) are ethanol from corn and coal-based Mobil gasoline. Hydrogen apparently is the most expensive, but it has major environmental attractions if it is produced with clean electricity.

TABLE 20 ESTIMATED PLANT-GATE PRODUCTION COST OF TRANSPORTATION FUELS (1986 $)

Feedstock	Fuel	$ Per Physical Gallon	$ Per Million Btu
Petroleum ($15/bbl)	Gasoline	0.40-0.50	3.70- 5.00
Petroleum ($30/bbl)	Gasoline	0.80-1.00	7.00- 9.00
Petroleum ($60/bbl)	Gasoline	1.50-2.00	13.00-17.40
Natural Gas (U.S.)	Methanol	0.50-0.80	9.10-14.50
Remote natural gas	Methanol	0.40-0.80	7.30-14.50
Remote natural gas	LNG	---	5.00- 7.00
Coal	Methanol	0.80-1.30	14.60-23.60
Coal	M-gasoline	2.00-3.25	16.30-28.30
Coal	Gasoline	1.80-2.60	15.70-22.60
Oil shale	Diesel	1.40-2.00	10.40-14.80
Wood	Methanol	0.80-1.35	14.50-24.50
Corn	Ethanol	1.20-1.70	15.80-22.40
Sugar cane (Brazil)	Ethanol	0.80-1.10	10.50-14.40
Water (and solar electricity)	Hydrogen	---	25.00-40.00

SOURCES: Standardized cost estimates of biomass fuels are presented in U.S. Congress, OTA, Energy from Biological Processes, Vol. II (Washington, D.C., GPO, 1980), p. 173. For Brazilian ethanol costs see chapter 4 of this book. Standardized cost estimates for mineral fuels and biomass methanol are presented in Jet Propulsion Laboratory, California Methanol Assessment (Pasadena, Calif.: JPL, 1983), chap. 4. JPL's estimates are based on the following sources: Acurex, Alternative Fuel Strategies for Stationary and Mobile Engines: Evaluation of Clean Coal Fuels, Volume III (Prepared for California Energy Commission, Mountain View, California: July 1981); Bechtel Corp., "Economic Feasibility of Synthetic Fuel Project," San Francisco, Calif., November 1981; Engineering Societies' Commission on Energy, Synthetic Fuels Summary (Prepared by A. K. Rogers et al. for U.S. DOE, Springfield, Va: NTIS, 1981, FE-2468-82). Other standardized estimates of mineral fuel production costs are included in ICF Inc., "Methanol from Coal," in U.S. National Alcohol Fuel Commission, Fuel Alcohol, Appendix (Washington, D.C.: GPO, 1981); Chevron, Inc., The Outlook for Use of Methanol as a Transportation Fuel (Richmond, Calif., 1985); M. F. Lawrence, J. Skolnik, and L. Lent, The Cost of Methanol During Transition (Bethesda, Maryland: J. Faucett and Assoc., 1986, 86-322-8/11). For assessment of hydrogen costs, see M. A. DeLuchi, D. Sperling, and R. A. Johnston, Comparative Analysis of Future Transportation Fuels (Berkeley, Calif.: Institute of Transportation Studies, 1987).

Before examining production costs more closely, it is instructive to examine capital costs independent of feedstock and other variable costs. Capital costs are defined here as the initial costs incurred in constructing and preparing a plant for operation, and in purchasing all ancillary equipment. Capital costs therefore include not only actual construction costs but also initial catalysts and chemicals, construction management services, and so on. By themselves, capital cost requirements have little significance in terms of an overview cost assessment—for instance, one feedstock-production option may have higher capital costs but lower total costs per unit of energy output than another option. On the other hand, the magnitude of capital costs is important because it influences the nature of investment participants. For example, because of the small amount of capital required, biomass plants could be built and owned by farmers and local entrepreneurs. A large minerals plant would require backing by major corporations (or government). The sizes of capital investments are important also because they indicate the sensitivity to changing economic, market, and political conditions. A plant with large capital costs (relative to variable costs)—and therefore large loans—is vulnerable to fluctuating interest rates and market conditions, while smaller bioconversion plants, which have relatively small capital costs relative to total costs, can shut down for part of the year without suffering onerous debt payments during the off-season. Capital-intensive investments are more adversely affected by labor strikes, inclement weather, breakdowns, and the like.

Capital cost estimates for a range of feedstock-production options are presented in table 21. These cost estimates, like those in table 20, are presented with a certain hesitation because many of the estimates are highly uncertain, and because costs will vary from one site to another. Again, these estimates should be treated as suggestive, not definitive.

One can observe from table 21 that the thermochemical processes of indirect liquefaction and of direct hydrogenation are the most capital intensive. Note that biomass conversion is not necessarily less capital intensive than coal conversion, though the smaller scale of wood biomass plants implies a smaller sum of capital for any one process plant.

The pyrolysis of oil shale requires somewhat less initial capital input than coal-fed liquid fuel processes, while bioconversion and natural gas processes have the lowest capital requirements. The apparent anomaly of a high proportion of capital costs being devoted to the production of

TABLE 21 MID-RANGE ESTIMATES OF CAPITAL REQUIREMENTS FOR FUEL PROCESS PLANTS, 1986$[a]

Conversion Route	Capital Charge as % of Total Cost (at plant gate)	Approximate Capital Cost per Daily Oil-Equivalent Barrel
Crude Petroleum → Gasoline	5-10	$ 11,000
Natural gas → LNG → CNG	65	36,000[b]
Natural gas → Methanol (land-based)	40-65[c]	36,000
Natural gas → Methanol (barge-mounted)	70	55,000
Coal → Methanol[d]	75	102,000
Coal → Methanol/SNG	75	82,000
Coal → M-Gasoline	75	122,000
Coal → Gasoline[e]	75	111,000
Oil shale → Gasoline	70	78,000
Wood → Methanol	60	107,000
Corn → Ethanol	20	62,000
Sugar cane → Ethanol (Brazil)	25	62,000

SOURCES: Based on Jet Propulsion Laboratory, _California Methanol Assessment_, Vol. III (Pasadena, Calif.: 1983): U.S. Congress, OTA, _Automobile Fuel Efficiency and Synthetic Fuels_ (Washington, D.C.: GPO, 1982, chap. 6; Chevron, Inc., _The Outlook for Use of Methanol as a Transportation Fuel_ (San Francisco, Calif.: 1985), attachment 5; H. Geller, "Ethanol Fuel from Sugar Cane in Brazil," _Ann. Rev. Energy_ 10 (1985): 135-164.

[a]Assumes about 20% return on 100% equity capitalization.

[b]Includes regasification facilities at destination and additional costs for cryogenic ships.

[c]Low value based on cost of $4 per million Btu for natural gas; high value based on (extraction) cost of $0.50 per million Btu.

[d]Costs are based on bituminous coal; would be about 10% greater for lignite.

[e]Costs are highly uncertain; based on pilot plants only.

methanol and CNG from natural gas is explained by the relatively low cost of remote natural gas feedstocks and the relatively low total cost of the resulting methanol and CNG output.

COST UNCERTAINTY

As suggested earlier, production costs are difficult to generalize because of the uniqueness of each process plant—its physical setting, financial circumstances, mix of products, etc. If the process has not been widely commercialized, then the future costs are even more uncertain.

Cost uncertainty is distinct from market uncertainty. Market uncertainty deals with consumer response to fuels, prices of competing products, and availability and cost of end-use technologies—it is addressed later. Cost uncertainty relates to the cost of constructing and operating a process plant; it is the misestimation of plant costs and performance.

Cost misestimations may be due to unforeseen changes in external factors such as government regulation, weather, labor strikes, inflation, and interest rates. When viewed over the long term, misestimation would also be due to fundamental changes in societal values, beliefs, and goals. For instance, heightened safety and environmental concerns may result in additional regulations and conditions being placed on LNG, oil shale, and direct liquefaction, causing those options to be unattractive to investors.

A study by Rand Corporation found that the major causes of cost and performance misestimation for innovative energy process plants were (1) the extent to which the plant's technology departs from that of prior plants, (2) the degree of definition of the project's site and related characteristics, and (3) the complexity of the plant.[2]

Cost and performance misestimates are best understood by distinguishing between precommercialization and postcommercialization. During precommercialization, costs are generally underestimated because the project has not been fully specified and because initial estimates are generally made by (optimistic) engineers who are proponents of the project. As engineering details are specified, cost estimates tend to increase. This phenomenon is illustrated in figure 18. The graph shows that as five mineral fuel process plants moved from initial conception to construction, their estimated costs increased at least twofold.[3] Precommercialization process technologies include water electrolysis, cellulose-to-alcohol processes, and direct liquefaction of coal. Those processes

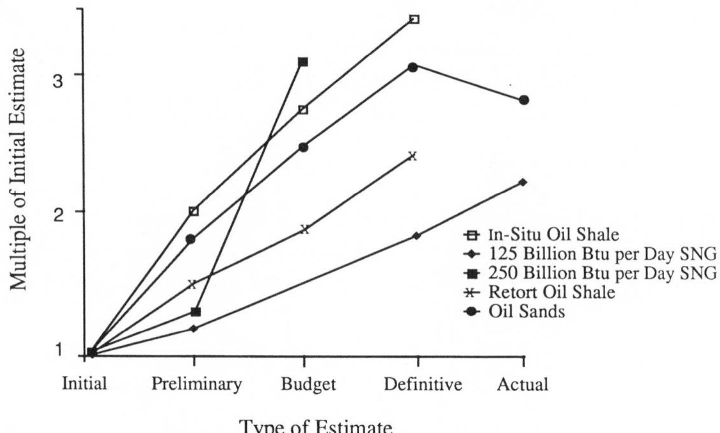

Figure 18. Empirical Cost Estimates of Pioneer Process Plants

Source: R. Hess (Rand Corp.), "Use of Cost Estimating Relationships in Energy Technologies," in R.E. Horvath, ed. *Selected Papers from the Energy Workshop: Industry Perspectives on Pioneer Process Plants* (Santa Monica, Calif: Rand Corp., 1981, N-1709-DOE), p. 90.

that were on the edge of commercialization in the late 1980s are in-situ oil shale and conversion of coal into methanol and high-Btu gases.

As conceptually represented in figure 19, the postcommercialization phase is one of *falling* cost estimates. As operating experience is gained with a particular process technology, the design is improved, operation is made more efficient, performance increases, and costs decrease. Again, it is difficult to anticipate the magnitude of these learning improvements. Some indication of the learning rate is indicated by historical experiences with process technologies having similar characteristics in other industries. Rand Corporation, in a study for the U.S. Synthetic Fuel Corporation, identified three industries that had production technologies similar to those of large, complex mineral fuel plants: low-density polyethylene (LDPE), ammonia, and alumina refining industries. LDPE is a petrochemical process. As shown in figure 20, the learning rate in all three industries was striking; as experience was gained and plant sizes increased, production costs dropped about 20 percent for each doubling of output.[4] For the first two industries, costs dropped steadily over time as output expanded; for the alumina industry costs dropped more rapidly initially and less so later.

All the energy conversion processes deployed in recent years show evidence of these learning benefits.[5] For instance, in South Africa it took

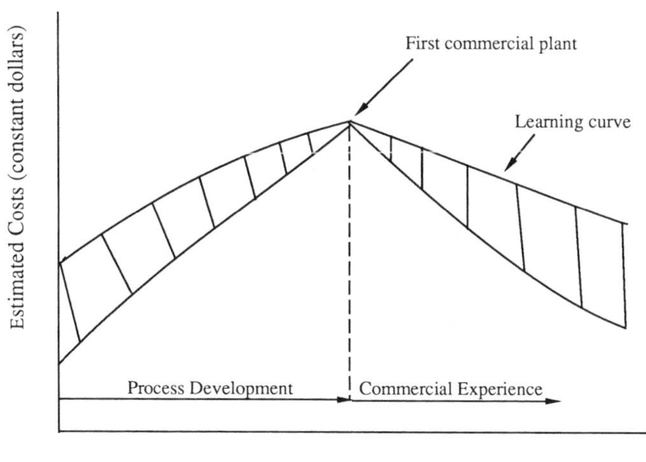

Figure 19. Conceptual Representation of Cost Estimates for New Process Technologies as a Function of Development Stage

five years for Sasol I to attain 75 percent of design capacity, but only two years for Sasol II and nine months for Sasol III. Moreover, the throughput of the original 1950s gasifiers was gradually increased to 45 percent over design capacity. Also, Sasol III, although identical to Sasol II, was operated 20 percent more efficiently.

Significant cost and efficiency improvements have been gained even with the well-known fermentation and distillation processes. In Brazil, for instance, operating improvements raised output in many distilleries to 20 to 30 percent above rated capacity, and small additional investments increased production to as much as 50 percent above capacity.[6]

After reviewing the performance of many energy-conversion technologies worldwide, the Rand report concluded that with constant attention to and improvements to plant operation, a single plant can be expected to achieve 5 to 40 percent improvements in production cost over its life.[7] These cost improvements are much more substantial between successive plants. Table 22 presents estimates by the U.S. Synthetic Fuel Corporation (based on the Rand Corp. studies cited above) of likely cost improvements in coal gasification, oil shale, and oil sands production. It estimated that third-generation plants would produce fuel for only one-fifth to one-half the cost of first generation pioneer plants.

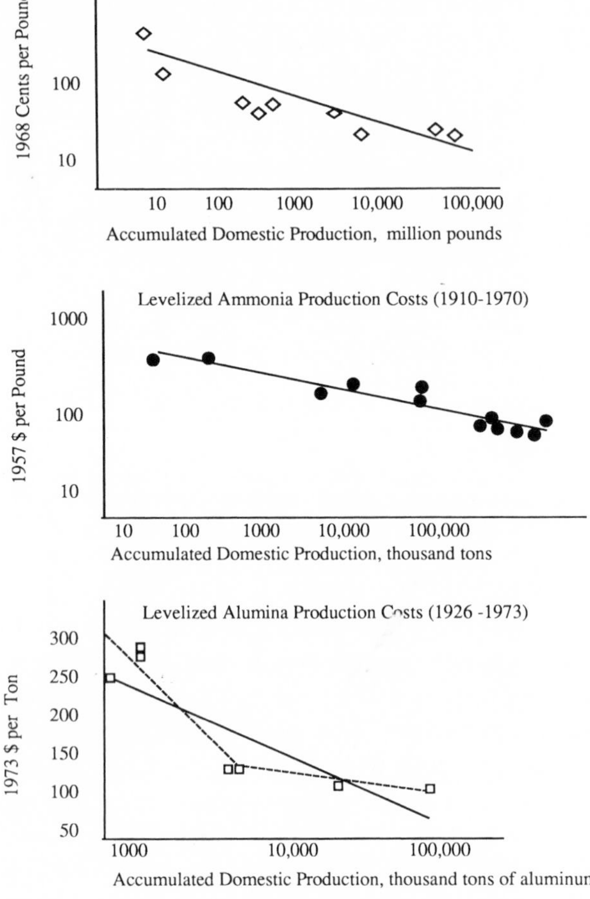

Figure 20. Levelized Production Costs for Three Industries

Source: U.S. Synthetic Fuel Corporation, *Comprehensive Strategy Report Appendices* (Washington, D.C.: 1985).

CASE STUDY

As indicated, cost estimates presented in this chapter are sensitive to a great many factors, some of which are highly uncertain and cannot even be specified accurately. To emphasize this point and to demonstrate how sensitive those estimates truly are, the actual cost analysis for a proposed fuel process plant is replicated here and used as the basis for testing the cost sensitivity of the project to various factors.

The proposed project was to be located adjacent to a major coal field

TABLE 22 COST CHARACTERISTICS OF PIONEER AND SECOND AND THIRD GENERATION PLANTS

	Pioneer Plant			Second Generation		Third Generation	
Plant Type	Plant Scale (BOED)[a]	Unit Capital Cost (per BOED)	Unassisted Levelized Cost[b,c] ($/BOED)	Plant Scale (BOED)	Levelized Cost[c] ($/BOED)	Plant Scale (BOED)	Levelized Cost[c] ($/BOED)
Oil shale[d]	14,500	$75,000	$110	30,000	$45-$50	45,000	$35-$40
Coal[e]	7,500	$85,000	$125	20,000	$40-$50	40,000	$25-$40
Oil sands[f]	5,500	$30,000	$ 90	10,000	$55-$60	40,000	$40-$50

SOURCE: U.S. Synthetic Fuel Corp., Comprehensive Strategy Report (Washington, D.C.: 1985), p. H-10.

[a]BOED: barrels of oil equivalent per day.

[b]Levelized costs for assisted projects are likely to be 20 to 30 percent lower because assistance improves availability and terms of financing.

[c]Includes total manufacturing costs plus capital charges.

[d]Includes upgrading of raw shale oil. Operation of twenty years.

[e]Includes only facilities to produce medium-Btu gas. Operation of twenty years.

[f]Does not include upgrading. Operation of twenty-five years.

in New Mexico. The process plant was to use Lurgi gasifiers in converting subbituminous coal into methanol and high-Btu substitute natural gas (SNG). At full capacity, it was designed to produce the energy equivalent of approximately 65,000 barrels per day of petroleum. When operating at 91 percent of design capacity as assumed by the project sponsors, output would be the equivalent of 59,000 b/d of oil, of which about 50,000 would be methanol and SNG. The remaining 9,000 b/d would include naphtha, ammonia, sulfur, phenols, carbon dioxide, and creosote, all of which have some commercial value.

The cost analyses for the proposed project were conducted by Bechtel, Lurgi, Texas Eastern Synfuels, Utah International, and others. The cost studies were based on detailed engineering designs. The feasibility study was funded by the U.S. Department of Energy and published for public review.[8] The project sponsors (Texas Eastern and Utah International) did not receive any further financial assistance from the government and did not construct the project.

Estimated costs of the proposed project have been replicated to specify a base case for the analysis which follows. Cost assumptions and financial parameters that were not explicitly stated in the study and

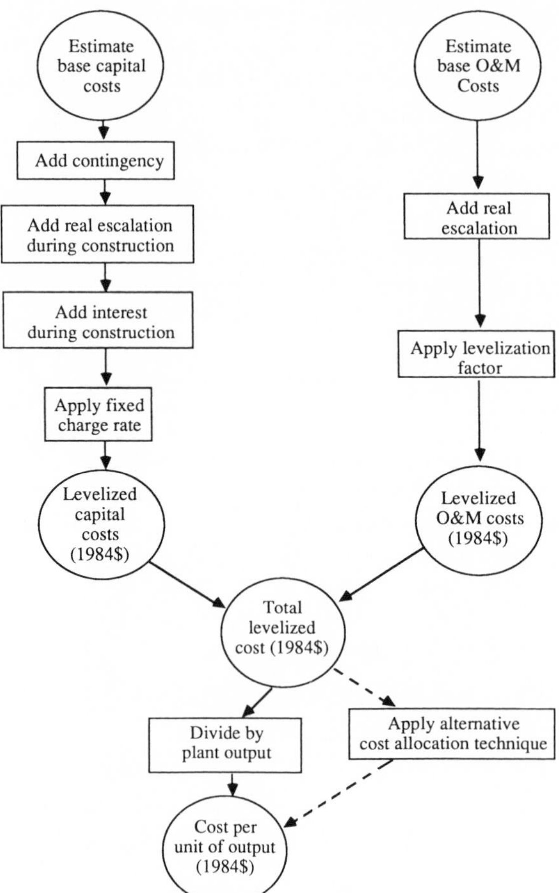

Figure 21. Flowchart of Cost Model

assumptions and parameters that were out of line with typical values used elsewhere were assigned new values (and noted in the text). After specifying the base case, a range of values, usually a high and a low value, was selected to test the sensitivity of various assumptions and parameters. The range in values was derived from a review of cost studies of other chemical and energy process plants and from discussions with experienced individuals in the industry. For the purposes of this analysis, plant construction is assumed to have begun in early 1985. All costs are expressed in 1984 dollars.[9]

The cost model used for the analysis is presented schematically in figure 21. That figure shows the step-by-step procedure used to reconstruct the estimated base case. The purpose of using this method of analysis was to enable the stream of costs to be levelized over the life of the plant so that costs may be expressed on a present-value basis, in this case in terms of 1984 dollars.[10] Capital costs were treated independently of operation and maintenance (O&M) costs and summed together at the end to determine the total cost per unit of output in 1984 dollars. The costs were first calculated per unit of Btu. Since multiple outputs were produced and each output required different processing and had different unit values, a second set of calculations was carried out using a cost allocation method.

BASE CASE

The base case capital costs are those for plant-related facilities, handling facilities for coal and water, land, and initial costs incurred before start-up, including administrative, labor, and chemical costs. The following capital costs are taken from the feasibility study for the New Mexico plant and escalated to 1984 dollars:

Plant-related facilities:	$2,730 million
Coal and water facilities:	$ 360 million
Other:	$ 409 million
	$3,499 million

These costs are spread over an eight-year period preceding start-up. Those eight years encompass time not only for construction but also for preliminary design work and feasibility studies. The assumed starting date is early 1985. The following construction expenditure cash flow is based on the expenditure profile over time of the proposed New Mexico plant:

Year 1	6.3%
2	7.2%
3	10.5%
4	16.8%
5	17.3%
6	17.2%
7	14.8%
Year 8	9.9%
	100.0%

Added to the \$3.5 billion capital cost are contingency funds to cover any unplanned capital expenditures. A contingency factor of 22 percent was used in the base case (the original study used 15 percent); this percentage was based on the analysis of cost growth in pioneer process plants by Rand Corp.[11] Another capital cost adjustment is for escalation of costs during construction. In recent years construction costs have typically increased at rates greater than the general rate of inflation. For example, from 1970 to 1979 the GNP price deflator rose at an annual rate of 7.2 percent, while the producer's price index for construction materials and components rose at a 9.4 percent rate and equipment and machinery prices rose at an 8 percent rate.[12] In the base case it is assumed that the *real* cost of construction increases at 1 percent per year during the construction period.

An additional cost is for financing of the project. It is assumed that 75 percent of the cost is financed externally by debt and 25 percent by equity and that the real interest rate on the debt is 5 percent. It is assumed that debt payments begin as soon as the project begins operation.

The capital costs of the project at the end of the construction period are summarized as follows:

$$TC = \sum_{n=1}^{N} [(C_n(1+x)\,(1+e)^j]\,[D(1+i)^{N-n} + (1-D)]$$

where:

TC = total capital cost at conclusion of the construction period
C_n = base capital costs expended in year n
x = contingency factor
e = real escalation rate
j = number of years by which year n exceeds base year of analysis
D = percentage of debt financing
i = interest rate during construction
N = number of years in construction period
n = number of years of construction that have been completed

The addition of contingency funds, real escalation costs during construction, and interest during construction raised the base case capital costs from \$3,499,300,000 to \$5,237,500,000 (1984 \$). All capital costs were levelized to 1984 dollars using a fixed charge rate. The fixed charge rate was calculated using a discount rate, various weighted cost of capital parameters (i.e., ratio of stock and debt for financing the

TABLE 23 ASSUMPTIONS AND METHOD FOR CALCULATING FIXED CHARGE RATE

(a) Fixed Charge Components:

1. Weighted Cost of Capital = (debt ratio x debt cost) + (preferred stock ratio x preferred stock cost) + (common stock ratio x common stock cost)

2. Sinking Fund Depreciation (sff) = $\frac{r}{(1+r)^N-1}$

   where: r = discount rate (weighted cost of capital)
   N = book life

3. Allowance for Retirement Dispersion = $\sum_{n=1}^{N} V^n S_n$

   where: N = book life
   V^n = present worth factor at year n at rate r
   S_n = percentage surviving in year n, calculated using the Iowa State University retirement dispersion curves
   r = discount rate (weighted cost of capital)
   sff = sinking fund depreciation factor

4. Levelized Annual Income Tax = (capital recovery factor + allowance for retirement dispersion - straight-line depreciation) x (tax rate/1-tax rate)) - (capital recovery factor + allowance for retirement dispersion - straight-line depreciation) x (debt ratio x debt cost/weighted cost of capital) x (tax rate/(1-tax rate))

   where the Capital Recovery Factor = $\frac{r(1+r)^N}{(1+r)^N-1}$

5. Property Tax and Insurance = .01 combined (by assumption)

(b) Assumptions (base case)

FC Component Assumptions	Assumption Values
Book Life	22 years
Tax Life	20 years
Debt Financing	.75
Equity Financing	.25
Weighted Cost of Capital (Discount Rate)	.05 (in real terms)
Investment Tax Basis/Investment Book Basis	.80
Federal and State Combined Marginal Tax Rate	.51
Investment Tax Credit Rate	.10 (not included in base case)
Property Tax Rate	.005
Insurance Rate	.005

SOURCE: For elaboration of methods used to calculate fixed charge rates, see Technekron Research, Inc., Cost Estimates and Cost Forecasting Methodologies for Selected Nonconventional Electrical Generation Technologies (Sacramento, Calif.: California Energy Commission, 1982, P300-82-006).

project), and income tax, property tax, and insurance factors (see table 23 for assumptions and calculation methods).

Operation and maintenance (O&M) costs are incurred once a plant is completed and operating. The plant's O&M costs are disaggregated into two categories: coal costs and all other costs.

The plant was assumed to use subbituminous coal from a nearby coal field with a 1982 coal price of $20.30 per ton. It was assumed coal prices would rise at a 2 percent real annual rate,[13] which gives an annualized cost of coal over the life of the project of $310 million in the base case.

Other O&M costs were those for labor, administration, material supplies, storage, and energy costs. Also included was the operation and maintenance of a short-distance railroad used to deliver coal from the nearby coal field. The total noncoal O&M cost, based on an escalation of the original cost estimates to 1984 dollars, was $166,400,000 per year. Since this analysis is being conducted in real terms, and because it is assumed that O&M costs excluding coal will escalate at the rate of general inflation, the total noncoal O&M cost above is also equal to the total annualized value of noncoal O&M costs.

The total annual O&M costs in 1984 dollars are summarized below:

O&M costs excluding coal:	$166.4
O&M coal costs:	$310.1
Total O&M Costs:	$476.5 million

The last important assumption that is needed to calculate the production cost of the fuel is to specify the nature of the output. For the base case, production costs were calculated per unit of energy output. In the sensitivity analysis that follows shortly, the different value- and revenue-generating abilities of methanol, SNG, and other incidental products were considered.

The total base case cost of fuel is summarized below in 1984 dollars per barrel of oil-equivalent output and per million Btu:

Capital costs	$47.00	$ 8.10
O&M costs	$38.90	$ 6.71
Total	$85.90/bbl	$14.81/mm Btu

In table 20 the range of coal-to-methanol production costs was estimated as $14.50 to $23.60 per million Btu in 1986 dollars, which

TABLE 24 SENSITIVITY OF COSTS TO CONTINGENCY AND CAPACITY FACTORS AND CONSTRUCTION SCHEDULES, DOLLARS PER BARREL OF OIL-EQUIVALENT FUEL

	Contingency Factor			Capacity Factor			Construction Schedule (years)		
	.08	.22	.37	.53	.72	.91	5	8	11
Capital Costs	$41.60	47.00	52.80	87.00	47.00	29.40	43.90	47.00	50.70
O and M Costs	$38.90	38.90	38.90	54.30	38.90	31.60	37.00	38.90	39.90
Totals	$80.50	85.90	91.70	141.30	85.90	61.00	80.90	85.90	90.60

roughly coincides at the lower end with the base case estimate here. Recognize that these costs are for finished (refined) products and are not comparable to crude oil costs. In any case, the precision of the cost estimate is not important here. The objective is to demonstrate the sensitivity of cost estimates, especially for new process technologies.

SENSITIVITY OF COSTS

In this section production costs are tested for their sensitivity to a number of factors related to construction, operation, financing, cost allocation, and government policies. Initially, all assumptions and parameters are held constant except for the single variable being tested.

First, consider factors related to construction and operations. As shown in table 24, changes in the construction period and in contingency requirements from the base case each have a significant effect, about plus or minus 6 percent, on production costs. The shorter construction period is probably more plausible than the longer period except in unusual situations in which there are significant technical problems or particularly strong public opposition (as is frequently the case with nuclear power plants).[14] Differences in capacity utilization have a much greater effect on costs. The New Mexico feasibility study assumed a capacity factor of 91 percent, which is an engineering rule of thumb for conventional industrial process plants, but this was lowered for the base case analysis because of findings by Rand Corp. that a more representative capacity factor for large new plants is 72 percent.[15] The range from 53 percent to 91 percent is suggested by that same Rand study. If the plant were to operate at 91 percent of capacity, production costs would be about $25 per barrel less than the base case value of $85.90.

Low utilization could come about as a result of a myriad of factors including technical operational problems and material supply problems. (Much higher operating capacities are also possible; the Great Plains coal gasification plant, for instance, was operating at 100 percent of design capacity within two years of start-up. This cost sensitivity analysis may therefore be excessively inflating costs.)

Factors related to financing also could significantly impact production costs. Key factors are the debt/equity ratio, interest rate, discount rate, and tax obligations. Effects of different debt/equity ratios were not tested here.

The interest rate affects only external financing, in this case 75 percent of the cost of the project. It is a measure of the cost of capital (see table 23, item a.l.). The interest rate is important because when a project is under construction for a long period of time, during which time costs are accruing, no revenue is being collected, and interest must be paid for money borrowed to cover construction costs. The prevailing interest rate is affected by government action and varies over time as a function of various macroeconomic factors such as money supply controls, deficit spending, inflation expectations, and world events. Real interest rates increased from about 2 to 3 percent in the 1960s and early 1970s to as much as 9 percent in the early 1980s.[16] Few economists had foreseen the rapid increase, and few pretend to have confidence in new forecasts.

By guaranteeing loans for a borrower, government may also alter interest rates on loans for particular projects. Since alternative fuel projects are risky, banks would normally require a borrower to pay a premium interest rate to compensate the bank for the risk it is assuming in making the loan. If the government pledges to honor the loan through a loan guarantee, banks will offer a considerably lower interest rate.

Loan guarantees were the principal subsidy tool employed by the U.S. Synthetic Fuel Corporation. In table 25 the effects of varying interest rates on production costs are shown. The 3 percent and 7 percent interest rates in that table are not extreme values, but are likely to represent the interest rates encountered by investors. The effect on production costs of this 3 to 7 percent range is relatively modest—a difference of only about 3 percent in production costs.

Another parameter that was tested was discount rates. The discount rate is used to convert costs from different time periods to equivalent values. Discount rates may be interpreted as the rate of return on an investment for output priced at actual production costs. A real rate of 5

TABLE 25 SENSITIVITY OF METHANOL PRODUCTION COST TO FINANCIAL ASSUMPTIONS

	Discount Rate (real)			Interest Rate (real)			Tax Incentives	
	.03	.05	.07	.03	.05	.07	with[a]	without[b]
Capital Costs	38.50	47.00	55.60	44.80	47.00	50.10	35.90	47.00
O and M Costs	38.80	38.90	39.00	38.90	38.90	38.90	38.90	38.90
Totals	77.30	85.90	94.60	83.70	85.90	89.00	74.80	85.90

[a]See text for explanation.

[b]Base case.

percent was used for the base case. As shown in table 23, varying the discount rate by 2 percent from the base case rate causes about a 10 percent change in production costs.

Another policy tool, one that is more readily available and more widely used, is tax incentives. Two commonly used techniques for altering costs through changes in taxes are changes in depreciation schedules and in credits for investment in capital facilities. In the base case it was assumed that a straight-line tax depreciation schedule was used and that no special tax benefits were available. A tax incentive scenario was created to illustrate the effect of tax changes in which a 10 percent investment tax credit and an accelerated depreciation schedule are allowed.[17] As shown in table 25, these two tax changes have a substantial effect, reducing production costs by 13 percent, from $85.90 per barrel to $74.80.

A goal that government is particularly vigorous in pursuing (especially in the U.S.) is preservation of the quality of the physical environment. Government actions to protect environmental quality in the form of rules, regulations, and taxes may have a significant effect on fuel production costs (see chap. 15 and 16 for detailed analysis of pollution impacts and control strategies). Objectives of government intervention might be to reduce acid rain, carbon dioxide emissions (to diminish the greenhouse effect), acid runoff from coal mines, emissions of toxic and carcinogenic materials to the air and into groundwater, and so on. Suppose, for example, that Congress enacts acid rain legislation that imposes stringent sulfur oxide restrictions on coal-burning plants. Since New Mexico has low-sulfur coal, one effect would be to increase the

TABLE 26 SENSITIVITY OF PRODUCTION COSTS TO ENVIRONMENTAL REGULATIONS

	Base Case[a]	Environmental Regulation Scenario[b]
Capital Costs	47.00	47.00
O and M Costs	38.90	51.40
Totals	85.90	98.40

[a]Assumes 2% real annual increase in coal prices.

[b]Assumes 4% real annual increase in coal prices (see text).

demand and price of New Mexico coal, causing an escalation in feedstock costs for the New Mexico synfuel plant. The effect on fuel production costs for the New Mexico plant, if real coal prices increased at a rate of 2 percent per year above the base case price, is shown in table 26. The effect is a 15 percent increase in fuel production costs (on a present-value basis). This analysis suggests the effect that a number of other environmental policies and regulations might have on production costs.

Production costs are also affected by how costs are allocated to different outputs. In the base case and in the previous sensitivity tests, all costs were allocated equally across all output, with output measured on an energy (e.g., Btu) basis. If the plant only produced one product, that combined-output cost allocation method would be appropriate. With multiple outputs it is not appropriate, because a particular piece of equipment in the plant may be used only in the production of one product—for instance, SNG—and not for the methanol coproduct.

Several different cost allocation methods may be used to determine the cost of producing the methanol portion of the output for this New Mexico plant. One is a chemical engineering method that uses the chemical heat of formation of different outputs; an alternative approach is based on the revenue streams of the outputs (or product prices of outputs). No single method is fully satisfactory. Different analysts use different methods, depending upon their expertise, which data are more accurate, and how the findings will be used.

The revenue-based cost allocation method is described here and the results are compared to the results of the previously calculated

TABLE 27 METHANOL PRODUCTION COSTS FOR COMBINED-OUTPUT AND REVENUE METHODS, 1984 $ PER OIL-EQUIVALENT BARREL

	Combined-Output Method (base case)	Revenue-Based Method*
Capital Costs	$47.00	$ 63.90
O and M	$38.90	$ 52.80
Totals	$85.00	$116.70

*Assumptions for estimating product revenues are listed below. These price assumptions are mostly based on a marketing analysis conducted by the Pace Co. in conjunction with the New Mexico Synfuel Project feasibility study.

Product	Unit	Annual Output(10^6)	Product Price in 1993 (1984 $)	Real Escalation	Annual Revenue (10^6 $)
SNG	cu. ft.	29.53	5.25 per MCF	1984-1992:0% 1993-2014:2%	194.47
Naphtha	barrels	.78	54.02	1984-2014:0%	42.12
Ammonia	tons	.04	360.47	1984-1992:0% 1993-2014:1%	17.59
Sulfur	tons	.05	189.64	1984-2014:1%	11.60
Phenols	tons	.02	490.04	1984-2014:0%	11.47
Carbon Dioxide	cu. ft.	52.61	2.43 per MCF	1984-2014:0%	127.83

combined-output method. According to this method, the estimated annualized revenue from each product, excluding the principal product in question (in this case methanol), is applied as a credit against the project's total revenue requirement (or total annualized cost); the residual dollar quantity represents the revenue (or cost) that is assigned to the product in question (or the revenue that must be generated by that product if plant costs are to be covered).

The anticipated revenue from each of the plant's outputs, excluding methanol, is estimated and applied as a credit against the project's total annualized cost (or revenue requirements).[18] The annualized revenue is then applied as a credit against the annualized cost components (capital and O&M costs) in proportion to each component's share of the total annualized cost. Methanol production costs calculated on a combined output basis (base case) and revenue basis are compared in table 27. Using the assumptions listed in that table, the revenue-based approach gives a methanol cost estimate 26 percent lower than for the combined output approach.

The cost estimate for the revenue approach is sensitive, of course, to the revenue assumptions. For example, if SNG prices escalated 2 percent per year faster than assumed in table 27, then methanol production costs would be $101.20 per oil-equivalent barrel rather than $116.70.

The final sensitivity-testing exercise is the creation of low-cost and high-cost scenarios to represent the aggregate effects of changes in a selected group of parameters. These two scenarios, along with the base case scenario, are presented in table 28. The variance represented by

TABLE 28 SENSITIVITY ANALYSIS FOR CHANGES IN MULTIPLE PARAMETERS, 1984 $ PER OIL-EQUIVALENT BARREL (COMBINED-OUTPUT BASIS)

	Low-Cost Scenario	Base Case	High-Cost Scenario
Capital Costs	$18.10	$47.00	$136.20
O and M	$23.60	$38.90	$ 73.40
Totals	$41.70	$85.90	$209.60

Assumptions for Scenarios:

Parameters	Low-Cost Scenario	Base Case	High-Cost Scenario
Capacity Factor	0.91	0.72	0.53
Contingency Factor	0.08	0.22	0.37
Construction Schedule	5 years	8 years	11 years
Coal Price Escalation	1984-2014:0%	1984-1992:0% 1993-2014:2%	1984-1992:2% 1993-2014:4%
Financing Rate	0.03	0.05	0.07
Discount Rate	0.03	0.05	0.07
Tax Preference Allowance	taken	not taken	not taken

the low- and high-cost cases is extreme and somewhat exaggerated—ranging from $41.70 to $206.60 per oil-equivalent barrel. Nonetheless, it demonstrates the sensitivity of cost estimates rather strikingly, and suggests the ease by which estimates can be manipulated, intentionally or otherwise.

SUMMARY

The range of cost estimates presented earlier in this chapter represents conventional wisdom. The estimates show, for instance, that the production of fuels from inexpensive natural gas is competitive with petroleum priced at $30 per barrel, and that other options are more expensive. However, the cost estimates are highly sensitive to a large number of uncertain future conditions and project-specific assumptions, as demonstrated in the second half of the chapter. The project-specific assumptions include the following: engineering details that are not always well understood; degree of project definition; biases by estimator or project proponent; changes in project design, product slate, or feedstock; management practices and capability; market for principal and secondary products; and bad luck—for instance, with regard to equipment delivery, weather, labor strikes, and labor shortages. Other uncertainties and variabilities are due to government intervention, changes in macroeconomic conditions and fuel markets, improvements in competing technologies, variability in the value of resources in competing uses, and resiliency or fragility of the physical environment.

An informed analyst, investor, or policy maker should be extremely cautious in accepting any cost estimate. In chapter 19 it will be shown that, in fact, many of these uncertainties are biased in a particular direction by past governmental policies, past investments (i.e., to preserve the value of sunk costs), and shifts in societal values and goals. It is argued that, in the long term, choices are based not so much on projected price trajectories as on past choices and fundamental societal values. When concerned with short time horizons, a price-determined approach is valid as a basis for making public and private investment decisions, but when the time horizon is pushed further out, social costs and fundamental values and goals should be considered in taking a more normative approach to specifying costs.

NINE

Further Assessment of Feedstock-Production Options

Production cost is the key criterion for evaluating the attractiveness of feedstock-production options. But as seen in the previous chapter, production cost estimates are often subject to considerable error, especially for long-term planning. This chapter expands the evaluation of feedstock-production options by considering other important factors, including the role of risk and uncertainty, investor participation, and commercial readiness. Another important factor, environmental effects, is discussed in chapters 15 and 16. To the extent that production costs are unreliable, these other factors take on greater importance.

NET ENERGY ANALYSIS: A MISLEADING MEASURE

One factor that is not of itself important is net energy output. It is addressed here because of the widespread attention it has received. Net energy analysis compares energy output with energy input. Using the first law of thermodynamics, one would compare energy content (i.e., heating value) of all inputs with energy content of outputs (see table 19 for first-law thermal efficiencies). In practice such a measurement is difficult to make. It becomes mired in questions of how to distinguish between renewable and nonrenewable feedstocks, how to incorporate energy required to power the process plant, and how to treat secondary outputs (coproducts and by-products). For instance, how does one compare the efficiency of converting wood, coal, or grass into alcohol?

Is one Btu of wood equivalent to one Btu of coal, petroleum, or natural gas? Certainly not! It would be a case of comparing "apples and oranges."

The question of net energy effect of fuel production options gained attention during the late 1970s. The shocking discovery was made that the conversion of corn to ethanol used as much or more hydrocarbon energy as it produced in the form of ethanol.[1] That is, it was found that more petroleum and natural gas was being consumed in growing, harvesting, fermenting, and distilling the corn (as fertilizer, fuel for tractors, and process heat for distillation) than was being displaced by the ethanol that was produced. The findings were valid but accurate only for those process plants that had formerly been beverage distilleries and corn processing facilities—plants that were built many years before without much concern for energy efficiency. Many of these plants used petroleum and natural gas for process heat. Process plants built in the 1980s are much more efficient and use mostly lower-cost coal.

While the case was overstated, it remains true that the production of ethanol from sugar and starch feedstocks is energy intensive. Depending upon a large number of assumptions, the production of ethanol from corn, even in efficient plants, consumes half or more of the high-quality hydrocarbons it displaces.[2] This is because corn production uses large amounts of hydrocarbon-based fertilizer and because distillery operations require large quantities of energy.

The concept of net energy balance and conversion efficiencies is important, but in practice it has been abused because of inattention to basic underlying assumptions and methodologies. Apart from the difficulty of using a common methodology, estimations of energy balances and conversion efficiencies are not critical because they can be subsumed into cost analysis, which, despite its shortcomings, is nevertheless a more robust measure. For instance, in the case of corn-to-ethanol, if more petroleum was consumed in production than was displaced by the alcohol, then that feedstock-production option would clearly be uneconomic and unwise. Simple economic analysis would have demonstrated that ethanol produced from corn was uneconomic—because of the high cost of corn, which in part results from the hydrocarbon-intensive farming methods used in growing corn, and because of the high cost of processing, which resulted in large part from the use of large amounts of petroleum and natural gas for process heat. Simple modifications in subsidy programs could be made to respond to these undesired energy effects, for instance, by not allowing oil-fueled distilleries to receive government subsidies. Normally, though, the market system by its very nature acts to prevent undesirable energy balances; if the cost of hydrocarbon inputs is greater than the value of the fuel itself then production of that fuel would be unprofitable.

The analysis of net energy balance presented in the previous paragraph represents a narrow application of the first law of thermodynamics. In the general case, using the second law of thermodynamics, one would compare the energy quality and quantity of the end product, the transportation fuel, with the energy quality and quantity of inputs. The energy content (heating value) of coal, corn, or whatever feedstock is used will always dissipate as it is converted into another form. The energy loss (in heat, gases, and other by-products) is accepted because the end product is in a more desired form than the input. For instance, gasoline is presumably more desirable than crude oil, electricity more desirable than coal, and alcohol more desirable than wood, otherwise the conversion process is not undertaken. The second law of thermodynamics addresses the quality and form of energy.

Ideally the market system will assign value to each commodity (based in part on the quality of the energy consumed and produced) and determine the attractiveness of any particular conversion process. In practice, energy balance analysis is useful only to the extent that market mechanisms are distorted and future prices of feedstock materials are uncertain. Reliance on energy balance estimates in terms of both the first and second laws of thermodynamics tends to lead one astray. Because fewer parameters are considered, energy balance estimates are a less powerful and less robust measure of attractiveness than cost analysis. Net energy measurements are useful indicators, but as suggested here (and demonstrated in the comparison of indirect and direct liquefaction processes in chap. 7), they should not be treated as primary or even important evaluation criteria.

REVENUE UNCERTAINTY

A much more critical consideration is uncertainty about the energy market and about future revenue streams of particular projects. In the previous chapter the various risks and uncertainties surrounding production costs were analyzed. While the risks associated with production costs indeed are large, an even greater risk from a producer's perspective is the future market value of the fuel itself. Thus investors face risk not only in production costs but also on the revenue side of the ledger.

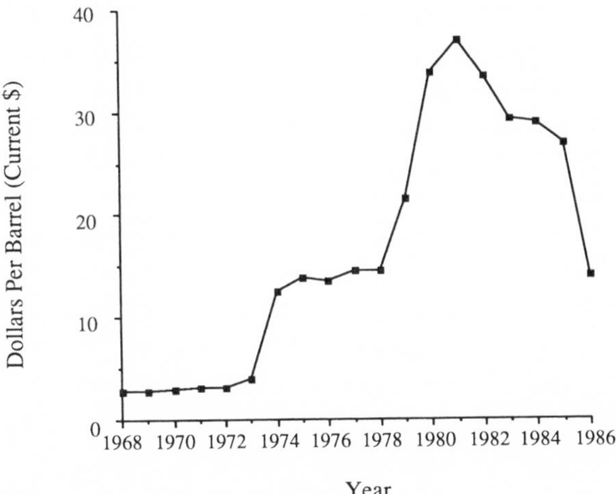

Figure 22. Historical Oil Prices

Source: U.S. Energy Information Agency, *Annual Energy Review 1986* (Washington, D.C.: GPO, 1987).

Note: Prices are refiner acquisition cost of crude oil imported into U.S.

The single most important yet unpredictable factor affecting market uncertainty, and the criterion against which nonpetroleum fuel investments are tested, is the price of petroleum. The price of petroleum will determine the value of the alternative fuel and, therefore, the revenue that an investor will collect from the sales of the fuel. But since there is a five- to ten-year lag from the time an investment decision is made until fuel production actually begins, fuel producers are vulnerable to unforeseen shifts in market conditions and fuel prices. The best illustration of the unpredictability of petroleum prices is a simple graph of recent petroleum prices (fig. 22). None of the major petroleum price changes, the fourfold increase in 1973–1974, the threefold increase in 1979–1980, or the 60 percent decline in 1985–1986, was widely expected or predicted by forecasters. Petroleum price forecasters have a poor track record. Forecasters do not and cannot anticipate political events such as collusion and price setting, government overthrows, massive sabotage, wars, embargoes, and other types of disruptions.

Market uncertainty is greatest for fuels dissimilar to petroleum. This market uncertainty is associated not only with petroleum prices but also with the development of a market for the product. How much market penetration can be expected for methanol or ethanol or CNG? Risk is

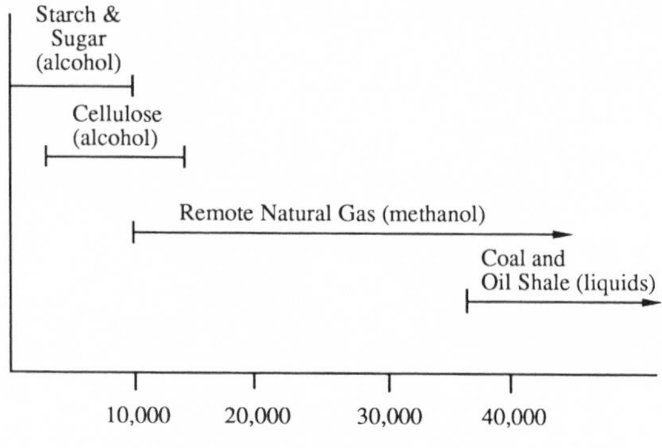

Figure 23. Economically Efficient Sizes of Process Plants

especially great for large process plants, because the investment is so large and because of the greater difficulty of marketing very large volumes of fuel. If markets do not develop as expected for dissimilar fuels, or if market prices are lower than expected, a large plant becomes a "white elephant." The more expensive the plant and the smaller the company, the greater the risk. An expanded investigation of market uncertainty is provided in chapters 13 and 14.

ECONOMIC PARTICIPATION

Because of the major role energy industries play in the economy, economic participation patterns will affect the long-term distribution of wealth and power in a society. Who will invest in what plants will be influenced by the size and nature of the process plant. At one extreme are small biomass fuel plants and at the other are multibillion-dollar mineral fuel plants. A decision to pursue one set of options or the other therefore directly affects the social and economic structure of an economy and society.

First, let us examine the magnitudes of these investments. The typical size of different types of process plants is summarized in figure 23. A rough approximation of the cost of individual process plants can be made by applying the capital cost requirements of table 21 to the determination of efficient plant size in figure 23. These costs, expressed as total capital costs (1986 $) per process plant, are as follows:

Bioconversion plants	\$20,000–\$170,000,000
Thermochemical biomass plants (mass-produced)	\$10,000,000–\$20,000,000
Thermochemical biomass plants (one-of-a-kind)	\$180,000,000–\$1,100,000,000
Remote natural gas plants (land or barge-borne)	\$250,000,000–\$1,400,000,000
Coal and oil shale plants	\$2,300,000,000 +

As shown, the capital cost for process plants ranges from about \$20,000 for small bioconversion plants that might be deployed on farms to multibillion-dollar megaprojects that would be owned by a handful of the largest corporations. Owners and investors in biomass projects would tend to be small and moderate-size industrial establishments that are located on or near the land where the biomass is grown. The symbiotic relationship between feedstock and process plant, and the small-to-moderate size of biomass process plants and therefore of production investments suggest that those farmers, cooperatives, and corporations that grow and own the biomass feedstock will also tend to play a major role in producing and marketing the alcohol fuel. Forest product companies and agricultural and food processing cooperatives and corporations, as well as individual farmers, would likely be the predominant participants in biomass fuel production. Farmers, small cooperatives, and other businesses would be capable of financing smaller fermentation and thermochemical plants, but only the largest agricultural, forest products, and food-processing firms would have the resources to invest in the largest biomass plants—plants that would cost several hundred million dollars or more.

The case of wood-based fuel activities is instructive. Roughly 28 percent of all commercial forestland in the U.S. is owned by the government,[3] and the prime areas for growing biomass for fuel east of the Mississippi River have dispersed land ownership and land management.[4] Land ownership is especially dispersed for forested areas of the U.S. Southeast, where the largest potential source of biomass is located (larger corporations of the forest product industry lease the land from the smaller landowners and manage wood production themselves). Thus the likely effect of woody biomass fuel production would be substantial land rents returned to government and ultimately to the general population, and widespread economic participation in production on private lands.

Economic participation in biomass fuel activities would be expanded by the creation of many new opportunities for aspiring entrepreneurs and local businesses—for instance, serving as intermediaries between landowners and plant operators in the collection and sale of feedstocks. These entrepreneurs would tend to be small and flexible, and have the local knowledge to respond to local opportunities.

A survey of planned (bioconversion) ethanol plants conducted in 1979 by the National Alcohol Fuel Commission indicates the potential for widespread economic participation in biomass fuel activities. Of 300 firms only five were oil and gas companies and none of those five ranked as a major energy firm. Almost all prospective producers were small companies, and most were involved in some agricultural activity.[5]

Although it was demonstrated earlier that the economics of biomass fuel production would push ownership toward atomistic patterns, there is evidence that very large economic enterprises would also participate. Already Texaco, Chevron, and Ashland Oil companies have entered joint ventures to produce ethanol from agricultural crops. Large energy companies could be expected to pursue this pattern of limited partnerships as a means of hedging their other energy investments, and to assist smaller firms in raising capital and marketing fuel. However, large energy companies are capital-intensive firms that have little experience with agricultural land management activities (though there are exceptions) and are therefore unlikely to dominate the production end of a biomass fuel industry.

The major participants in biomass fuel production are likely to be large companies in the forest product and agricultural industries. Initially, one food-processing firm (ADM) has dominated ethanol fuel production in the U.S., accounting for over 50 percent of ethanol fuel production in 1985. When (or if) biomass fuel production expands, ADM's market share will certainly diminish. It is uncertain what mix of food, agricultural, and wood products companies would characterize a large-scale biomass fuel industry—whether the industry would be dominated by relatively large firms such as ADM or whether participation would be more dispersed. It should be kept in mind, however, that even the largest agricultural and forest products firms are much smaller in size than the energy companies that would be the predominant investors in mineral fuel production. In 1984 in the U.S. the largest industrial company with any food or wood-related activities was ranked thirtieth (by sales), while the highest ranked forest products firm was sixty-second.[6] ADM was ranked seventy-fourth.

Oil, gas, and petrochemical companies, however, are among the largest corporations in the world and are the most likely investors in mineral fuels plants. Petroleum and chemical manufacturing companies accounted for fourteen of the largest twenty-one, and twenty-three of the largest fifty companies in the U.S. in 1984, and twenty-six of the top fifty industrial companies in the world.[7] Those companies with large coal reserves and otherwise unused oil shale, oil sands, and RNG resources would seek to exploit their holdings. Those firms that refine and market gasoline can readily provide markets and marketing expertise for these new fuels. Most important, these energy companies have the financial resources to pursue multibillion-dollar investments.

The squeezing out of small businesses and investors in large mineral fuel projects will be a result of the large risk involved. Even the largest energy companies are reluctant to undertake the full risk of a large process plant. Almost all the projects that sought financial support from the U.S. Synthetic Fuel Corporation were joint ventures, with some projects including as many as seven major oil companies as sponsors. When sponsors of the Great Plains SNG project defaulted in August 1985, the financial impact was large—even though five investors shared in the costs, the companies were large and the plant not very large for a commercial mineral fuels plant, and even though the government absorbed most of the loss because of loan guarantees. One company (Transco) had a loss of $91 million, which represented 70 percent of the company's total earnings of the previous year.[8]

The nature of hydrogen plants and investments is the least certain. However, there appear to be few if any economies of scale in photolytic and photovoltaic hydrogen production. The production plants would be built in modules. The only important economy of scale would come in pipeline distribution, but those economies would be tiny compared to the production cost. Since these plants would be using solar energy, they would also be land intensive. For instance, a very large photovoltaic plant (or set of plants) that produced about 1 percent of U.S. energy demand (0.6 quadrillion Btu) would cover 800 square miles. The small-scale nature of the production plants lends itself, as does biomass production, to decentralized energy production systems and broad ownership patterns.

In summary, mineral-based fuels production would be dominated by large energy companies seeking to exploit their resource holdings and to diversify their energy activities, while biomass fuel (and perhaps hydrogen) activities would be dominated by a wider mix of small and middle-size businesses.

COMMERCIALIZATION SCHEDULE OF FEEDSTOCK-PRODUCTION OPTIONS

The array of opportunities for producing alternative fuels will now be brought into sharper focus by considering the following three factors:

Development status of production technologies and lead time required to bring a new facility on line

Attractiveness of competing opportunities for feedstocks

Production cost.

The feedstock-production systems rate very differently according to these three criteria. The first limiting feature is the commercial readiness of the technology (see table 29). The state of technology of different processes varies greatly. Most advanced are gas-to-methanol technologies and the simple bioconversion systems that use starch and sugar feedstocks, both of which are already widely commercialized. Some thermochemical systems are nearly ready for commercial production, while others require more development at the pilot-testing stage. The

TABLE 29 DEVELOPMENT STATUS OF CONVERSION PROCESSES

Feedstock-Production Process	Pilot Stage	Demonstration Stage	Technologically Ready	Commercially Established
Biological conversion (ethanol)				X
Oil sands (pyrolysis)				X (Canada)
Barge-borne methanol (RNG)			X	
Indirect coal liquefaction			X (U.S.)	X (S. Africa)
Oil Shale (pyrolysis)		X (Union Oil)		
Direct coal liquefaction	X			
Indirect biomass liquefaction	X			
Hydrogen	X			

greatest research effort has been devoted to the conversion of coal and oil shale feedstocks. Wood and municipal waste feedstocks have received much less research; small efforts have been nurtured at several universities and nonpetroleum firms (e.g., Union Carbide and the company spun off from International Harvester). Other potential cellulosic feedstocks have received almost no attention from process engineers; basic process technology now available would have to be modified to handle those feedstocks (such as grasses). Those modifications require time and resources for redesign and testing. The lead time needed to bring these technologies to a state of commercial readiness and actual commercial production depends in part on costs and market factors, and in part on level of R&D effort. For instance, because political pressures favor ethanol bioconversion, cellulose-to-alcohol processes have received very little funding from the U.S. government during the 1980s, and as a result their development is lagging.

The time between investment decision and the actual start-up of plants is measured in years. Start-up is delayed by the time required to acquire permits, complete site-specific engineering designs and land preparation work, and time to construct the production facility. Lead time is crucial, as will be demonstrated in later chapters, because it affects the ability of investors to respond to incentives and market opportunities. It also influences the extent to which end-use, distribution, and production investments can be coordinated, a critical consideration in the case of alcohols and other fuels that differ from petroleum.

The longest lead times are required by the largest production facilities. Actual construction times are about three to five years for coal and oil shale plants, about two years for barge-borne and large biomass plants, and somewhat less time for small biomass plants. Preparation for construction may take as much time as construction, but it is difficult to predict exactly how long the permit application and final design and site preparation activities might take. Generally, though, the preliminary preparation phase should take about one year for small plants (and barge-borne plants) and, as shown in figure 24, about two to five years for large projects. Small plants are less controversial and are not likely to be delayed; large projects are more complex and more disruptive of local communities and physical environments and therefore more susceptible to delay in the permitting process. Overall, then, the largest coal and oil shale projects would require five to ten years from the prefeasibility stage to actual operation, while biomass and barge-borne plants would require less than four years. There are two excep-

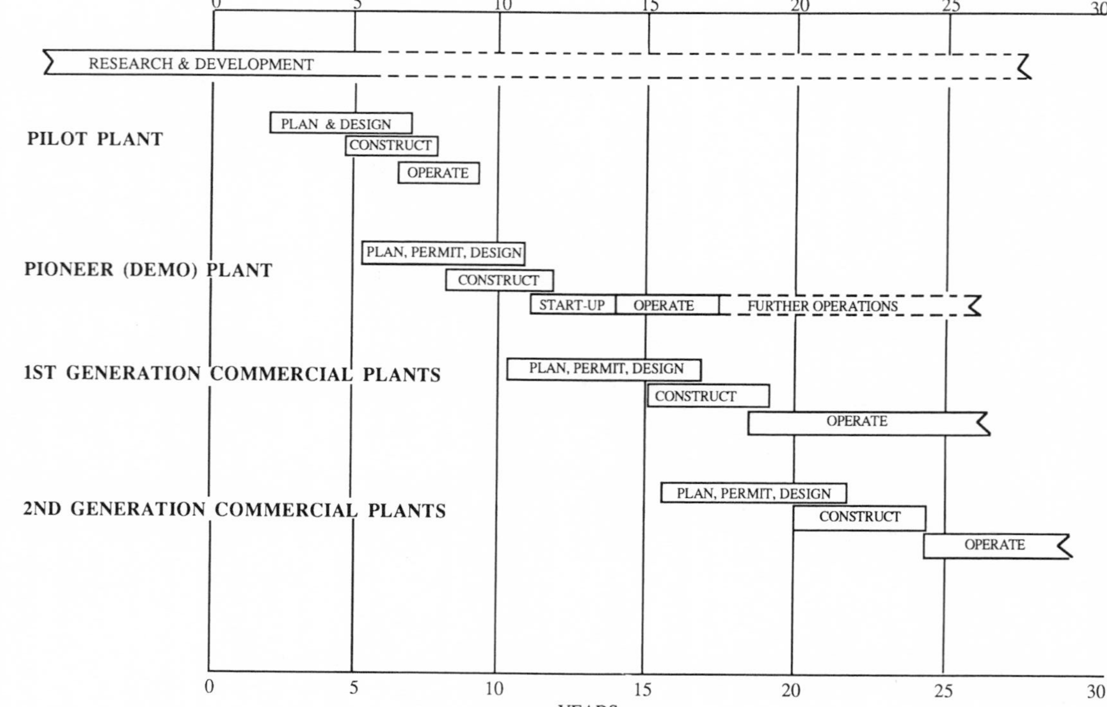

Figure 24. Timescale for Commercialization of Coal and Shale Conversion Technologies

Source: U.S. Synthetic Fuels Corporation (Distributed at Conference on Transportation Fuels in the 21st Century, SAE, Washington, D.C., November 1985).

tions to these generalizations: first, mass produced plants would require much less time than one-of-a-kind plants, and silvicultural farms would require five to fifteen years because of major site preparation activities and the growth period of wood biomass.

The lead time for developing efficient process technologies is much longer than indicated by the previous paragraph, since successive generations of the technology must be designed, built, tested, and operated—each succeeding design showing improvement. For direct coal liquefaction processes, industry lead time would be at least twenty-five years, assuming continued commitment, or more if development work wanes.

A factor that could slow the development of a particular feedstock-production option is competition for the feedstock by other uses (see fig. 25). An optimization model developed in a Ph.D. dissertation to test the relative cost attractiveness of some major biomass options showed that if natural gas prices do not increase relative to oil, the most profitable energy use for biomass feedstocks in the U.S., apart from firewood, is alcohol fuels.[9] Using 1978 feedstock prices and production cost, and ignoring technology readiness considerations, that research found that residues available in concentrated locations, such as at food-processing plants and at lumber mills, could be economically converted to alcohol fuel. With a moderate increase in real oil prices, the model indicated that logging residues would become profitable. With a doubling of real oil prices, crop residues and silvicultural energy production on excess land also became profitable. Only if oil prices quadrupled would energy production be attractive enough to divert traditional agricultural commodities into alcohol production and to divert agricultural land to wood biomass for energy production.

Although the assumptions are now out of date and it is difficult to verify the accuracy of feedstock prices and production cost functions used in that study's linear programming model, the results give a sense of the relative opportunities for introducing alternative fuels. The important finding from that analysis is that alcohol fuels are a highly attractive option relative to other biomass energy applications.

The production of low-Btu producer gases from cellulosic biomass is not a major competitor with biomass alcohol even though the production process is much simpler and less expensive; that is because the gases are expensive to transport (because of their low energy density) and therefore can only be used near the production site. Thus even if these low-Btu gases were to be attractive (which was found in the Ph.D.

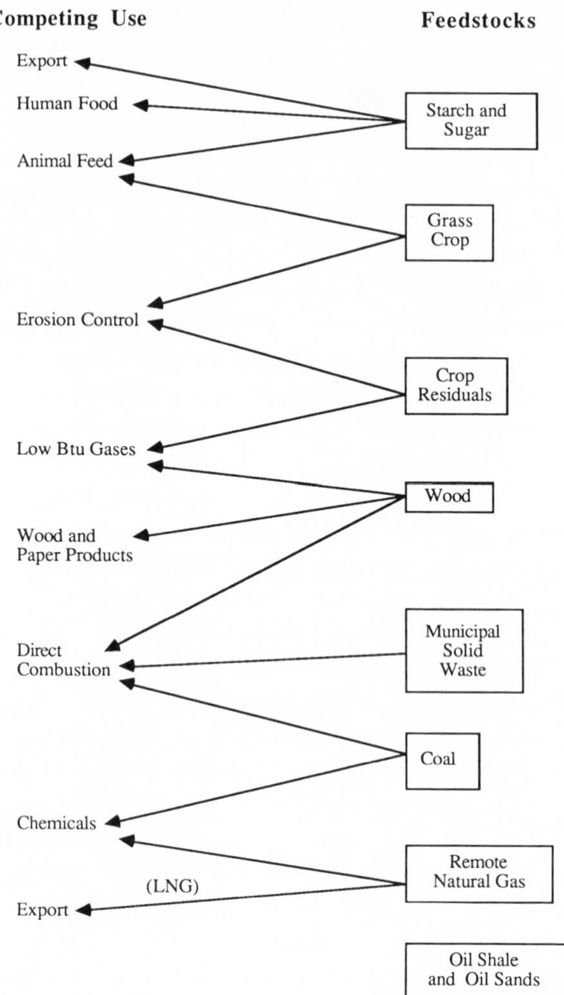

Figure 25. Principal Nontransportation Uses For Energy Feedstocks

study to occur only in the unlikely situation that natural gas prices rose to levels comparable to those for oil), their market is sharply limited.[10]

CONCLUSION

As a means of summarizing the four chapters on feedstock-production options, a list has been created of the relative attractiveness of the many options based on the three criteria of technology readiness, attractiveness of transportation fuel relative to other end-use applications, and production costs. It is not based on any normative or long-term considerations and ignores for now fuel distribution and end-use considerations, environmental and social factors, and government subsidies. The ranking of feedstock-production options according to the three criteria is provided below:

1. Compressed natural gas from remote natural gas
2. Coal gases for coproduction of methanol and combined cycle electricity
3. Methanol from remote natural gas
4. Methanol and high-Btu gas from coal
5. Petroleum-like liquids from oil shale
6. Methanol from concentrated biomass residues
7. Gasoline from methanol (from coal)
8. Methanol from forest residues and intensively managed wood production
9. Petroleum-like liquids from coal (direct liquefaction)
10. Ethanol from corn and grains
11. Methanol from grass crops
12. Methanol from silvicultural farms
13. Long-term and exotic options (e.g., aquaculture, hydrogen, fuel cells).

This list is suggestive and is not intended as a definitive evaluation. Indeed, a theme of this book is the fallacy of making judgments based on so-called average conditions and the desirability of examining energy options in the context of their societal setting. The ranking would vary tremendously from one region (and site) to another and one time period to another. Thus, in Brazil, ethanol from sugar feedstocks probably ranks ahead of all options except petroleum. Similarly, in remote areas with large natural gas reserves, compressed natural gas would rank at or near the top of the list. Governments, of course, often intervene in such a way as to disrupt this rank ordering; the continuing expansion of corn-to-ethanol investments in the U.S. is a vivid example.

In later chapters this fairly narrow initial analysis of feedstock-production options is expanded and made more robust by including social and environmental factors, and fuel distribution and end-use considerations. The evaluation will take on a more strategic and normative orientation as additional information and insight is incorporated.

PART IV

Fuel Distribution

Fuel distribution is the movement of fuel from the production plant to the end user. Since most nonpetroleum fuels are and will be produced in remote locations far from end-user markets, fuel distribution considerations often play an important role in determining when and where the fuel will be produced and marketed. As has been so vividly illustrated by past debates over the building of the Alaskan oil and gas pipelines, LNG terminals, coal ports, and other major distribution facilities, fuel distribution considerations are not trivial. Few attempts have been made, however, to deal with the strategic and systems questions of developing distribution systems for new fuels, especially for those fuels dissimilar to petroleum.

Fuel distribution activities have been virtually ignored in academic and policy circles and by government regulators, not only for new fuels but also for petroleum products. The reason for this neglect is that fuel distribution in the U.S. has been almost totally a private sector activity that has been performed efficiently with no catastrophes or "white elephant" embarrassments. As a result, fuel distribution activities have received little regulatory attention and are poorly understood outside industry. A well-respected and otherwise knowledgeable congressman who was serving on an energy subcommittee demonstrated the extent of that ignorance at a 1980 hearing on alternative fuels with the query: "we don't, at the present time, ship gasoline in pipelines, do we?"[1] The answer is yes—virtually all gasoline is pipelined.

Because there is a huge sunk investment in the existing petroleum product distribution system, investments and activities that will not use those existing facilities are at a disadvantage. The extent to which petroleum products and natural gas can share their highly efficient system with new fuels will critically influence when and where new fuels will be produced and used.

Compatibility of new fuels with the existing petroleum and natural gas distribution systems may be categorized in three groups: material, operational, and logistical compatibility. Material compatibility may be specified on strictly technical grounds. But operational and logistical compatibility are far more complex phenomena; they deal with size and location of new process plants, size of fuel shipments, and the operational practices of pipelines and other system components.

The primary focus of these two chapters on fuel distribution is alcohol fuels, since they face much greater compatibility problems than natural gas and nonpetroleum fuels similar to petroleum (e.g., shale oil). The cost and difficulty of establishing a network of natural gas fuel outlets, the costs and problems with LNG transport, and compatibility of hydrogen with an existing natural gas pipeline network are other nonalcohol issues addressed herein.

The key questions in these next two chapters are not technical, but economic and strategic. Even though alcohol and gasoline have almost identical viscosity and flow characteristics, there are major impediments to a shared system. The concern is that if it were perceived that the costs and barriers to change are high and that alcohol could not be effectively integrated into the gasoline system, then fuel investments would be directed toward fuels that are more compatible with the existing system but possibly inferior to alcohol. An example might be a decision to produce synthetic gasoline from methanol, even though the synthetic gasoline would be more costly to produce and of lower quality than the methanol it is produced from.

These next two chapters investigate the difficulty and cost of introducing even small changes to large long-standing systems. In chapter 10 the existing petroleum product and natural gas distribution systems are examined. A functional analysis of the various transport modes is conducted in which the salient attributes of each mode are specified. Chapter 11 analyzes the material, operational, and logistical compatibilities of alcohol and petroleum products, and the factors influencing the establishment of retail outlets for alcohol and gaseous fuels.

TEN

Current Fuel Distribution Systems

PETROLEUM PRODUCT DISTRIBUTION

The first bulk transport of liquid fuels began in 1859 at the site of Colonel Drake's oil well in Pennsylvania.[1] Horse-drawn carriages loaded with large barrels soon gave way to barrel-loading on rail cars and ships and small cast-iron pipes. Cast-iron pipes were eventually replaced by steel pipes, and specialized rail tank cars and tanker ships replaced portable barrels. Later, in the twentieth century, the development of motorized tanker trucks facilitated the dispersal of fuel stations to locations not served by railroad. In summary, the basic technologies for transporting liquid fuels as we now know them were in place in the early 1900s. Their basic characteristics have not changed since then, nor have the liquids they carry.

The productivity of liquid transport has increased many times in the intervening years. These productivity increases are derived almost entirely from technical improvements that increased economies of scale. Those transport technologies with the greatest inherent potential for expanding capacity—ships, barges, and pipelines—have come to dominate liquid fuel transport, mostly at the expense of the railroad system. Figure 26 shows that pipelines have greatly expanded their market share over the last six decades while railroads have seen their market share practically disappear.

The emergence of pipelines as the dominant mode for high-volume shipments eroded the market shares of other transport modes.[2] Other

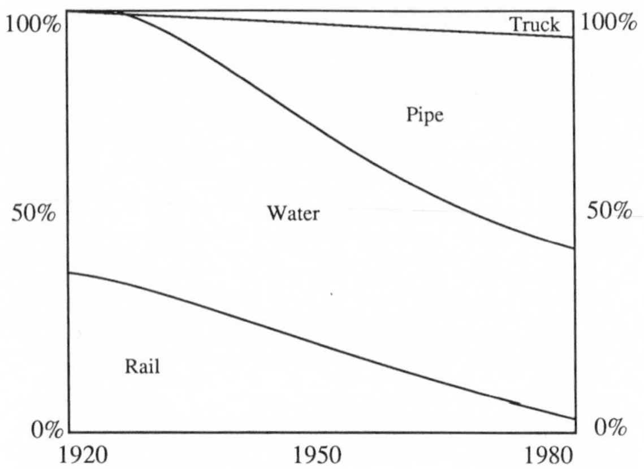

Figure 26. Market Share of Petroleum Products Transported Domestically (by ton-miles), U.S., 1920–1980

Sources: Adapted from U.S. Congress, Congressional Research Service, *National Energy Transportation* (Washington, D.C.: GPO, 1977); and U.S. Department of Transportation, *National Transportation Statistics* (Washington, D.C.: GPO, annual).

modes gradually withdrew to the market for which they were best suited. Coastal tankers and barges now carry heavy liquids, rail serves remote and specialized markets (e.g., small-volume chemicals), and trucks make short local deliveries. Only at the edges of their market niches are transport modes competitive with each other. In summary, each mode is suited to a particular type of transport service. The relative attributes of each mode are presented in table 30 and elaborated upon in this chapter.

Specialization of freight modes to particular market niches has been the result of the cost, carrying capacity, and guideway characteristics of those modes. But over time these modes have developed other attributes that are not necessarily inherent to the technology of that mode. For instance, the structure of the petroleum product distribution system reflects the supply and demand patterns of petroleum products as well as the physical and chemical properties of petroleum. The existing structure and operation reflect the peculiarities of the petroleum industry; but this is not a function of some invariant laws of technology.

Petroleum products and alcohol fuels have similar physical and chemical properties; thus an ideal future fuel distribution system for alcohols would have similarities with today's petroleum product distribution system. In particular, the transport and storage technologies

TABLE 30 ATTRIBUTES OF LIQUID TRANSPORT MODES

Modal Attributes	Tank Truck	Rail Car	Unit Train	Existing Pipe	New Pipe	Water
Unit Shipping Cost*	H	HM	M	L	L	L
Capital Cost*	L	L	M	L	H	L
Ease of Deployment	H	H	M	M	L	M
Flexibility	H	HM	M	L	L	H
Shipment Size	L	L	H	M	HM	M
Service Reliability	H	L	HM	H	H	M

*Costs based on full utilization over lifetime of equipment/vehicle. Some of these attributes have a larger range than indicated, depending upon various circumstances. These relative rankings are illustrative, not definitive, and are subject to some variation.

H = high, large; HM = medium high; M = moderate; ML = medium low; L = low, small.

would be similar. However, network structure would be very different because feedstock sources and process plants of most biomass and non-petroleum mineral fuels are not located near petroleum refineries. The challenge, then, is to discern the rigidity of the existing system and the extent to which it is flexible enough to accommodate change. Low tolerance to change may place such overwhelming start-up costs on certain incipient fuel options that their growth might be stunted for decades.

The petroleum product distribution system includes transportation technologies and storage terminals. The configuration of current distribution systems for transportation fuels is depicted in figure 27. Gasoline and diesel and jet fuel are usually transported from oil refineries to

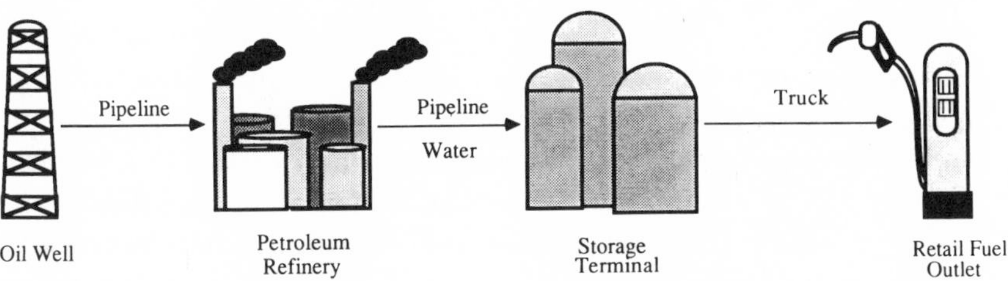

Figure 27. Configuration of Transportation Fuel Distribution System

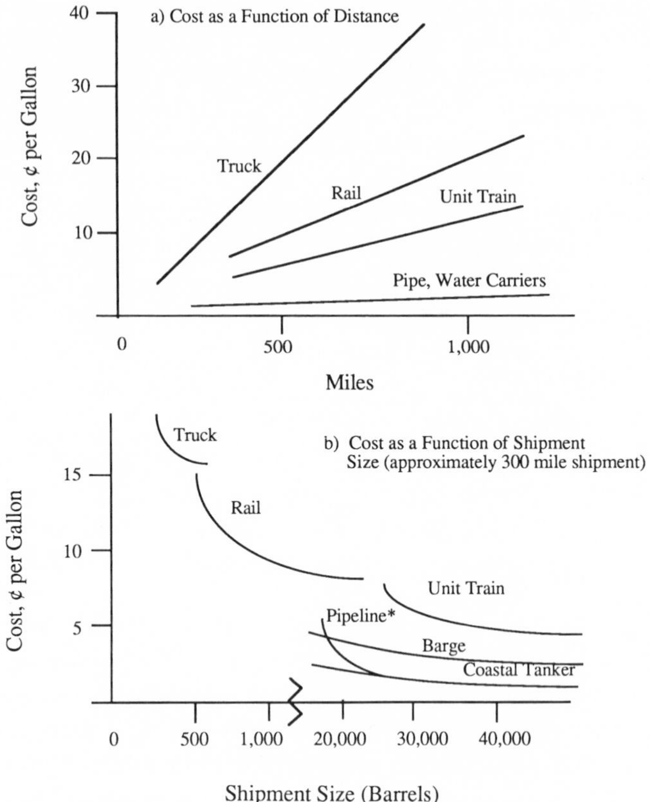

Figure 28. Liquid Commodity Transport Costs as Function of Distance and Shipment Size (early 1980s)

Source: Rate quotations by Southern Pacific (rail and unit trains), Archer Daniel Midlands (rail and truck), Bulk Carrier Conference (truck); published tariffs by Federal Regulatory Commission (pipelines); U.S. DOE, *U.S. Petroleum Pipelines: An Empirical Analysis of Pipeline Sizing* (all modes); and G. Wolbert, *U.S. Petroleum Pipelines* (Washington, D.C.: American Petroleum Institute, 1979), Appendix A (all modes).

bulk storage terminals by pipeline (and occasionally by water transport). Bulk storage terminals are located in or near major metropolitan areas. The fuel is then shipped by small pipeline or truck to a smaller terminal located at smaller urban centers *or* shipped by truck directly to end users (mostly retail fuel outlets). Gasoline deliveries to end users are almost always by truck.

Cost characteristics of transport modes are presented in figure 28.

The cost functions in the two graphs are intended to be representative, not definitive. In practice those costs will vary according to various factors such as topography, type of liquid, pipeline diameter, and operational characteristics. The costs in figure 28(a) are estimated for the large size of each transport technology, e.g., long unit trains and large-diameter pipes. Considerable evidence exists that the structure of those cost functions in terms of trip distance and shipment volume is relatively stable and that the *relative* costs (though not the numerical values) represented in figure 28 will hold over space and time.

The graphs illustrate the considerable difference in cost and carrying capacity between modes. Generally speaking, truck is the preferred mode for small shipments under 200 miles, and barges, where accessible, are the preferred mode for volumes under 150,000 barrels per day and for dense products that do not flow easily in pipes. Elsewhere, pipelines are preferred. In the U.S. pipelines carry about 80 percent of all light- and middle-distillate petroleum products.

PIPELINES

Pipelines dominate in the transportation of light petroleum products because overall they are the least expensive and most reliable mode of liquid transportation. But these benefits are accessible only for large and regularized shipments. The scale economies of pipelines are shown in figure 28(b). As shipments between two points are consolidated and shipped in ever-larger pipes, unit transport costs drop substantially.

The large scale economies of pipelines have led to an unusually high level of joint ownership. In order to boost throughput in a line and therefore to decrease per gallon shipping costs, oil companies and their subsidiaries have tended to consolidate their traffic into fewer but larger pipelines. They have formed joint venture companies and "systems" in which each company has a specified interest; in the latter case of joint venture "systems," a specific volume is allotted to each participating company. In 1975, seventy-eight product pipelines were owned by joint ventures representing 56 percent of all petroleum product pipelines.[3] Pipelines not owned by joint ventures are generally smaller and are mostly owned by petroleum company subsidiaries.

The joint venture arrangement may prove to be a strong deterrent to fuels dissimilar to petroleum products, since acceptance of a new fuel would require the consent of all owners. Although pipelines are common carriers and are required by law to serve the public, there are many ways to avoid compliance. In the case of alcohol there may be some justification for noncompliance, since the long-term material effects of alcohol in pipelines might not be detected for many years, creating a certain amount of risk for the pipeline owner.

Material incompatibility takes several forms: corrosion, other forms of material damage, scouring, and stripping away of corrosion inhibitors. Alcohols affect materials in different ways than petroleum. Methanol, which has more severe effects than ethanol, has been found to adversely affect aluminium, zinc, lead, magnesium, some plastics (e.g., polyurethane), fiberglass, and most important, carbon steel, which is used for pipes, valves, pumps, and storage tanks.[4] Before methanol can be used in pipelines and their associated storage facilities, considerable modification or replacement of materials will have to take place; for instance, by coating surfaces with methanol-resistant substances or by replacing the original material.

In one sense the material compatibility problem in pipelines is not severe. Most of the materials incompatible with alcohol are not widely used. In the case of steel, which is of course widely used, the corrosion rate is extremely slow, and, as suggested above, corrosion-resistant materials could be placed on the pipes. But corrosion is exacerbated by two other properties of alcohol: unlike petroleum, it dissolves and scours rust, gum, and dirt deposits that accumulate on steel, thereby exposing pipes (and tanks) to new corrosion and accelerating the corrosion process, especially when water is present. Alcohols also strip away internal corrosion inhibitors used in petroleum pipes, again permitting corrosion to occur. Possible solutions are to thoroughly clean pipes before alcohol shipments are made, to add corrosion inhibitors to alcohol (as has been done in Brazil with ethanol), and to apply protective coatings to pipes and tanks.[5]

The fact that carbon steel is commonly used with methanol in the chemical methanol industry suggests that the problems can be solved without excessive cost.[6] Still, some industry representatives emphasize privately that they would be reluctant to ship methanol in their lines without convincing evidence that methanol will not cause long-term deterioration. Such evidence would only be forthcoming after many years of experience and testing.

In terms of operations compatibility, a major impediment to shipping new fuels in petroleum product pipelines, especially those from small biomass plants, is the requirement that shipments meet minimum tender requirements. A minimum tender is the minimum amount that a pipeline will accept for shipment.[7] Pipelines set minimum tenders to assure that product contamination of adjacent batches is kept at acceptable levels.[8] The minimum tender that a pipeline needs to maintain product integrity will depend on the diameter of the pipeline: the larger the diameter of the pipeline, the larger the needed minimum tender.

In the past, oil companies that owned pipelines have set high minimum tender requirements to restrict access to petroleum pipelines and thereby to restrict competition.[9] To combat this anticompetitive behavior, the U.S. Interstate Commerce Commission (ICC) ruled in an important decision in 1940 that a minimum tender requirement of over 10,000 barrels for crude oil or 25,000 barrels for petroleum products was excessive and therefore unreasonable.[10] Although this policy rule has not been well enforced either by the ICC or by its successor in petroleum pipeline regulation (FERC), relatively few pipelines today have minimum tenders that are higher than that stipulated by the ICC rule.[11] A 1979 study by the U.S. General Accounting Office found thirteen pipelines out of a total of 110 with minimum tender requirements set at levels higher than the 1940 ICC rules.[12]

The highest minimum tender, 75,000 barrels, is required by the Colonial Pipeline Company. However, most petroleum product pipelines have minimum tenders of around 10,000 barrels, while some have minimum tenders as low as 4,000 barrels. To put the minimum tender requirements of petroleum pipelines into perspective, recall that a typical coal-based plant would have a capacity of around 50,000 barrels per day.

It is plausible that if pipeline companies do accept alcohol shipments, they would impose large batch-size requirements so as to minimize interface with other products. If they intended to keep alcohol out for any reason, pipeline owners could use a number of tactics such as inconvenient scheduling, lack of storage service, and higher tariffs.

Clearly, there are no overwhelming technical or operational problems that would keep alcohol out of pipelines. Indeed, there already are numerous experiences with alcohol in pipelines. For instance, a small quantity of methanol moves via a six-inch 2.5-mile methanol-dedicated pipeline in Lake Charles, Louisiana;[13] methanol also routinely moves through pipes at process plants where methanol is produced and used. There have also been five reported trial shipments of methanol in pipelines by Atlantic Richfield Chemical Company, in which there were no reported problems with product quality or with the pipeline itself.[14] Also, Brazil has been routinely shipping ethanol in gasoline pipelines, reportedly with no unusual problems, although too short a time has passed to determine long-term effects.[15] The fact that methanol and ethanol have been successfully handled suggests that there are workable permanent solutions.

Since there are no overwhelming technical or operational problems that will keep alcohol out of pipelines, the determining factors will be of a strategic and economic nature. From a business perspective alcohol will be attractive to those pipelines with low utilization. Many pipelines will likely fall into this category because of declining and shifting traffic. In 1977, when the use of light petroleum products was near its peak, an extensive industry survey showed that pipelines still had a large amount of unused capacity.[16] Of forty-six integrated systems, twenty-nine did not have to "proration" entry at any point on their lines during the survey period.[17] Of the seventeen that did proration, only two had substantial prorationing. Underutilization was therefore significant even in the late 1970s. Since the usage of light and middle distillates is not expected to expand significantly (see chap. 14), and since refinery dislocations will continue to occur as a result of shifting crude oil supplies, one would expect even more underutilization of pipelines in the future.

Probably the strongest factor acting to deter alcohol transport in petroleum pipelines is a mismatch in supply-demand patterns. Observe the map of product pipelines in figure 29. It is a stylized map representing major network links. Some links represent multiple parallel pipelines; many smaller flows are not included. The network radiates from the Gulf Coast area toward the Great Plains, the Midwest, and the Northeast. This network structure evolved during the early and middle part of this century as petroleum production shifted toward the Gulf Coast area and as market demand continued to grow along the eastern coast and midwestern parts of the country. In figure 30 one observes the concentration of oil refineries on the Gulf Coast, a response to large but now dwindling oil production in Texas, Oklahoma, and Louisiana.

This pipeline network structure is mostly incompatible with the needs for new fuels. The existing network would not serve a coal and shale liquids industry located in the Great Plains and Rocky Mountain areas, although it could serve coal liquid plants in the Appalachian and Midwest coal regions and biomass plants in the wooded areas of the southern United States by connecting them to product pipelines passing nearby. Apart from material compatibility problems, the viability of those connections would depend on fuel marketing locations and the potential for assembling large batches.

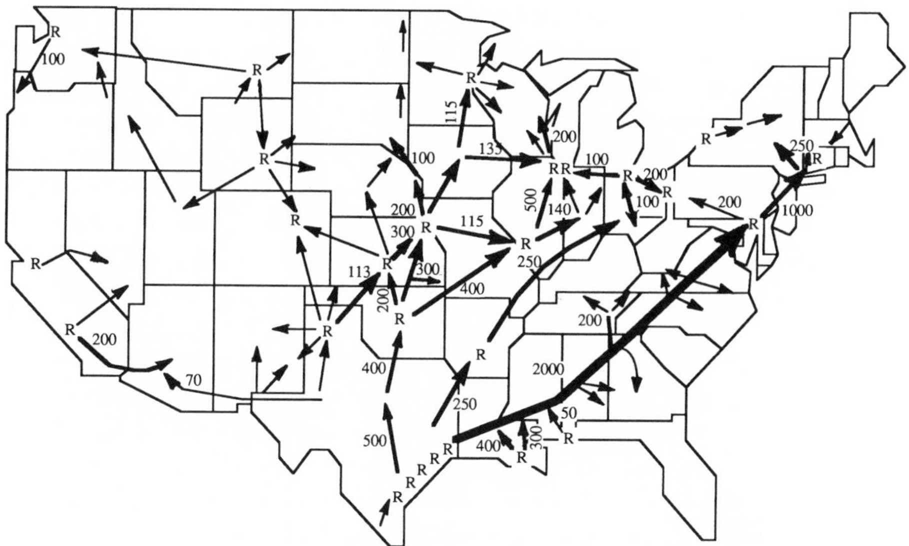

Numbers in map refer to flow in thousand barrels per day.

R = Petroleum Refineries

Figure 29. Petroleum Product Pipeline Network, Major Links and Nodes

Source: National Petroleum Council, *Storage and Transportation Capacities* (Washington, D.C.: Vol. III, 1979), p. D-3.

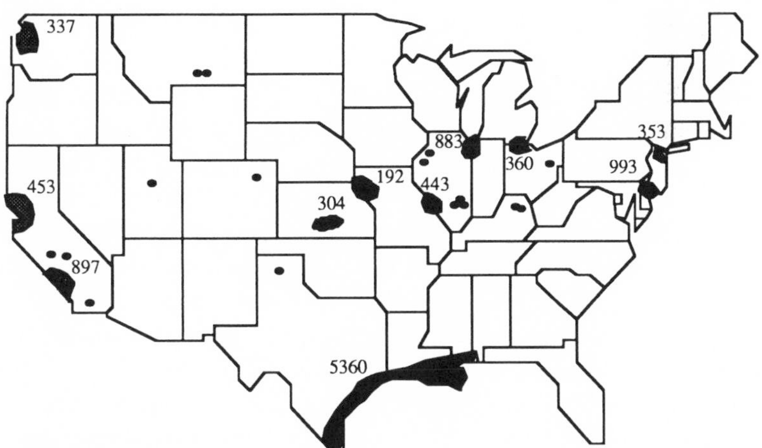

• ■ Refining Area

Figure 30. Refining Areas in the United States (Thousand barrels per day)

Source: U.S. Bureau of Mines, compiled in Piercey, *The Petroleum Pipeline Segment of the Domestic Petroleum Industry*, Ph.D. dissertation, University of Oaklahoma, Norman, Oklahoma, 1978, p. 85.

Other options for pipelining alternative fuels do not rely on the existing product pipeline network. The first option is to build new pipelines dedicated to single fuels—i.e., methanol, ethanol, coal liquids, and shale liquids. Another option is for new fuel producers in the area west of the Mississippi River—those who will be producing liquids from coal, oil shale, and petroleum (from the new fields in the Overthrust Belt)—to cooperate in jointly designing, building, and operating multiproduct pipelines that are compatible with a large range of liquid fuels. A third option is the use of coal slurry pipelines that use methanol (instead of water) as the liquid for the slurry. Since conventional water-based coal slurry pipelines have failed to gain eminent domain powers (so that they can override the refusal of railroads to provide rights of way), it is likely that methanol-coal pipelines will also fail. A fourth opportunity that has received little attention but may be practical in some situations is the insertion of alcohol into crude oil pipelines.[18] The two materials, oil and alcohol, would be separated at the destination in a preprocessing unit located at or near the oil refinery. Studies indicate that crude oil and methanol separate better than would be expected and that methanol cleans the pipeline and improves its flow characteristics. Arguments against this option, apart from possible corrosion problems, are the contamination that would result and the cost of separating the crude oil and alcohol.

WATER-BORNE TRANSPORT

Water transport is price competitive with smaller pipelines and much less expensive than truck or rail (see fig. 28). Generally, barges are considered price competitive with pipelines of less than a twenty-inch diameter, while the moderate-size tankers handled by U.S. ports are roughly competitive with all but the largest pipelines.[19] Absolutely no technical or operational constraints hinder the shipment of alcohol or any other liquid fuel over water. Indeed, chemical alcohol is currently shipped by water. In fact, most alcohol-producing and alcohol-consuming plants are located on waterways, in large part because water transport is relatively inexpensive and requires relatively small capital outlays.

Virtually all existing methanol plants rely on water transport for distribution. The feasibility of fuel methanol plants proposed in the 1970s and early 1980s—a peat-to-methanol plant on the North Carolina coast, a coal-to-methanol plant in southern Alaska, and barge-borne methanol plants—were all premised upon the availability of low-cost water transport.

Where the opportunity exists—as with some peat deposits along the east coast, forested areas in the southeastern and northwestern parts of the U.S., and corn fields along the Mississippi-Ohio River systems—water transport would generally be the preferred transport mode unless shipment volumes became very large. Except for alcohol plants located in northern climates where ports are closed during winter months (such as along the Great Lakes and the upper Mississippi-Ohio River system), water transport is highly attractive, especially during the early years of an alcohol fuel industry when cost, scheduling flexibility, and freedom from large initial investments are key considerations.

Water carriers have a wide range of shipment sizes.[20] Barge capacity ranges upward from 4,000 barrels, to as much as 100,000 barrels when joined in flotillas. Coastal petroleum tankers commonly carry between 100,000 and 300,000 barrels. Ocean tankers carrying chemical methanol in the mid-1980s typically carried about 380,000 barrels, although the largest U.S. coastal tanker carrying methanol had a capacity of only 80,000 barrels.

RAIL TANK CARS

Rail transport is moderately attractive as a liquid fuel carrier for many reasons, but it does not have any single feature that makes it clearly superior to alternative modes. The rail network is fairly ubiquitous, but less so than highways; rail transport is less expensive than truck transport, but more so than pipelines or water carriers; minimum size shipments are smaller than for pipe and water modes, but larger than for trucks. As a result of its middling performance, rail's early dominance of petroleum products markets has eroded to almost nothing. In the U.S. it now carries only 2 to 3 percent of the petroleum product traffic (see fig. 26).

While rail transport is not valued by petroleum shippers, it undoubtedly will be by alcohol shippers; shippers of both biomass and mineral-based alcohol are likely to make rail a widely used transport mode during the early years of an incipient alcohol fuels industry, and in the case of smaller biomass plants, far into the future. Rail is attractive because it is flexibile. The rail system can handle shipments anywhere from about 550 barrels in a single rail car up to about 50,000 barrels in a unit train. This flexibility allows producers to forgo large initial investments in fixed pipelines until their cash flow improves, demand patterns stabilize, and capacity expansion decisions are finalized.

Unit trains may be the key to large rail shipments. A unit train used for liquid transport consists of a string of interconnected tank cars that can be quickly loaded or unloaded. Such trains carry only one commodity from one origin to one destination, and they have shorter transit times (they do not need to be broken and coupled at intermediate terminals), quick loading and unloading times (because they would be highly automated, using only a few connections), and the flexibility to carry varying shipment volumes (by coupling and uncoupling extra rail cars).[21] Because of these efficiencies, unit trains also offer lower rates to customers than conventional rail service.

A large producer and leaser of rail cars claims that unit tank trains may be cost competitive with pipelines in some situations.[22] Generally, though, unit trains are more expensive, although there is no question that they offer greater flexibility and lower initial investment than pipelines. This flexibility is somewhat restricted, however, by the ownership and purchasing practices of the rail industry. Rail tank cars are usually owned not by the railroads or even by the shippers, but by leasing companies—who own about 80 percent of all rail tank cars in the U.S.[23] Tank cars have about a twenty-five-year life and therefore represent a long-term capital investment, especially because the resale market is highly erratic.[24] Leasing companies and fuel producers themselves will have to be convinced that the rail cars will be usefully employed for many years before they invest in the purchase of new cars.

The attractiveness of unit trains would be greatly enhanced if the initiation of new fuel industries induced a booming market in rail transport of liquid fuel; as tank car traffic increased there would be greater demand for used tank cars. Unit trains could then be deployed on a short-term basis, greatly increasing the flexibility of producers and shippers.

TRUCKS

Although trucks can go almost anywhere and are the most flexible of all modes, their unit transport costs tend to be significantly greater than other modes. Tanker trucks, therefore, are used principally for local deliveries of petroleum fuels; most trips are shorter than fifteen miles.

Tanker trucks range in size from the single-unit straight trucks carrying 2,600 gallons (62 barrels) to trailers carrying up to 10,000 gallons

TABLE 31 BULK STORAGE CAPITAL COSTS (1986 $)

Storage Tank Size	Cost
125,000	$ 378,000
250,000	574,000
500,000	870,000
1,000,000	1,319,000

SOURCE: U.S. Environmental Protection Agency, Distribution of Methanol as a Transportation Fuel, by R. D. Atkinson (Springfield, Va.: NTIS, 1982, PB83-116822). Costs inflated to 1986.

(238 barrels). Newer trucks, especially those owned by "for-hire" firms (common carriers), generally have stainless steel tanks because they are more versatile. They can be used for a wide variety of liquids, including alcohol.

For-hire carriers own about 60 percent of all petroleum product tank trucks in the U.S.[25] For-hire trucking companies provide an efficient alternative to maintaining an expensive fleet for intermittent use.

STORAGE TANKS

Storage tanks serve as buffers in the movement of liquids from process plant (refinery) to end user. They are also used for inventory control, since production is not always constant over an entire year (as would be the case with some biomass process plants) and because fuel demand fluctuates seasonally. Storage tanks are an integral part of fuel distribution systems.

Petroleum product storage tanks are ubiquitous. Most such tanks, except those at retail fuel outlets (service stations), are above ground. Generally, alcohol may be stored in the same tanks now used for petroleum products once the tanks have been thoroughly cleaned. Typical costs for bulk storage tanks are presented in table 31.

Above-ground tanks have a floating roof, a rigid cone roof, or both, but no survey has been taken to determine how many are in each category. A floating roof is used in pollution-sensitive areas to curtail the evaporation of emissions from unburned fuel, and a cone roof is used in areas of heavy rainfall and snowfall to deter water intrusion.

Modifications of above-ground storage facilities to accommodate alcohol are expected to be minimal. No modifications are needed for petroleum-like fuels. Storage centers play an important role in fuel distribution, but are not a critical factor in developing a distribution system for new fuels.

RETAIL FUEL DISTRIBUTION

The last distribution activity is delivery to the end user. Some transportation fuel is delivered in bulk to large vehicle fleet operators, but the great majority is sold through retail fuel outlets. One of the most prominent factors impeding the introduction of new transportation fuels is the absence of retail outlets selling those fuels. While operators of gasoline vehicles give little thought to availability of gasoline in their vehicle purchase and trip-making decisions, prospective purchasers of vehicles that operate on dissimilar fuels such as methanol, ethanol, compressed natural gas, or hydrogen would weigh fuel availability considerations heavily in their vehicle-purchase decisions. Refueling considerations and concern about the availability of fuel at retail outlets are increased by the fact that vehicles that operate on alternative fuels tend to have shorter driving ranges between refueling than gasoline or diesel vehicles do because most nonpetroleum fuels have lower energy densities than petroleum.

Creation of a minimal network is not a simple process. In 1985 there were about 130,000 primary retail outlets in the U.S. that were owned and controlled by a large number of companies of various sizes. A critical initial task for any prospective fuel producer is to gain commitments from some core group of retailers to market the new fuel.

While a large number of firms own and operate retail fuel outlets in the U.S., the majority of the outlets are owned or controlled by a few large, vertically integrated companies. The fourteen largest oil firms operating in the U.S. account for 80 percent of domestic gasoline production; of this, about 8 percent is sold through company stores and another 24 percent is sold through outlets leased to individuals and small corporations.[26] Thus these fourteen vertically integrated oil firms directly control 32 percent of all gasoline sales. They also retain control over another 27 percent of sales, which are sold to independent marketers but marketed under the brand name of one of the fourteen refiners.

The structure of distribution marketing networks varies quite markedly from one area to another partly because of the tendency of marketers, even large oil refiners, to concentrate on specific regions of the country. In 1978 the share of the gasoline market held by company stores owned and operated by major refiners varied from 38 percent in Ohio to less than 3 percent in some states. Likewise, outlets leased from oil refiners accounted for as much as 47 percent of gasoline sales in several large states and as little as 9 percent in others.

A possibly important consideration in the development of biomass fuels is the already existing participation of agriculture-based firms and cooperatives in the fuel marketing business. Firms and cooperatives such as Agway, Farmer's Union, and Farmland have sizable chains of fuel outlets in rural farmland areas. Farm cooperatives have been especially active in distributing liquefied petroleum gases (LPG) as fuels in rural areas.[27] To some extent these farm cooperatives and agriculture-related firms are likely to participate in biomass alcohol production. Their expertise and local orientation would match the relatively small-scale nature of biomass alcohol producers.

No technical barriers inhibit the dispensing of alcohol from existing service stations, but as indicated in the next chapter, the cost penalty may be substantial. In the past, all underground tanks at retail fuel outlets were made of steel and were compatible with alcohol. Since the late 1970s, however, fiberglass tanks have come to be preferred because they are resistant to corrosion and are expected to have a much longer life (over thirty years, compared to ten to fifteen years for steel tanks).

Fiberglass tanks are *not* compatible with alcohol, however; or more accurately, the fiberglass tanks being sold as of 1987 are not compatible with petroleum fuels containing more than 10 percent ethanol. The largest manufacturer of underground tanks has already developed a fiberglass tank that handles straight methanol, and charges only a small surcharge for them, but almost none of their sales are of that type. While most existing underground tanks are still made of steel (as of 1987), that proportion is dropping. The effect of this transition to fiberglass tanks that are incompatible with alcohol is to inhibit any fuel-station operator who might in the future consider switching to alcohol fuels.

An additional disincentive to alcohol fuels is a regulatory requirement that was adopted in California in 1985 (and possibly elsewhere) and is expected to be adopted nationwide shortly: a requirement that new underground tanks have a double wall so as to reduce the risk of contamination of groundwater by leaking fuel. Since generally it will be necessary to install new tanks wherever straight alcohol is to be marketed, there will thus be an additional cost beyond that of a standard

TABLE 32 SUMMARY OF TYPICAL GASOLINE DISTRIBUTION COSTS IN U.S. (1986 $)

Cost Item	Cents per Gallon
Bulk transport (pipeline)	5
Bulk terminaling	5
Local distribution (truck)	2-5
TOTAL	12-15
Retail markup	20%[a]
Excise taxes	19-24[b]

SOURCE: DHR, Inc., Methanol Use Options Study (Springfield, Va: NTIS, 1981), table 2.12. Costs were inflated in this table to 1986 $.

[a]Based on delivered wholesale price.

[b]In 1986 the Federal gasoline tax was 9¢ and state taxes were about 10¢ to 15¢ per gallon. Taxes on alternative fuels will likely be calculated on an energy-equivalent basis, and thus the per-gallon tax will be different for those fuels.

tank—the extra $10,000 or so to purchase the more expensive double-walled tank. While this rule is not meant to be discriminatory (and will be very important in slowing the contamination of groundwater in many areas), it is another example of a small hidden obstacle to change.

SUMMARY OF PETROLEUM PRODUCT DISTRIBUTION SYSTEM

The total cost for moving gasoline from the refinery to the fuel station pump is summarized in table 32. The total cost, excluding taxes, is about twelve to fifteen cents per gallon plus approximately 20 percent retail markup by fuel station operators. This cost estimate provides a baseline value for comparisons with nonpetroleum fuels.

In general, dissimilar nonpetroleum fuels will face very different distribution situations than gasoline. The choice of transport mode and distribution option will depend on market and production conditions and locations. Biomass fuels will tend to depend on truck and rail, while liquid fuels produced in large facilities will depend on water-borne and pipeline transport.

NATURAL GAS DISTRIBUTION

Pipeline is by far the least expensive mode for transportation of natural gas (or any gas). Pipeline's relative superiority to other modes is much greater with gases than with petroleum transport. Because of the low energy density of natural gas—less than 2 percent that of petroleum at ambient temperatures—natural gas is expensive to carry in an enclosed tank, whether on a truck, railroad, or boat. Thus, unlike the situation for petroleum, pipelines are used to transport even very small volumes of gas, for instance, to individual houses. In the U.S. there are over 1 million miles of natural gas pipelines, 6.5 times as many as crude petroleum and petroleum product pipelines combined.[28] About 250,000 miles of these are for long-distance transmission, about 700,000 miles for local distribution, and less than 100,000 are gathering lines in gas fields. No other country in the world has such an extensive network of gas pipelines.

When pipelines are impractical, the alternatives are to use high-pressure or low-temperature cryogenic tanks that may be carried on trucks, rail cars, and ships. Pressurized tanks generally have an economically constrained limit of 3,000 pounds per square inch (psi), about the same as CNG tanks used in CNG automobiles, but they could be built to withstand higher pressure. However, the energy cost of compressing the gas increases exponentially and soon becomes prohibitive. At 3,000 psi the size and weight of the containers is so great as to make transportation prohibitively expensive except for very small volumes such as would be carried in an automobile.

Transport of gas in liquid form at very low temperatures (−260° F) in cryogenic tanks is much more economical. Intercontinental transport of natural gas is always in cryogenic LNG ocean tankers. These specially constructed tankers are thermally insulated, big and bulky, and expensive. Because LNG is very light even in liquid form, large LNG tankers have much less capacity than similar-size oil tankers—the standard-size LNG ship carries the energy-equivalent of only 132,000 barrels of oil. One estimate is that LNG ships are about ten times as expensive as oil tankers of comparable energy transport capacity;[29] another estimate is that the total transport cost by LNG ship is about $1.74 per million Btu ($0.20 per gasoline-equivalent gallons) for transportation from shipments from the Middle East to Western Europe.[30] About seventy LNG ships exist in the world, half of which serve the Japanese market.

LNG transport is also economically practical in truck and rail tankers. The gas may either be transferred directly in liquefied form from ships to other tanks or it may be liquefied in a special facility for overland LNG transport. Small liquefaction facilities already exist in the U.S. and elsewhere and are considered economically feasible at a scale of one facility per large truck fleet.

An LNG truck tanker carries about one million cubic feet of gas, which is equivalent to 8,700 gallons of gasoline, about the quantity carried by a gasoline tanker truck. According to cost figures provided by industry sources, the cost of transporting LNG by truck is about the same as that for transporting gasoline. That seems unlikely, but it suggests that perhaps the costs are not much greater.

The cost of transporting gas by pipeline, although much less expensive than transporting it by any other mode, is still about twice as much as pipeline transport of petroleum on an energy-equivalent basis.[31] The higher cost results from the low energy and mass density of the gas; more pipeline capacity is needed to move the same quantity of energy, and the energy costs of pumping gas are greater than for pumping gasoline.

An LNG transport system requires a liquefaction plant at the trip origin, specialized cryogenic ships, and a receiving terminal with regasification facilities. The cost of a modest-size system of LNG facilities that handles the energy equivalent of 80,000 barrels per day of petroleum is about \$2.6 to \$3.8 billion.[32] These investments in LNG are risky because they are so large and so dependent on stable market demand at the destination. If natural gas demand diminshes in the region served by LNG, the owners of the LNG facilities are stuck with a financial disaster because there is no way to recover the huge investment. As a result, investors in LNG usually require long-term take-or-pay contracts of up to twenty years with gas purchasers before they are willing to build the system of LNG facilities. Because natural gas and petroleum prices did not increase as expected in the 1980s, huge LNG-related financial losses were incurred during that time. Spain reportedly incurred take-or-pay losses of \$800 million through 1985 in an LNG contract with Algeria, and still had ten years remaining in the contract.[33]

The establishment of a network of refueling networks for natural gas would not present significant technical problems, although, as shown in the next chapter, substantial cost is involved. CNG stations are simpler than LNG stations. A CNG station would generally be located along a natural gas pipeline. The station would require a compressor to pressurize the gas (to 2,400–3,000 psi) into fuel cylinders in vehicles. The refueling facilities could be located at petroleum fuel outlets or at a dedicated CNG outlet.

Two different refueling methods are used: slow-fill and fast-fill. In the slow-fill method, gas from the pipeline is compressed to either 2,400 or 3,000 psi and fed directly over a period of several hours into the vehicle's storage cylinder. Slow filling is suitable for large fleet operators who can centrally refuel their vehicles overnight.

Fast-fill systems would be used at retail outlets. They deliver CNG in two to five minutes from a rack of large bottles prefilled to about 3,600 psi.

Another refueling option is at home, since many homes are connected to the natural gas pipeline distribution system. The problem is the high cost of compressors. As of 1987 the cost of a small home compressor that slow-fills a vehicle was about $3,000.

LNG refueling stations are somewhat more sophisticated. The safest and most practical LNG station would probably use on-site self-powered liquefiers. Several companies already build skid-mounted natural gas liquefiers with capacities up to 1 billion Btu per day (equivalent to almost 9,000 gallons of gasoline per day) that are ready to operate. Although they are considered safe and easy to use on delivery, they are also rather expensive.

The next chapter investigates the specific compatibility problems and distribution costs associated with the introduction of alcohol, natural gas, and hydrogen.

ELEVEN

Fuel Distribution Costs and Problems for New Fuels

Alcohol fuels face the greatest distribution barriers of all the alternative fuels because they are at least partially incompatible with the existing fuel distribution systems. Natural gas fuels face fewer distribution problems because they can use the existing natural gas pipeline network (except for intercontinental transport), although they face the major cost of establishing a new network of fuel outlets. The fewest problems are faced by petroleum-like fuels from coal and oil shale which are confronted only with the cost of providing connecting pipelines between the fuel plants and the existing petroleum product pipeline network.

This chapter focuses on the distribution barriers and costs that would face alcohol fuels during the early phase of a transition and the costs of establishing fuel networks for natural gas and alcohol. First, the problem of water intrusion for fuel blends is analyzed. Next, the additional costs incurred in transporting alcohols in petroleum product pipelines are specified, followed by an analysis of the costs and other criteria for building new dedicated pipelines. Then, the problems of establishing a network of retail outlets for alcohol and gaseous fuels are considered, and, finally, the overall distribution costs and risks inhibiting the introduction of gaseous and alcohol fuels are assessed.

ALCOHOL FUELS

The transition from gasoline directly to straight alcohol would be disruptive from both distribution and end-use perspectives. Timing the start-up of alcohol production capacity so that it coincides with the establishment of an extensive network of retail fuel outlets and with the appearance of hundreds of thousands of alcohol vehicles is unrealistic to expect, even with strong government involvement.

The principal option for mitigating the disruptiveness of an abrupt switch to straight alcohol is to introduce alcohol first as a blend component with gasoline. Blending is an attractive strategy because it requires little or no modification of motor vehicles and no new retail fuel outlets or fuel dispensing facilities. But a blending strategy confronts a new and potentially formidable barrier: phase separation.

WATER INTRUSION AND PHASE SEPARATION

A fundamental problem for blends of alcohol and gasoline is that alcohol is completely soluble in water while petroleum is not. Current practice allows large amounts of water to enter the petroleum product distribution system—at the refinery, in pipelines, at terminals, and at service stations. Water is used in processing at the refineries and is periodically inserted into pipelines to test the pressure. Since it is heavier than gasoline, water simply collects at the bottom of storage tanks and is drained off periodically. Water intrusion, therefore, is not a significant problem with gasoline and petroleum materials. Indeed, the petroleum product distribution system is for the most part allowed to be a "wet" system.

In contrast to gasoline, alcohol is infinitely soluble in water and immediately adsorbs water on contact. If alcohol is being shipped and used as a distinct product, there is no problem. In fact, alcohol burns more efficiently with some water in it. But when gasoline and alcohol are blended and come in contact with water, the alcohol adsorbs the water, and as a result separates from the gasoline. No longer is there an alcohol/gasoline blend; instead there are two phases of liquids, an upper layer of gasoline and a lower layer of mostly alcohol and water. The extent of separation depends upon the quantity of water that intrudes, the temperature of the air and liquids, the type of alcohol, the composition of the gasoline, and the proportion of alcohol in the mixture.

Phase separation occurs even when small quantities of water are present. Depending upon gasoline composition and temperature, as little as 0.05 to 0.2 percent water will cause a 10/90 blend of methanol and gasoline to separate, while about 0.5 percent will similarly affect a

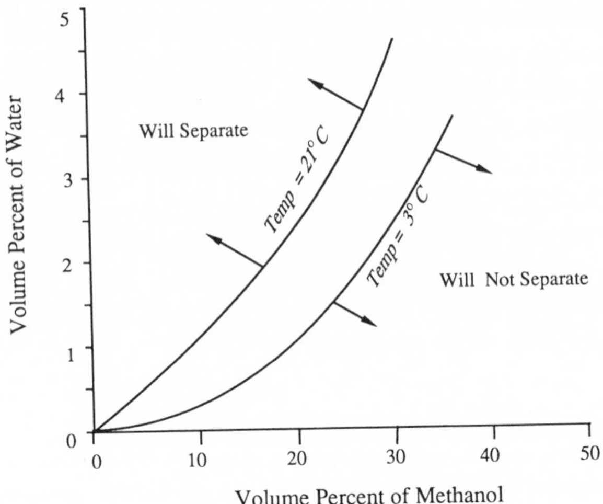

Figure 31. Phase Separation of Methanol and Gasoline

Source: American Petroleum Institute, *Alcohols: A Technical Assessment of their Application as Fuels* (Washington, D.C.: 1976, No. 4261).

10/90 blend of ethanol and gasoline.[1] Methanol blends are more susceptible to phase separation than ethanol blends.

The sensitivity of blends to water decreases substantially at higher temperatures and as the proportion of alcohol increases beyond 10 percent. Figure 31 illustrates water sensitivity of methanol blends. It shows that by increasing the methanol content from 10 percent to 30 percent, an almost twofold increase in water volume could occur without causing phase separation.

Water intrusion is greatest in marine vessels and bulk storage terminals. A survey by Union Oil Company found that tanks served by marine vessels contained 0.13 to 1.07 percent water; another survey of seventy-nine bulk storage tanks in the western United States found that half the tanks contained more than 0.9 percent water. One tank contained 8.42 percent water (measured by volume and calculated for half-filled tanks).[2]

The same surveys found water levels to be much lower in pipelines and underground service stations tanks; only 0.02 percent water was found in 95 percent of the gasoline transported in a major pipeline, and the median level for service stations tanks was 0.1 percent. These survey results indicate that ethanol and possibly methanol blends may be safely carried in pipelines, but that service station tanks and bulk storage tanks would need to be dehydrated before blends could be stored in them. More recent tests in Europe confirm those findings.[3]

The relatively high water levels found in storage tanks are a result of lax attitudes toward water ingress. While some of the water found in the storage tanks results from condensation, most of it is thought to result from rain that passes the seals of bulk "floating roof" tanks. In one fairly typical case 5 percent of the rain falling on a tank actually entered.[4]

Technically speaking, there are no obstacles to dehydrating the system—what might be required are "desiccators" on the air vents of the storage tanks, dry gas "blankets" to cover the surface of the liquid, the use of cosolvents such as higher-order alcohols, and generally tighter operational controls.[5] Dehydration is already selectively practiced in fuel distribution systems that handle water-sensitive liquids such as liquefied petroleum gases (LPG) and aviation fuel. Specialized dehydration units using various chemical drying agents are deployed in LPG pipelines.[6] These particular drying agents are not suited to alcohol blends because they remove alcohol along with the water; presumably other chemicals could be found that would be effective. Removal of water from pipelines has already been accomplished in tests by using scrapers ("pigs") that would remove not only water but also petroleum residues.[7] The cost of eliminating water from the storage tanks of the Colonial Pipeline (which has a capacity of over 2 million b/d and connects the Gulf Coast with northeastern U.S.) and upgrading the pipeline for use in transporting methanol blends is estimated at $50 to $100 million.[8] In conclusion, technical solutions are readily available to handle the water intrusion problem, but at a cost.

The ready availability of technical remedies does not mean, however, that phase separation will not be a problem. Intentional and unintentional violation of rules and quality-control procedures are inevitable. For instance, in 1984 the U.S. Environmental Protection Agency fined twenty-four fuel distributors in Michigan and Ohio for blending illegally large proportions of alcohol in gasoline.[9] The media also reported numerous cases in Ohio during 1983 and 1984 of alcohol-gasoline blends that were partly separated. In this latter case, the problem was probably lack of knowledge and diligence regarding water intrusion controls.

The success of new fuels will require strict adherence by industry to quality-control procedures. Failure to do so tarnishes the fuel's image in the marketplace, as was the case with methanol blends in Ohio and Michigan.

TABLE 33 ESTIMATED COST OF TRANSPORTING METHANOL BY PIPELINE, DOLLARS PER BARREL (1984 $)

Distance (miles)	10 inch dia. (50,000 b/d)	14 inch dia. (100,000 b/d)	20 inch dia. (200,000 b/d)
50	.22- .40	.18- .32	.11- .19
200	.50- .90	.37- .66	.23- .42
500	1.05-1.89	.74-1.33	.49- .88
1000	1.97-3.54	1.36-2.45	.90-1.63

SOURCE: Adapted from U.S. Department of Transportation, The Transport of Methanol by Pipeline (Washington, D.C.: 1985), p. 5-17.

Low cost: Same as cost of transporting gasoline; no cost penalty for corrosion or dewatering (see text).

High cost: "Low cost" + dewatering cost penalty (40% of gasoline transport cost) + corrosion prevention cost penalty (40% of gasoline transport cost) (see text).

INCREMENTAL COSTS FOR METHANOL FOR SHARING DISTRIBUTION INFRASTRUCTURE

Insertion of a new fuel into the existing petroleum product distribution system incurs additional costs beyond those that would be normal for petroleum product shipments of the same volume and distance. Those incremental costs would be minimal for truck, rail, barge, and ocean tanker transport, since those modes already transport many diverse liquids and use tanks that are compatible with alcohol. Thus costs would be low both for blends and for straight alcohol.

However, the additional costs incurred in using existing pipelines and some storage tanks are not negligible. Table 33 presents an estimate of the incremental cost for shipping methanol by pipeline. The "low costs" in the table represent the situation in which the cost of shipping methanol is equal to the cost of shipping gasoline (on a volumetric basis). The high cost includes two additional cost components, dewatering and corrosion prevention costs. The cost estimates for dewatering are based on a 1974 report by Exxon[10] that compared the costs of transporting aviation fuel and gasoline. Aviation fuel must be kept water-free in order to minimize the possibility of ice crystals forming in the fuel systems of aircrafts during flight. Exxon concluded that keeping aviation fuel "water-free" resulted in a 20 percent increase in pipeline transport costs. If alcohol were to be used in a blend, the entire system would have to be kept even drier than it is for aviation fuel. In table 33 it is therefore postulated that dehydration of a pipeline system for methanol use would result in a 40 percent increase in costs over the low "baseline" (i.e., low-cost) case. Estimating the cost of corrosion prevention is even more difficult, since no laboratory or field tests have been conducted. The Transportation Systems Center, research arm of the U.S. Department of Transportation, estimated that corrosion prevention costs would be roughly similar to the costs of dehydrating a pipeline.[11] The high estimates in table 33 therefore include the 40 percent increment for dehydration and another 40 percent cost increment for corrosion prevention (thus high costs = 1.8 × low costs).

The cost estimates in table 33 are somewhat speculative and are based on sparse data. Nonetheless, they provide some sense of the magnitude of the cost penalties to be incurred. They are somewhat deceptive in that they are measured on a volumetric basis. Since methanol has half the embodied energy of gasoline, roughly two gallons of methanol would have to be shipped for each gallon of gasoline. Making this adjustment, the cost of shipping one barrel of gasoline 1,000 miles in a fourteen-inch pipeline would cost $1.36 (or 3.2 cents per gallon); the cost for an energy-equivalent volume of methanol (using the high estimate) would be $4.90 (11.7 cents per gallon). There are other adjustments to be made in interpreting table 33. First, if the methanol is to be used unblended, then there would be no need to dewater it, and the cost penalty would be much less. Second, the estimates in table 33 are based on equal sharing of the incremental costs across all shipments; but some of the shipments might continue to be gasoline and so the dewatering and/or corrosion prevention costs should be borne only by alcohol shipments. Third, the costs in table 33 include all capital and operating costs. It may be reasonable to argue, as does the DOT methanol pipeline study, that only operating costs should be considered, since the capital costs are sunk. In this case, paying only the incremental costs (i.e., operating cost and cost penalties for alcohol) would be sufficient to induce existing pipeline operators to accept methanol shipments. This incremental cost analysis would be most relevant for pipelines with unused capacity.

Another important cost element necessary to make the existing petroleum product distribution compatible with alcohol fuels is modification of fuel-dispensing stations. If the alcohol is to be used in blends, then it would be necessary to replace certain components in the fuel pumps to make them alcohol-compatible, and to clean water out of the existing tanks.

If a new tank were to be installed to replace an incompatible fiberglass tank or leaking steel tank, the total cost of retrofitting a fuel station would be about $40,000. This estimate includes material replacement costs of about $1,000 to $3,000 and about $30,000 to $40,000 for purchase and installation of a double-walled tank and fuel lines (see analysis later in this chapter).

The cost of modifying bulk storage tanks is an additional consideration, but not a major one. The same DOT report that studied methanol pipeline costs also estimated the cost of providing bulk storage of methanol. Using the same analysis of low and high costs, they estimated the minimum additional storage cost at about ten cents per barrel if no extra corrosion prevention or dehydration/dewatering measures are used, and a maximum of forty-one cents (one cent per gallon) if extreme dewatering and corrosion prevention actions are taken.[12]

A 1986 study[13] for the U.S. DOE and EPA estimated that the total capital cost of establishing a complete methanol fuel distribution system within the Los Angeles area to handle 250 million gallons per year (which represents less than 10 percent of gasoline demand for that area) would be $5.1 to $28.4 million. That estimate includes all investments needed to get the methanol from the port to the fuel pump at the station, including a short pipeline, tanker trucks, and storage facilities. The lower cost range reflects the possibilities of using existing facilities. The cost is equivalent to 0.4 to 2.2 cents per gallon. The actual extra cost for local distribution of methanol (from port to retail outlets) would be much greater, however, because roughly two gallons of methanol replace each gallon of gasoline. Thus the per gallon costs cited above must be doubled, and one must add in operating costs and retail markup for the second methanol gallon. The total additional cost for local distribution of methanol might be as much as ten cents per gasoline-equivalent gallon, and for the total trip from the plant gate to end user, as much as fifteen cents. This cost penalty would not decrease much for larger volumes, but could be significantly higher for smaller volumes and more remote areas.

In conclusion, the costs for adapting the existing petroleum product distribution system are significant, but they are not by themselves overwhelming barriers to change. What may prove to be a greater barrier is the large number of firms that participate in fuel distribution activities. Every gallon of fuel is handled by a large number of firms as it passes through one or more pipelines, is stored in a bulk terminal, is perhaps transported to a local terminal, and then is trucked to a service station where it is placed in underground tanks. Only a very small proportion of gasoline that flows through this chain of activities is under the control of a single firm. The major concern of brand name producers in the case of blended fuels would be their inability to enforce quality control and therefore to assure that the fuel delivered to the customer will meet their standards.

INCREMENTAL COSTS FOR GASEOUS FUELS

The existing natural gas distribution system, like the petroleum products distribution system, is very extensive in the U.S., Canada, much of Europe, Japan, and several other countries. Using the existing system for CNG or LNG transportation fuels poses no problem except that of capacity. Using the pipeline system for hydrogen does, however, create some problems, although they appear to be of a relatively minor nature.

Three problems associated with putting hydrogen in natural gas pipelines are leakage, higher pumping costs, and embrittlement of the steel in the pipes.[14] Hydrogen gas leaks because it is lighter and therefore less viscous than natural gas. In tests, hydrogen leaked at about three times the rate of natural gas, but on an energy basis the leakage was equivalent. On the other hand, the need to use higher pressures to move the hydrogen gas through the pipeline network increases the transmission cost by a factor of 2.3 to 2.7 per unit of energy.

The third problem is hydrogen's tendency to react at the pipeline surface with certain steels, causing embrittlement and accelerating the growth of fatigue cracks in the walls. Recent experiments by Brookhaven National Laboratory have identified several compounds such as sulfur dioxide that completely eliminate embrittlement.[15] So far the identified compounds are too explosive or too toxic, but the limited research suggests that the embrittlement problem is solvable.

In conclusion, the only apparent deterrent to placing hydrogen in natural gas pipelines is a minor cost penalty for compression and pumping.

BUILDING A NEW PIPELINE SYSTEM

New fuels will require at least some new distribution facilities. To some extent these new fuels will be able to share existing distribution equipment and facilities, but in some cases existing facilities will not be physically available (e.g., storage tanks, pipelines), will have material compatibility problems, or will not have excess capacity. The situation is most severe with pipelines. There will be a fundamental mismatch in the existing flows of petroleum products and the future flow pattern of nonpetroleum fuels. End-use markets are identical, but feedstock and process plants are located in different regions. The existing pipeline network would not match the supply-demand pattern of nonpetroleum fuels.

For large-volume shipments, the most inexpensive transport modes are barges, ocean tankers, unit trains and pipelines. Where water transport is available it will be the superior choice because it requires relatively small investments (the barges or ships may often be hired or leased), and its unit shipping costs are much lower than those for truck and rail. Unit trains offer the advantage of less sunk investment and greater flexibility in that they may be operated over the large existing network of railroad tracks. If markets grow or shrink, routings and shipment volumes are easily altered. The cost of three sets of one hundred tank cars would be about $15 to $24 million,[16] but these rail cars may be sold if shipments are diverted to another mode or if the plant closes prematurely. Despite these advantages, unit trains are not likely to be cost competitive with pipelines in shipping large volumes. One analysis indicates that unit trains would be less expensive than pipelines for shipments of less than about 27,000 b/d when transported a distance of fifty miles.[17] For longer distances of 1,000 miles, the threshold at which pipelines become superior was estimated to be 38,000 b/d.[18] In any case, pipelines will be the most efficient mode available for large shipments (see table 34).

Pipelines have huge economies of scale in construction. The capital for acquiring and preparing right of way, and acquiring and installing steel pipe, pumping stations, and other equipment accounts for about 75 percent of total costs. Compared to other modes, operating costs are very low: pipelines are highly automated and have few employees, and energy and labor costs are small relative to other transport modes. Since capital costs dominate, economies of scale in construction are reflected in total costs (see tables 34 and 35.) A twenty-inch line has four times the capacity of a ten-inch line and about one-half the unit transport costs. These strong scale economies are generic to all pipelines. Since new pipelines would be designed and built specifically for the new fuels, there would not be any incremental cost to resolve material compatibility problems.

TABLE 34 COSTS FOR TRANSPORTING LIQUIDS BY TRUCK, UNIT TRAIN, AND PIPELINE, CENTS PER GALLON (1986 $)

Shipment distance	Tank Truck (6900 gallon capacity)	Unit Train (1,000,000 gallon capacity)	Pipeline (10 inch, 50,000 b/d, 75% utilization)	Pipeline (20 inch, 2,000,000 b/d, 75% utilization)
50	2.0¢	1.2¢	0.9¢	0.5¢
100	3.1¢	1.6¢	1.3¢	0.6¢
500	-	4.5¢	4.4¢	2.0¢
1000	-	8.1¢	8.2¢	3.8¢

SOURCE: Adapted from U.S. DOT, The Transport of Methanol by Pipeline (Washington, D.C.:1985), tables 5.4 and 5.6.

NOTE: Cost estimates include full capital and operating costs. Pipeline cost estimates are mid-range. 1,000,000 gallons = 23,800 barrels.

TABLE 35 SCALED RELATIONSHIP BETWEEN PIPELINE CAPACITY AND COST

Pipeline Diameter	Capacity	Cost per Unit of Throughput
1	1	1
2	5	.36
4	20	.20

SOURCES: Computed from annual pipeline cost summaries published in August issues of Oil and Gas Journal.

Unlike the case for other transport modes, there is a strong incentive to design and construct pipelines for expected future flows even though the line may be lightly used for several years. Large producer-shippers are faced with decisions whether to use unit trains temporarily and wait until markets stabilize to build a pipeline, whether to build a small pipeline immediately, or whether to delay or relocate plant construction (or port facilities) in order to aggregate flows into large joint venture pipelines.[19] Each decision will have to be made on the basis of the specific circumstances of that situation. For now, note that pipelines are attractive because they are very efficient, but they are also inflexible and burdened with large sunk costs.

COST OF ESTABLISHING FUEL OUTLETS

The cost of establishing a network of fuel outlets is considerably easier and less expensive for alcohol than for gaseous fuels. The cost per fuel outlet could be as little as $2,000 for alcohol if steel tanks in good condition are in place. The $2,000 cost would be for replacement of materials in the fuel-dispensing pumps that are incompatible with alcohol.[20] If a new tank is needed, because of the presence of an incompatible fiberglass tank or requirements that double-walled tanks be used, then the cost would be about $30,000 to $40,000.[21] This higher cost is equivalent to 6–7 cents per gasoline-equivalent gallon (amortized over 15 years at 12 percent interest rate and 15,000 methanol gallons sold per month).

The cost for establishing a network of natural gas outlets is much greater because the gas must be compressed or liquefied. The major cost item for a CNG station is a compressor pump—an industrial-scale pump that can be used for fast filling motor vehicles costs over $250,000. The total cost of a station, excluding retail profit margin but including all capital and operating costs, has been estimated at $1.65 to $3.81 per million Btu (19 cents to 44 cents per gasoline-equivalent gallon).[22] These costs are for fairly low-volume stations such as in New Zealand, Canada, and Italy; at higher volumes the costs would decrease considerably. These costs are much higher than the marginal costs calculated earlier for alcohol because unlike the calculations for alcohol, the CNG calculations include costs for station attendants, land, etc. Note, however, that even with these high refueling costs, retail CNG prices in New Zealand and Canada are still only about half those of gasoline.

For small volumes of fuel, the cost of LNG refueling is also fairly high. For example, skid-mounted self-powered liquefiers with an output capacity of 1 billion Btu per day (9,000 gasoline-equivalent gallons) were sold in 1987 for $2.5 million. The company that markets those self-contained liquefiers estimates the total cost, including all operating and capital costs, to be about 32 cents per gasoline-equivalent gallon.[23] Other estimates for LNG refueling stations are lower.[24] Considerable economies of scale exist with liquifier costs, but these economies cannot be taken advantage of because of the low volumes of fuel handled at retail fuel outlets.

THREE MODELS

There are three policy approaches for establishing initial networks of fuel outlets. One model is based on a laissez-faire free-market approach, as illustrated by the experiences with CNG, LPG, and diesel cars in the United States. This approach results in very little market penetration even when there are large economic benefits for consumers. For example, gaseous fuels have been considerably less expensive (on an energy basis) than gasoline and diesel fuel for many years, yet their only significant market penetration has been with a few vehicle fleets that have their own refueling facilities.

A second model is one of government incentives. In New Zealand and Canada, offers of grants and attractive financing and tax benefits were made to retail fuel outlets to supply CNG and to consumers to retrofit or purchase CNG vehicles. In both New Zealand and Canada, the incentives have been substantial enough to achieve some market penetration of the retail fuel market.

A third and last model is based on an authoritarian approach that uses mandates, such as was the case with unleaded gasoline in the U.S. In this government-orchestrated transition to unleaded gasoline, the federal government required all fuel outlets that sold at least 200,000 gallons of gasoline per year (which included most outlets) to also provide unleaded gasoline.[25] Thus uncertainty was avoided by the imposition of government mandates.

If alcohol and gaseous fuel sales at retail outlets are not mandated by government, the process of setting up the network will be slow in part because of the substantial initial investment (especially for gaseous fuels). The difficulties and obstacles would be reduced significantly in the case of alcohol if major oil companies were supportive of these fuels, something that would come about more readily if they were financially involved—for instance, in the ownership of the production facilities.

RATIONAL DESIGN VERSUS UNCERTAINTY

A rational fuel distribution system is one that exploits the features of its technologic components to minimize the cost of delivering fuel from process plant to end user. Today's petroleum product distribution system is often cited as an example of a highly rational and efficient system, and in some sense it is. It is argued that each transport mode has been deployed in a manner that best exploits its cost and performance characteristics: pipelines for long distance point-to-point bulk shipments; water transport for heavy liquids unsuited to pipelines; and trucks for small, short shipments.

The present configuration of the petroleum product distribution system is not necessarily appropriate for new fuels. A rational system for biomass fuels, for instance, would be very different from a rational system for mineral fuels. Biomass fuels would not rely on pipelines because output from any one biomass fuel process plant would generally be too small to justify building a pipeline or even to consider shipping batches of fuel in existing petroleum pipelines (even if the pipeline operators proved willing to accept nonpetroleum liquids). The very largest biomass fuel plant would produce around 150 million gallons per year, equivalent to less than 10,000 barrels per day, which is just over the threshold of minimum volume needed to justify building a pipeline. But since a single plant will not necessarily serve a single market location, it cannot be assumed that all the fuel output would be shipped on a single route (or single pipeline). Thus even a very large biomass plant will often find pipelines inappropriate for its needs.

If pipelines are not utilized by biomass producers, railways and trucks are left as transport options. Unit trains, with capacities of over 24,000 barrels (1 million gallons), may be attractive for some large biomass plants, but their large volume requirements and their delivery to a single destination point (and not to multiple market locations) would still be onerous for most biomass plants. Tanker trucks and railcar-size shipments provide the best match with biomass plants: they are suited to small shipments, they provide dense spatial coverage and, by leasing or hiring transport services, require little or no commitment of capital. Trucks and railways perform as simple and highly flexible links that respond easily to changing market and production conditions, a desirable feature for the irregular and seasonal production characteristics of biomass plants. Almost all shipments of biomass alcohol in the U.S. and most shipments in Brazil have been by truck and railcar.

Truck and rail transport have high tariffs and costs, however, relative to pipeline, unit trains, barges, and ocean tankers. The effect of using truck and rail transport is to sharply curtail shipping distances. The market area for each biomass process plant would be restricted in order not to incur large transport costs. Therefore, the most efficient distribution network would have a hub-and-spoke structure, with the biomass process plants located at the hub.

In contrast, distribution systems for mineral-based fuels would be very different from those for biomass fuels. Process plants for these fuels would be very large, ranging from 30,000 b/d or so to over 100,000 b/d. Producers of fuel from coal, oil shale, and oil sands would be tempted to deploy pipelines, either by clustering their plants and sharing a pipeline or by shrinking their marketing region to a single area to consolidate their shipments. Importers of methanol produced from remote natural gas face similar decisions about where to market their fuel inland. Should they market the fuel primarily in major port cities, or should they expand their markets inland using pipelines?

Deployment of pipelines is risky, however, as suggested earlier. If a producer-shipper had the foresight or foreknowledge of how markets and prices would evolve and of the effect new competition would have on those markets and prices, then a low-risk and highly rational pipeline deployment strategy could be formulated. But that foresight and foreknowledge do not exist. Lacking that foreknowledge, producers and shippers must choose between a large initial investment in pipelines and the potential cost efficiencies that implies, or smaller investments in unit trains or small pipelines that have higher unit transport costs but are less risky.

It is difficult to determine if and when pipeline deployment is rational. For producers of petroleum-like fuels and distributors of natural gas this is less of a problem than it is for producers of entirely new fuels such as alcohol. During the early years of the alcohol fuel industry, market uncertainty would be particularly great. Large investments in fixed facilities such as pipelines would tend to limit future marketing options and lock producers into certain decisions that might be in neither their own nor the country's long-term interest. While pipelines have the advantage of making virtually any large market location accessible because of their very low unit costs (when amortized over long time periods and large shipment volumes), the flip side of that relationship is that a prospective market must be stable, as well as large, or else the efficient pipeline will become a permanently underutilized financial disaster, a white elephant.

A similar dilemma faces LNG investors. They invest billions of dollars for a single set of facilities with the hope that demand will be stable over the life of the project. History indicates that that hope is unlikely to be realized because of the exceptionally erratic nature of the world energy market.

The possibility of financial disasters is very real. The oil and gas industries, for instance, have made many disastrous investments in pipelines, offshore ports, supertankers, oil refineries, and LNG terminals in recent years. They came close to investing in the more than $40 billion Alaskan Gas Pipeline, which may truly have become a financial disaster, and have abandoned or sold for scrap metal many supertankers and at least one LNG terminal, absorbing losses in the many billions of dollars. These investments were made in large part by huge, mature, vertically integrated corporations. The likelihood of incipient alternative fuel industries making wrong decisions, given the uncertainty that exists, is probably many times greater.

At this point it may be fruitful to step back for a moment to scrutinize the costs and risks of alternative distribution and marketing strategies from the perspective of an individual investor and society as a whole. From an individual investor's perspective, the options are highly constrained. Given that a large fuel distribution system is already in place, the tendency is to rely on the existing pipeline and storage infrastructure for regional and national distribution. Other options are available—for example, unit trains and water transport for long-distance shipments—but the choices are strongly influenced by the existing fuel distribution system.

Individual choices may not, however, be in the long-term interest of the economy or society. Indeed, the petroleum product distribution system may be growing incrementally in a direction that is contrary to the best interests of the overall economy and that perversely shuts off the most desirable long-term energy production options. It was stated earlier that the petroleum product distribution system is viewed as an example of a highly rational and efficient system. That is true only when it is viewed in a narrow suboptimal manner. In a broader sense, the distribution system evolved over time in a manner designed to preserve sunk investments made elsewhere in the chain of petroleum activities.

The existing system is efficient and rational only when it is considered in isolation from these other activities. For instance, consider that the huge fixed investment in petroleum refineries along the Gulf Coast and in the pipelines that deliver their output to much of the rest of the country (see figs. 29 and 30) compels the oil industry to preserve those investments. Construction of a multibillion dollar deep-water port off the Louisiana coast was begun in the 1970s to attract imported oil to those Gulf Coast refineries. In turn, those Gulf Coast refineries have been continuously rebuilt and expanded to preserve the investments in ports and pipelines. Now the deep-water port is unlikely ever to be fully completed, refineries are being closed, and some pipelines are prorationed while others are underutilized.

This distorted distribution system is a historical artifact to which new fuels must conform, at least to some extent. It is another example of how situations and conditions peculiar to the past alter today's price and cost estimates, even though those conditions no longer prevail. To the extent that this is true, government must be particularly attentive in guiding private investment decisions to make sure long-term social goals are not being ignored or contravened because of narrow attention to self-interest or failure to grasp an understanding of fundamental shifts in energy demand and supply.

CONCLUSIONS

The distribution of fuels that are physically and chemically different from petroleum fuels does not face any technical barriers, but those fuels do face significant economic and, to some extent, institutional barriers. The size of these barriers is reduced when vertically integrated petroleum companies are involved in alternative liquid fuels and when natural gas distribution companies are involved in alternative gaseous fuels. These companies would benefit by integrating these new fuels into existing facilities and activities, in effect reducing initial economic and institutional barriers.

But evidence indicates that energy companies are resistant to taking the risk of introducing new types of fuels. Petrobras resisted ethanol (chap. 4) and almost all U.S. petroleum companies resisted gasohol and methanol blends (chap. 3). Likewise, natural gas utilities have resisted converting their own small fleets of vehicles to natural gas, even though the technical risks are negligible and the economic benefits are substantial (in the U.S. only about 11 percent of gas industry vehicles had been converted to natural gas as of 1987). Each company sees itself as just one player in a complex interconnected system. They see only risks and costs in purchasing compatible underground storage tanks, establishing nonpetroleum fuel outlets, and designing their pipeline, storage, and refinery facilities to accommodate the introduction of new fuels. That should not come as a surprise, since they are accountable to shareholders to maintain short-term profit levels and because market sales of new fuels are so uncertain. Without government intervention, the likelihood of a timely transition to new fuels is remote.

PART V

End Use

A growing proportion of petroleum is being used as transportation fuel (see table 36). By 1986, 63 percent of all petroleum used in the U.S. was used for transportation. That percentage has increased over the years because the electric utilities and the industrial, commercial, and residential sectors have replaced much of their petroleum use with other sources of energy, while the transportation sector remained almost completely dependent on petroleum. In fact, for decades petroleum has provided over 95 percent of propulsion energy for the transportation sector and over 99 percent of the energy for motor vehicles. To a large extent, the petroleum issue is therefore a transportation issue.

The underlying assumption of the next three chapters (and the book) is that personal motor vehicles will continue to dominate passenger transportation into the foreseeable future because motorists are willing to pay a very high price for personal mobility. As land-use patterns become less dense in response to the mobility offered by the automobile (and truck), dependence on personal transport becomes greater. Moreover, it is unlikely that new engine technologies will supplant the internal combustion engine; it is more likely that continuing improvements in combustion mechanics and material substitution (e.g., ceramics) and modifications to accommodate new fuels will allow the internal combustion engine to remain the dominant power plant for motor vehicles. Given the continuing dependence on personal transport, the ability of manufacturers to increase fuel efficiency considerably higher than the levels prevailing in the 1980s, and the unlikelihood of a dra-

TABLE 36 ENERGY AND PETROLEUM CONSUMPTION IN U.S. IN MILLIONS OF OIL-EQUIVALENT BARRELS PER DAY, 1970 to 1985

Year	Total Energy Consumption	Petroleum Consumption by Non-Transport Sectors	Petroleum Consumption in Transport Sector	Transportation Petroleum Consumption as % of Total Petroleum Consumption	Gasoline Consumption
1970	33.3	6.94	7.76	52.8	5.45
1973	37.2	8.28	9.03	52.1	6.79
1974	36.4	7.83	8.82	52.8	6.53
1975	35.4	7.39	8.93	54.6	6.72
1976	37.3	8.11	9.35	53.6	7.09
1977	38.2	8.69	9.74	52.7	7.29
1978	39.2	8.72	10.13	53.9	7.55
1979	39.6	8.52	9.99	54.0	7.29
1980	38.1	7.51	9.54	55.9	6.84
1981	37.1	6.57	9.47	59.0	6.73
1982	35.5	5.99	9.30	60.8	6.64
1983	35.4	5.82	9.41	61.8	6.50
1984	37.2	5.98	9.72	62.0	6.55
1985	37.1	5.87	9.88	62.2	6.67
1986	37.1	5.96	10.18	63.1	6.85

SOURCE: U.S. Energy Information Administration, Annual Energy Review (Washington, D.C.: GPO), published annually.

matically different engine technology supplanting today's spark and compression ignition engines, liquid and gaseous fuels will continue to be the principal transportation fuels into the foreseeable future. Perhaps some dramatically more efficient transit or electrified roadway technology will come into being in the late twenty-first century, but that is far off and highly speculative.

The next three chapters investigate the prospects and opportunities for replacing petroleum in motor vehicles and related end-use markets. The emphasis is on fuels significantly different from petroleum—in particular, alcohol and gaseous fuels—since no changes in motor vehicles would be necessary to accommodate fuels made from oil shale, oil sands, and directly hydrogenated coal that are essentially identical to and indistinguishable from petroleum products.

The objective of the next three chapters is to determine the flexibility of end-use technologies and institutions in responding to and accepting new fuels dissimilar to gasoline and diesel fuel. Natural gas, hydrogen, petroleum, and alcohol fuels will be compared. It will be shown that alcohol vehicles tend to face fewer obstacles during the initial years of a transition because a gasoline-alcohol multifuel vehicle is less compromised than a multifuel vehicle that uses a gas and liquid. Generally, liquid fuel vehicles tend to have somewhat more power, fewer on-board storage difficulties, and a longer driving range than gaseous fuel vehicles. But in the near to medium term, gaseous fuels—particularly natural gas—will be considerably less expensive to use in motor vehicles. The question, then, is how important to consumers are reduced costs. The answer is that some consumers will prefer lower costs while others would be willing to pay more for other attributes. Initial experiences with fuels indicate that more affluent societies may tend to prefer methanol, while less affluent countries may find natural gas fuels more attractive. These preferences will be explored from the perspective of individuals and organizations in chapters 13 and 14, and from a societal perspective in later chapters.

TWELVE

Motor Vehicle Technologies

(WITH MARK DELUCHI)

This chapter will focus on the use of alcohol, natural gas, and hydrogen fuels in spark and compression ignition engines. As explained below, liquefied petroleum gases (LPG), combustion (gas) turbines, and, to a lesser extent, electric engines are not likely to play major roles in future motor vehicles.

The thrust of this and the following two chapters is to demonstrate that while alcohol and gaseous fuels face large barriers in gaining acceptance in motor vehicles, those barriers are not because the fuels and engines are technically or even economically inferior, but primarily because they are different from existing vehicles and fuels. Current engine and vehicle technologies have been refined and modified over many decades to accommodate and exploit the particular characteristics of gasoline and diesel fuel. The introduction of a new fuel, even if it is superior, requires additional R&D efforts to learn how to accommodate and exploit fully that fuel's unique properties, requires new investments in manufacturing facilities, and depends upon consumer acceptance of the different attributes of the new fuels and vehicles.

ELECTRIC ENGINES

Electric engines can be designed to receive their electric power from batteries, fuel cells, or from outside the vehicle. Although they are quiet, virtually pollution-free, and highly efficient, they are unlikely, as explained below, to supplant internal combustion engines in the foreseeable future.

Fuel cells use electrolysis processes to convert a solid, liquid, or gaseous fuel into electricity.[1] They are potentially attractive because they generate almost no emissions and have extremely high efficiencies, much higher than those of internal combustion engines, thus promising low operating costs. Fuel cell technology was greatly improved in the 1960s and 1970s for use in the U.S. space program, and its development has continued through the 1980s.

A fuel cell operates much like a battery but, rather than being charged electrically, it is powered by fuel. The fuel is oxidized with the aid of a catalyst and converted directly into electric current, which is used to run the electric engine. Unlike conventional batteries, fuel cells will continue to operate as long as fuel (and oxidant) are available.

The most common fuel cell is powered by hydrogen and oxygen. The oxygen can be easily obtained from the air; the hydrogen can be stored on board as liquid or gaseous hydrogen or in the form of methanol (or any hydrocarbon). For motor vehicle applications fuel cells would probably use hydrogen stored in methanol. The methanol is dissociated on board into hydrogen. The reason for using methanol is the relatively high energy density compared to hydrogen. This is a critical factor because fuel cells are very bulky and by themselves take up a large amount of space. Because bulk is not so much a problem for buses and trucks, fuel cells may be used in buses and trucks, but even with major improvements, fuel cells are not likely to become a viable technology for motor vehicles until well into the next century, if ever.

The possibility of electric vehicles powered by externally provided electricity is also highly speculative. The concept is attractive: electric circuits would be imbedded in the pavement so that vehicles could pick up the electricity without needing on-board electricity generation or storage equipment.[2] The disadvantages of electrified roadways are the huge capital cost of retrofitting roads with electric circuits, the need to have well-maintained flat surfaces, the need to carry batteries on board where roads are not electrified, the difficulty in billing users, and exacerbated peaking problems for electricity suppliers.

The peaking problem is probably the most critical; at present electricity suppliers are attempting to flatten electricity demand over the period of a day because it is very expensive to maintain generating capacity when it will only be needed for a few hours per day. Because the afternoon peak of motor vehicle travel overlaps with peak periods of electricity demand, it would be necessary to increase electricity-generating capacity, which would raise the price of electricity (because the marginal cost is greater than the average cost). Limited applications of this technology—for instance, along heavily used bus routes—might be attractive, but it is unlikely that electrified roadways will be widely introduced in the foreseeable future.

The third electricity option is battery vehicles. This option is likely to gain more market acceptance than the other two electricity options, but will probably be used only in several small market niches. The problem is the long recharging time and low energy density of batteries, which sharply limits the driving range and power of the vehicle. Various innovative battery technologies have been developed that have high energy densities, but they are far too expensive for use in motor vehicles. Intensive research efforts have been ongoing for decades and have resulted in incremental improvements in batteries. Those improvements are likely to continue, but only with order of magnitude improvements will battery vehicles become competitive. Such dramatic improvements are not likely or expected. An Argonne National Laboratory review concluded in 1985, "It is very doubtful that any advanced electric vehicle containing the hoped-for super battery could ever be competitive, either in cost or performance, with today's internal combustion vehicles at present gasoline prices."[3]

Battery vehicles' only future role is likely to be in those market niches where the vehicles do not have to travel at freeway speeds or take long daily trips. Even with much higher energy prices and the depletion of oil, it is doubtful that battery vehicles will play more than a limited role in our transportation energy future.

Petroleum-like fuels, those fuels that are essentially indistinguishable from gasoline and diesel fuel, face no significant end-use barriers and therefore are not considered in this chapter. Petroleum-like fuels may be manufactured from oil shale, oil sands, coal, or other materials. They tend to have somewhat more hazardous emissions characteristics—for instance, shale liquids contain some additional impurities such as arsenic, and coal liquids have suspected carcinogens—but other than these air quality and health effects, petroleum-like fuels have no discernible effect on end-use activities.

LPG

Liquefied petroleum gases (LPG) are also not considered. While LPG is significantly different from gasoline and diesel fuel, it is not considered here because of limited supply potential. LPG consists mostly of propane (about 90 percent), and is derived from the lighter hydrocarbon fractions produced during petroleum refining and the heavier parts of natural gas removed before the gas is placed in pipelines. LPG represents about 3 percent of natural gas and a similar proportion of petroleum, but the proportion can vary by several percent, depending on the characteristics of the raw feedstock and the design and operating parameters of the oil refinery. LPG is an attractive fuel for spark ignition engines because it has a higher octane and is cleaner burning than gasoline. When stored under low pressure (about 160 pounds per square inch), it is in liquid form and has about 80 percent of the energy density of gasoline (92,000 Btu/gallon versus a lower heating value of 115,000 Btu/gallon for gasoline). LPG is burned in the engine as a gas. It is more similar to gasoline than compressed natural gas (addressed below), which explains its greater popularity. LPG has gained modest usage in several countries principally because its price is lower than gasoline's and because of the generally low cost of retrofitting gasoline vehicles for LPG—about $900 to $1,600.[4] In the early 1980s, LPG was consumed in about 200,000 vehicles in the Netherlands, 100,000 in Canada, 500,000 in the U.S., and in limited numbers elsewhere.[5] However, LPG does not represent an important long-term fuel because, as explained above, its production potential will always be limited to a small proportion of oil and gas production. LPG is currently used mostly for heating rural homes and businesses and for other uses, such as indoor crop drying and indoor industrial vehicles, where its clean-burning qualities are highly valued. It is not considered further here as a transportation fuel.

COMBUSTION TURBINES

Combustion turbine engines are used in the transportation sector to power jet planes and as stationary engines to generate electricity. With very little modification, combustion turbines can burn a large range of liquid and gaseous fuels. Fuels with low mass ratios of hydrogen, such as those petroleum-like fuels made from oil shale and via direct hydrogenation of coal, are possible candidates for use in combustion turbines because they can be burned with little or no upgrading. The liquid output of the Union Oil Company oil shale plant, for example, is being sold to the military, some of it presumably for their jet planes. But other alternative fuels with higher mass ratios of hydrogen—ethanol, methanol, and natural gas—are unattractive for planes because of their low volumetric energy density. To travel the same distance with alcohol as (petroleum-based) jet fuel, planes would have to carry a much greater volume and weight of fuel. Since alcohols and natural gas are unsuited to jet planes, combustion turbines are not considered further in this chapter.[6]

PHYSICAL AND CHEMICAL PROPERTIES OF FUELS

Gasoline and diesel fuels are not necessarily ideal or inherently superior fuels for spark ignition and compression ignition engines. Other fuels have superior attributes. However, gasoline and diesel fuel prevailed because of the abundance and low cost of the crude oil from which they are derived. As petroleum becomes scarcer and more expensive, the properties of alternative fuels should be examined to determine their relative attractiveness and to determine how well those fuels match existing engine technology.

Natural gas is one of the two fuels addressed in detail in this chapter. Like petroleum, natural gas is a hydrocarbon. It is about 85 percent methane (CH_4). Since natural gas has a substantially higher octane rating than gasoline (RON = 120 to 130, vs. 90 to 95 for gasoline) it is suited to spark ignition engines; because it has a very low cetane rating (a measure of self-ignitability), it is not well suited to compression ignition engines. Natural gas fuels also have broad flammability limits that allow engines to operate with leaner air-fuel mixtures than gasoline and to burn more completely, resulting in much lower carbon monoxide emissions—near zero, in some cases. Nitrogen oxide emissions are similar to those of gasoline, and higher than those of alcohol. In addition, combustion of natural gas generates fewer reactive hydrocarbon emissions. As a fuel, however, natural gas has several important disadvantages. At typical pressures in the U.S. (2,400 psi, 70° F; Italy and Canada use 3,000 psi), compressed natural gas (CNG) has only about one-fifth the energy density of gasoline (equivalent to about 20,000 Btu/gallon), thus requiring large storage containers in the vehicle. Like any gaseous fuel, it also suffers an inherent power loss compared to liquid fuels because it occupies a much larger volume of the charge mixture (it displaces air from the induction, unlike a mixture of suspended fuel droplets). This loss can be compensated for, however, by designing the engine to take advantage of the higher octane number.

A second major fuel is alcohol, which is a generic label that encom-

TABLE 37 SELECTED PROPERTIES OF METHANOL, ETHANOL, GASOLINE, AND CNG

	Methanol	Ethanol	Typical Gasoline	Compressed Natural Gas
Chemical Formula	CH_3OH	C_2H_5OH	Mixture of C_4 to C_{14} hydrocarbons	Primarily CH_4
Density, lb/gallon	6.63	6.61	6.20	1.08
Energy Content, Btu/gallon[a]	56,560	75,670	115,400	19,760[b]
Boiling Temperature, °C	149	172	80-437	-
Vapor Pressure, 33°C (Kg/cm^2)	0.32	0.21	0.6-0.84	-
Water Solubility	infinite	infinite	nil	-
Stoichiometric Air/Fuel Ratio	6.4	9.0	14.6	-
Research Octane Number (R)	112	111	94	120-130
Motor Octane Number (M)	91	92	84	120-130
(R + M)/2	101.5	101.5	89	120-130

SOURCES: American Petroleum Institute, _Alcohols: A Technical Assessment of Their Application as Fuels_ (Washington, D.C.: API 1976, 4261); R. Fleming and R. L. Bechtold, "Natural Gas (Methane), Synthetic Natural Gas and Liquefied Petroleum Gases as Fuels for Transportation," _SAE_ 820959 (1982).

[a] Lower heating value.

[b] At 2400 psi, 70°F.

passes methanol (CH_3OH), ethanol (C_2H_5OH), and other higher-order alcohols with longer molecular chains. Alcohols are represented by the chemical formula $C_nH_{2n+1}OH$. Higher-order alcohols are produced in small quantities and are of interest only as cosolvents. The physical and chemical properties of alcohol differ somewhat from those of gasoline and diesel—as indicated earlier, alcohols are in some ways superior, in others inferior, and in still others just different. Two of the most important differences are the octane and cetane ratings. Alcohols, like natural gas, have higher octane ratings and lower cetane ratings than gasoline, which makes them attractive for spark ignition engines but generally unattractive for compression ignition engines. Other properties of alcohol that differ from those of petroleum products are lower energy density, cooler burning temperature, faster burn rate, poorer lubricating characteristics, high solubility (miscibility) in water, and lower vapor pressure (see table 37).

Hydrogen is a third fuel dissimilar to petroleum. Hydrogen molecules comprise only two hydrogen atoms. As a result, hydrogen has a very low energy density at atmospheric conditions. To be used as a transportation fuel it must be stored at a much greater density; this may be accomplished by liquefying or pressurizing it, or combining it with certain metals to form hydrides. This low density is a principal deterrent to hydrogen's use as a fuel. Its major attraction is that it burns very cleanly. All pollutant emissions, including carbon dioxide, are eliminated or greatly reduced. The implications of differences between the various fuels are explored in the remainder of the chapter. Next, the use of alternative fuels in diesel engines and then spark ignition engines is explored.

DIESEL ENGINES

Rudolph Diesel patented the diesel cycle in 1892. Diesel engines differ from spark ignition engines in that compression heat—rather than sparks—is used to ignite the fuel. Initially, compression ignition (diesel) engines were very large engines used in stationary applications and in railroad locomotives and ships.

Since World War II diesel engines have been improved substantially. They have been downsized and made reliable for operating at variable speeds and have almost completely replaced spark ignition engines in heavy- and medium-duty trucks, farm equipment, and transit buses.

Diesel engines are valued for their high fuel efficiency, low maintenance, and long life. However, they are difficult to start in cold weather and have poor acceleration. Because of these negative attributes, diesel engines were not widely used in automobiles until the late 1970s (although Mercedes-Benz has been selling diesel cars since the 1930s). Recent efforts to downsize and increase engine speeds of diesel engines for use in automobiles has had mixed technical and economic success, less in the U.S. than elsewhere. General Motors, the first U.S. manufacturer in recent times to manufacture diesel cars, began to market them in 1978, but the vehicles suffered from persistent mechanical difficulties. In 1984 GM decided to terminate the production of small diesel engines. In the U.S. penetration of the new car market by diesel cars peaked at 6.1 percent in 1981.[7] In other countries market penetration is much higher—close to 20 percent of the new car market in some countries in Western Europe.[8]

The popularity of diesel cars and light trucks in the rest of the world is due almost entirely to the low price of diesel fuel. The historically low price of diesel fuels outside the U.S. is due to differential fuel taxes. Most countries place large taxes on gasoline but not diesel fuel, on the premise that diesel fuel is used by large trucks for freight transportation and that subsidizing freight transport will stimulate economic activity. Gasoline, on the other hand, is used for personal transportation, often by more affluent individuals, and is considered a luxury in the sense that expanded automobile use does not directly stimulate economic activity. In Western Europe the average gasoline price including taxes was $2.24 per gallon in 1984, while the average price for diesel fuel was only $1.51.[9] In the U.S. diesel fuel and gasoline were taxed at the same rate until 1984; since then diesel fuel has been taxed 6 cents higher per gallon.

Until the early 1980s, diesel fuel prices in the U.S. were close to those for gasoline, generally a few cents cheaper. That price relationship shifted slightly as diesel fuel demand increased relative to gasoline. Unlike the situations in other countries, the demand for diesel fuel was much lower than for gasoline. As a result, oil refineries had developed severe, energy-intensive processing techniques to increase the proportion of gasoline extracted from each barrel of crude oil, thereby raising the cost of gasoline relative to diesel fuel and other petroleum products. As diesel engines became more popular in agricultural equipment, medium-duty trucks, and even cars and light trucks (due in large part to the low diesel fuel prices), and as jet travel increased, the demand for middle distillates increased. Diesel fuel prices rose until they finally exceeded regular gasoline prices in 1982 and then unleaded gasoline prices in 1983.[10] This shift in relative prices was a major cause of the collapse of the U.S. diesel car market in the 1980s.[11]

Alcohols and natural gas are not well suited to diesel engines. Like gasoline, they do not easily self-ignite when compressed. Their cetane rating is close to zero (though some alcohols have cetane ratings between five and ten), while diesel fuels usually have ratings between forty and fifty. The minimum acceptable cetane number set by most diesel engine manufacturers is forty. Because of their low cetane rating, it was commonly believed until recently that alcohol and natural gas were not feasible fuels for diesel engines.[12]

Recent R&D efforts and engine tests, however, suggest that diesel engines can be redesigned to operate efficiently with alcohol and natural gas fuels. These redesigned engines depart somewhat from conventional design; in some ways they are a hybrid of the spark and compression ignition engine technologies. The motivating force for adapting diesel engine technology to alcohol and natural gas has been the high level of pollution generated by diesel-powered transit buses and trucks in urban areas. Regulations have been imposed in the U.S. that require substantially reduced emissions from diesel engines in the 1990s (see chap. 15). In some applications, such as urban transit buses, switching to alcohol or natural gas may be more attractive than using a diesel engine encumbered by pollution control equipment (see chap. 13).

In the early 1980s General Motors (Detroit Allison division), Mercedes-Benz, and MAN (a German bus manufacturer) began developing promising methanol-powered prototype buses. Each manufacturer has followed a somewhat different technological route. The Mercedes bus uses a modified spark ignition engine, the GM bus uses a turbocharged two-stroke diesel engine with a glow plug to assist in starting and cold operation, and the MAN bus uses a four-stroke diesel engine with spark plugs added. Each of these manufacturers has prototype buses in operation in demonstration programs around the world (MAN since 1980, Mercedes-Benz since 1981, and GM since 1983).

The performance of these prototypes is impressive given the perception in the early 1980s that alcohol was unsuited to diesel engines, and given the relatively small amount of effort that has been devoted to modifying diesel engine technology.[13] When operated on typical urban routes, methanol buses of all three manufacturers had 13 to 17 percent lower energy efficiency (e.g., miles per Btu) than production diesel buses. On test tracks more advanced versions of the MAN and Mercedes-Benz buses had the same energy efficiency as production diesel buses; the GM bus was 14 percent less efficient. A 1984 report by the U.S. Environmental Protection Agency makes three comments about these early methanol diesel bus tests:

> First, methanol bus engine design is at an early stage and it would be expected that significant improvements will be possible. Second, the prototype GM, MAN, and Mercedes methanol bus engines have all proven to be capable of higher power and torque levels than their diesel counterparts, especially at low loads. This means that the methanol buses can achieve steeper accelerations during stop-and-go bus operation, with concurrent reductions in fuel economy. Third, it appears that both MAN and Mercedes have found ways to improve methanol bus fuel economy. It is anticipated that future GM engines will also show improved fuel economy. The consensus of people within the industry is that methanol buses will ultimately equal, and possibly exceed, the energy efficiencies of diesel buses.[14]

With further development, alcohol should have about the same energy efficiency as diesel fuel in compression ignition engines. One reason for the high efficiencies with methanol is the high compression ratio of diesel engines, which allows alcohol's high octane to be exploited.

As previously mentioned, the motivation in the U.S. for using alcohol in diesel engines is to reduce emissions, especially particulates and NO_x. The data, especially for actual in-service tests, are incomplete and uncertain, but are nonetheless highly promising (see chap. 15). The EPA has concluded that advanced methanol buses would emit lower levels of particulate matter (i.e., smoke) and nitrogen oxides than diesel buses and, when equipped with catalytic converters, would reduce carbon monoxide and reactive organic emissions as well.[15]

General Motors estimates that at low production volumes of about 250 to 300 buses annually, the cost of manufacturing a methanol bus would be about $6,000 to $7,000 more than for a comparable diesel bus;[16] this additional cost would be much less, about $2,000 per bus, if annual sales reached 5,000 new buses or engines.[17] Compared with the typical $160,000 to $220,000 cost of new transit buses, these additional production costs are modest.

Other developmental work and laboratory testing with alcohol fuels in diesel engines has been carried out as well, including the production of ethanol-powered diesel tractors by Ford in Brazil.[18] Various techniques such as fumigation, dual injection, and use of blends and additives have been developed to use ethanol and methanol in diesel engines. Some show promise, but the methanol bus work is the most advanced and highly developed for highway vehicle applications.

The use of natural gas in diesel engines is more problematic than that of methanol, although this assessment might be due in part to the fact that less R&D has been carried out.[19] Four approaches may be taken to use natural gas in diesel engines: dual-fueling, fumigation, direct injection, or full conversion to an otto cycle. Dual-fueling is the injection of natural gas into the combustion chamber with small amounts of diesel. In fumigation the gas enters the chamber with an appropriate amount of air only when the engine is operating roughly at half load or greater. In direct injection the natural gas enters the chamber with the diesel fuel (not with air as in fumigation); the proportion of natural gas is increased at higher RPMs. A full otto cycle conversion involves adding a carburetor and spark ignition system, and lowering the compression ratio. At present, conversion kits for buses and trucks are expensive—$6,000 or more for dual fueling—and performance, efficiency, emissions, and engine wear have all been poor.

No work has been done with hydrogen in diesel engines, but the results are likely to be similar to those with natural gas.

SPARK IGNITION (GASOLINE) ENGINES

The third and most prevalent transportation end-use technology is the spark ignition (otto cycle) internal combustion engine. It is well suited to alcohol fuels, hydrogen, and natural gas. Spark ignition engines usually burn gasoline, but they may be easily retrofitted to burn alcohol, hydrogen, and natural gas (as well as LPG). They power most automobiles, light-duty trucks, motorcycles, school buses, many off-road vehicles, and some construction and agricultural equipment. This analysis of spark-ignited engines will be limited to highway vehicles, however, because they account for about 96 percent of all gasoline use.

Alcohol, hydrogen, and natural gas (and LPG) are similar enough to gasoline to be used in the basic spark ignition engine technology currently employed in most autos. However, differences between gasoline and these alternative fuels require that some modifications be made. Differences between alcohol and gasoline are presented in table 38, along with a list of the effects of using alcohol in gasoline engines, possible remedies for mitigating undesired effects, and modifications for capturing desirable effects.

The technical problems and incompatibilities with alcohol, natural gas, and hydrogen are all solvable—and perceived to be so by automakers —although further research and development is required. Referring to alcohol and natural gas fuels, Donald Peterson, then president and later chief executive officer of Ford Motor Company, stated in 1984:

> After years of modifying existing powertrains or creating new ones to run on a variety of alternative fuels, of field trials and field tests, of emissions testing, and of comprehensive fuel, material, component and systems research, we are confident that the transition to alternative fuels can be made on a large scale without major engineering or hardware problems related to vehicles. Of course there are more questions to be answered, such as the impact of new fuels on the environment. But it appears to us that they can be solved without the need for new invention, new technology or dramatic technical break-throughs.[20]

ALCOHOL

Indeed, the auto industry already has had considerable experience with alcohol fuels; for instance, General Motors, Fiat, Saab, Scavia, Volvo, and Mercedes-Benz have been manufacturing ethanol vehicles in

TABLE 38 EFFECTS OF ALCOHOL FUEL USE IN GASOLINE ENGINES

Desirable Attributes of Alcohol	Effect with Current Vehicle Technology	Possible Vehicle Adjustments
Higher octane rating	Reduces "knock"	Higher compression ratio gives greater fuel efficiency and higher torque
High heat of vaporization and lower flame temp.	Burns cooler and lower NO_x emissions; denser air-fuel mixture and more power	None needed (but increased power allows downsizing of engine)
Faster burn rate	Greater power per unit of energy	Cylinder redesign and spark advance to capture potential efficiency benefits
Greater tolerance to lean combustion	Generally lower overall emissions and higher energy efficiency	Engine redesign to capture potential efficiency benefits

Undesirable Attributes of Alcohol	Effect with Current Vehicle Technology	Possible Vehicle Adjustments
Lower energy density	Reduced range	Increase fuel tank volume
Poorer lubricating characteristics	May contribute to greater engine wear (though less important than formic acid formation problem)	Use of more compatible lubricating oils and use of corrosion inhibitor in methanol
Incompatibility with some materials	Ethanol damages certain materials in fuel lines; methanol corrodes fuel tanks and other materials	Coat or replace fuel tank; replace other materials
Lower vapor pressure and single boiling point	Poor starting below about 45° F (see Table 37)	Hydrocarbon additives, automatic heating of fuel line, dual fuel system for engine starting, fuel dissociation

Brazil since 1980. Both Volkswagen and Ford have been actively developing alcohol engines since the early 1970s. Most of the Japanese manufacturers also have active alcohol fuel R&D programs as well.

Experience with Brazilian alcohol cars has proved valuable, but in order to serve the U.S. market, that technology will need considerable modification. The principal differences between the two countries from a vehicle design perspective are the greater extremes of topography and weather in the United States, and the stringent emission standards in the U.S.

The major technical challenge in introducing alcohol vehicles is to eliminate the "cold start" problem.[21] The low vapor pressure and single boiling point of alcohol makes it difficult to start engines at less than about 10° C. Various means exist to overcome the cold start problem, such as the following technical options: providing an additional small gasoline (or propane) tank to be used only for starting, automatic heating of fuel to increase volatility, blending small quantities of hydrocarbon additives into alcohol to assist starting, and dissociation of alcohol in order to use the resulting hydrogen during the start-up phase. In Brazil a dual fuel system, using an auxiliary gasoline tank for starting, has been used to solve the cold start problem. In the U.S. and other colder climates, the auxiliary gasoline tank may not be a satisfactory solution because of the inconvenience of needing frequent refills. Another problem may be high emissions during cold starts. However, the cold start problem should not be a major impediment except in extremely cold climates.

One difference between alcohol and gasoline that makes alcohol relatively attractive involves emission characteristics. As elaborated upon in chapter 15, alcohol emits fewer nitrogen oxides and less total reactive organic matter (i.e., hydrocarbons and aldehydes) than gasoline; carbon monoxide emissions are about the same. An emission problem does arise, however, when alcohol and gasoline are blended. Blending leads to the formation of azeotropes (light hydrocarbon molecules) and high vapor pressures, which leads to higher evaporation rates. Improvements in evaporative emissions controls in vehicles and adjustments of the blends at the refinery may help reduce those increases to levels that comply with emission standards, but blended fuel will also evaporate in storage tanks and at fuel pumps. Increases in evaporative emissions are one factor militating against the use of alcohol/gasoline blends, although they will become less of a problem as fuel injection systems replace carburetors in motor vehicles and vapor recovery systems are installed on vehicles and/or service station pumps.

NATURAL GAS

Natural gas is an inexpensive and attractive fuel that is well suited to spark ignition engines. The greatest problem is that as a gas it cannot be stored in the same tank as a liquid. For a vehicle designed specifically for natural gas and dedicated only to natural gas use, this requirement for separate storage does not create an overwhelming problem, but during a transition period natural gas vehicles will have to accept the burden of carrying two fuel systems—one for natural gas and the other for gasoline—until a large network of natural gas refueling stations is in place. Unfortunately, these dual fuel vehicles will also be greatly inferior to dedicated natural gas vehicles—much more so, as will be explained later, than are dual-fuel alcohol vehicles relative to dedicated alcohol vehicles.

The inherent advantages of natural gas relative to gasoline are higher octane, broader flammability limits (which allows leaner air-fuel ratios), and generally lower emissions. Carbon monoxide emissions tend to be lower because of leaner engine operation and more complete burning, and hydrocarbon emissions are largely nonreactive (see chap. 15). Being a gas at atmospheric temperatures, natural gas does not present cold start problems.

Natural gas also has some inherent disadvantages, but they can be mitigated. The principal disadvantage of CNG is inherently less power than gasoline engines because the gas displaces more air than gasoline, thereby reducing the amount of oxygen available for burning. This disadvantage can be mitigated or eliminated by incorporating a turbocharger, increasing the compression ratio, adjusting the spark timing, and making other design changes. Natural gas fuels also have a much lower volumetric energy density, about 20 percent of that of gasoline (when the gas is stored in compressed form as CNG), thereby restricting vehicle range. Additional fuel cylinders can be placed in the vehicle, but they take space from other uses and add weight to the vehicle. Current steel tanks take about 1.5 cubic feet of space and weigh over 100 pounds. Three cylinders would take considerably more space than a gasoline tank and would hold the equivalent of only about 7.8 gallons of gasoline. Recent development of aluminum tanks wrapped in fiberglass or "Kevlar" has reduced the weight penalty by 50 percent or more, however. At high production volumes, lightweight aluminum tanks should cost no more than steel tanks.

In summary, CNG vehicles face no insurmountable technical challenges—indeed, they are already widely used in New Zealand and Italy, and to a lesser extent in Canada. CNG fuel has some attributes that are superior to gasoline (and alcohol) and others that are inferior. But their market acceptance has been limited; the problem is that dual-fuel CNG vehicles have redundant fuel tanks (which take up space and are expensive) and have degraded performance, efficiency, and emission characteristics relative to that of a dedicated vehicle designed specifically for CNG. All CNG vehicles now in existence, except for a few prototypes developed by Ford,[22] are gasoline vehicles that were retrofitted to operate on CNG.

Three options may be pursued to overcome the storage and vehicle range problems resulting from CNG's low energy density: higher pressure tanks, low pressure absorbents, or LNG. The first option, using higher pressure tanks, is least attractive. By using tanks designed for 10,000 pounds per square inch (psi) instead of the 2,400 psi tanks used in the U.S. or the 3,000 psi tanks used in New Zealand and Canada, considerably more fuel can be stored in the same volume of space. However, on an energy unit basis, these tanks may cost more, weigh more, and take longer to refill (the data do not permit a precise comparison).

These disadvantages would have to be balanced against the space saved. In any case, more R&D would be worthwhile.

An option receiving considerable attention from research engineers is lower pressure tanks (1,500 to 2,000 psi) that have honeycombed surfaces inside. Gas is pumped in and absorbed into the interstices in such a manner as to provide greater storage capacity per unit of volume than tanks pressurized to 2,300 psi. This option has the attraction of being suited to low-pressure compressors such as could be installed in homes. This storage technology is still far from commercialization, however.

A third option for increasing vehicle range and volumetric energy density is to liquefy the gas. As indicated in the previous chapter, gas may be liquefied and placed in storage tanks suited to motor vehicles at relatively small stations. The tanks that hold the liquefied natural gas (LNG) are highly insulated to maintain the very cold temperature of the fuel. The LNG enters the engine as a gas, just as CNG does, except that it is colder, which is an advantage because it cools the inlet air, providing greater volumetric efficiency and therefore more engine power.

A major problem with LNG from a consumer perspective is the "boil-off" of LNG vapors from the storage tank. This boil-off occurs through a valve that is installed so that as the LNG fuel warms, venting of vapors can occur so as to prevent a build-up of pressure (above about 60 psi). The release of LNG vapor will not necessarily present safety problems (see chap. 16), unless it occurs in confined, unventilated spaces, but it could be annoying and costly to some consumers. This boil-off would begin after about one week if a vehicle is not used. Virtually all of the LNG would be vented in three to ten weeks (depending on various design features). Enough LNG would be left to drive only about five miles or so. However, if the vehicle is driven at least once a week, the vapor build-up is consumed by the engine and no venting would occur. The boil-off problem can be mitigated by building the tanks to withstand greater pressures and to vent at a higher pressure, but this is more expensive.

HYDROGEN

The last fuel option considered in this chapter is hydrogen. Numerous hydrogen and dual-fuel hydrogen-gasoline vehicles have been built since the 1930s. All used retrofitted or modified gasoline engines, however, and therefore precise determinations cannot be made of the likely performance, emissions, and efficiency of vehicles optimized for hydrogen. Performance, emissions, and efficiency would also depend in part on how the fuel is stored on board.

Several general conclusions can be drawn. On the positive side, hydrogen vehicles will probably be more efficient than gasoline vehicles, since hydrogen mixes better with air and can burn at leaner fuel-air mixtures. Also, hydrogen will burn much more cleanly than any other fuel and, unlike all other fuels (except biomass crop fuels and electricity generated from solar, water, wind or nuclear power), will not generate carbon dioxide. On the negative side, because hydrogen ignites at much lower temperatures than gasoline, hot gases in the chamber may preignite the charge and cause backfiring. There are a number of ways to prevent this. Also, as with any gaseous fuel, hydrogen will supply less power, although as with LNG, this may be compensated for when the fuel is stored as a cold liquid. Because it does not contain carbon and is a homogeneous fuel, hydrogen will also probably result in smoother-running engines that require less maintenance, although because the explosion is of greater force, more stress will be placed on the engine, affecting the maintenance and life of the engine.

The critical factor is hydrogen's energy density, which is even lower than that of methane, and its constraining effect on vehicle range. As with natural gas, the fuel could be stored under greater pressure, under extreme cold as a liquid, or bound with certain materials (metals in this case) in a low-pressure tank. The latter two, liquid hydrogen and metal hydrides, are the most attractive.

Metal hydride storage systems are usually long, thin, hollow cylinders, tightly bundled, and pressurized with the hydrogen gas to about 500 psi. The hydrogen, when pumped into the cylinders, becomes bound to the metal lattices. The hydrogen is released when heat is added. The heat is supplied by the combustion process. Various metals have been tested and used, but the central problem is that hydrides contain a maximum of about 8 percent hydrogen by weight, and thus vehicular hydride storage systems are very heavy and large. Current hydride storage system weighing 1,000 pounds and displacing 10 cubic feet would provide a vehicle range of only about 100 miles. Considerable improvements are needed to make metal hydrides viable.

Liquid hydrogen storage tanks also need considerable improvement before they will be acceptable to consumers. Most technical problems, such as adequate temperature control in the tank and fuel lines and establishment of safe and easy refueling techniques, are being dealt with and should not be insuperable, but making the vehicle attractive to consumers might be more difficult. The main obstacles to consumer acceptance of liquid hydrogen vehicles are, again, boil-off and the bulkiness and cost of the storage tanks. Liquid hydrogen tanks that hold the equivalent of twenty gallons of gasoline would be six to eight times larger than gasoline tanks, while boil-off for liquid hydrogen would be somewhat more severe than for LNG.

TRANSITION STRATEGIES

The following three strategies could be pursued in introducing alcohol, natural gas, and hydrogen vehicles.

After-market *retrofitting* of gasoline vehicles

Providing new vehicles with a *multifuel* capability

Building *fuel-specific* vehicles exclusively for alcohol, natural gas, or hydrogen.

Each strategy reflects tradeoffs between fuel efficiency (see table 39), flexibility of use, and production cost. Strategic decisions must be made as to which of these approaches should be followed. Under conditions of extreme market uncertainty, automakers in principle would prefer the retrofit approach, in which they would play little or no role, or second, the multifuel approach, in which their product is adaptable to a wide range of market conditions. In practice, however, as was so vividly illustrated in Brazil, aftermarket retrofit conversions are often poorly done. As a result, automakers would probably actively oppose a retrofit strategy so as to preserve the quality image of their vehicles. The ideal situation for automakers would be to eliminate uncertainty and to spe-

TABLE 39 APPROXIMATE THERMAL EFFICIENCY OF ALCOHOL AND CNG COMPARED TO GASOLINE

	Engine/Vehicle Type	
Fuel	Multifuel or Retrofitted Gasoline Engine	Fuel-Specific Engine
Ethanol	0 to +5%	+10 to +15%
Methanol	0 to +5%	+10 to +20%
CNG	-20 to 0%	0 to +10%

Note: These relative efficiencies are sensitive to various factors: for instance, retrofitted engines would be more efficient if the compression ratio were increased; efficiency of multifuel engines depends on what fuel (or range of fuels) the engine is optimized for; and efficiency of fuel-specific engine depends on what grade of gasoline and what level of engine technology the new engine is compared to. See Jet Propulsion Laboratory, California Methanol Assessment (Pasadena, California: JPL, 1983), chap. 4, for a conservative analysis of thermal efficiencies in alcohol cars.

cify a single fuel (or fuel blend); automakers could then design and build all their vehicles for that single specified fuel.[23]

The inclination of fuel producers, however, is to produce and market the fuels that are least expensive and generate the largest profit. They would, for instance, prefer to blend ethanol, methanol, hydrogenated coal liquids and oil shale liquids in a manner most cost efficient to them, or to market natural gas if they had an inexpensive supply source. Thus specification of a single fuel would be assured only through a government mandate or arrangements made between the fuel production industry and automakers.

RETROFITTED VEHICLES

Vehicle manufacturers play little or no role in this strategy. They would produce and sell vehicles, and only afterward would modifications be made. For alcohols and CNG the preferred engine technology for conversion is the spark ignition engine. The sophistication and cost of vehicle conversions depend on the fuel used and the fuel efficiency desired. With no modification, most spark ignition engines designed for gasoline use should run well on alcohol/gasoline blends that contain up to about 10 percent ethanol or 5 percent methanol.[24] Beyond those levels, modifications must be made. For some vehicles conversion is simple and inexpensive. Gasoline-powered trucks and autos without fuel injection and without microprocessor-controlled fuel intake components (mostly pre-1981 cars in U.S.) may be converted to ethanol use by simply adjusting the fuel flow mechanism in the carburetor and replacing some plastic and rubber components that come in contact with the ethanol. This most basic retrofit could be accomplished by most mechanics for about $300,[25] although the long-term durability of such a vehicle would be questionable.

Other retrofits are more difficult and costly. For example, methanol conversions are more costly because methanol is more corrosive than ethanol and requires more material changes. To retrofit vehicles with microprocessor-controlled fuel flow would be beyond the capabilities of most repair shops and presumably would be very expensive. Still greater costs are incurred to capture potential fuel efficiency benefits of alcohol. The cost just for increasing the compression ratio is $600 to $800 per engine; the upper limit for a high-quality conversion to methanol would be about $2,000 per vehicle.[26] The Bank of America estimated in the early 1980s that their cost for retrofitting thirty-five GM cars to burn methanol, without changing the compression ratio, was $270 for parts plus four to six hours of labor (for a total cost of about $400); their cost for retrofitting 252 Ford cars, in which the compression ratio was raised, was $1,850 per vehicle.[27]

If the compression ratio is not changed, a retrofitted vehicle running on alcohol will have about the same energy efficiency (miles per Btu) as if it had been using gasoline. Because of alcohol's lower energy content (Btu per gallon), the volumetric fuel economy (miles per gallon) for ethanol and methanol would only be about 67 percent and 50 percent that of gasoline, respectively. Generally, retrofitted vehicles will have the same energy efficiency with alcohol as with gasoline,[28] but will have a shorter driving range per tankful of fuel. (New larger tanks would usually not be possible since they would not fit the existing vehicle con-configuration and might create safety hazards.)

These retrofit costs are substantial and discourage the introduction of alcohol fuels. California attacked this cost barrier by providing tax credits for 55 percent of the conversion cost, up to a maximum credit of $1,000, but even with this subsidy, the cost to the consumer is still substantial.

Retrofitting vehicles to run on alcohol is not only costly but may also have unforeseen effects on performance and durability. In Brazil problems with alcohol cars were primarily attributable to retrofitted cars and not to production-line vehicles (see chap. 4). Alcohol's corrosiveness and poor lubricity may cause unforeseen damage and engine wear in some vehicle models. Additional problems could result from inferior conversions by mechanics inexperienced with alcohol. Because of unforeseeable effects that may arise and the possibility of poor-quality conversions, vehicle manufacturers would generally recommend against converting gasoline vehicles to alcohol.

The situation with CNG is similar in many respects. The attraction of CNG retrofits is that they provide dual-fuel capability—the vehicle operator can change from one fuel to the other simply by flipping a switch.

Vehicles with CNG retrofits, however, like those with alcohol retrofits, have a number of major disadvantages, the foremost being the initial cost. Conversion kits provided by various companies around the world include one or more storage cylinders, new fuel hoses and control valves, fuel regulators, and a gas-air "mixer" to maintain the proper air-fuel ratio for combustion. For conversion of spark ignition engines to CNG, the minimal changes necessary are replacement of the air filter with a gas-air mixer (which during operation on gasoline serves as the air filter), and installation of distributors with two spark ignition schedules to improve engine power. The cost for purchase and installation of conversion kits in the early 1980s (U.S. dollars) was about $1,500 per vehicle in the U.S. and Canada,[29] and $1,000 in New Zealand.[30] The cost is somewhat higher for engines with electronic fuel injection.

Other disadvantages of retrofitted CNG vehicles are less storage space, reduced range when using CNG, less power, extra weight, and possibly diminished vehicle handling (due to added weight of the cylinders). The exact magnitude of differences in performance and fuel efficiency is difficult to specify because of the variety of vehicles involved, the use of nonstandardized testing procedures, availability of different types of retrofits, and the sparseness of test data. The differences are relatively small, however, except for impacts on luggage space.[31] Of course, these differences would be negligible if the CNG vehicles were designed especially for CNG.

Advantages of CNG in retrofitted CNG vehicles are reduced corrosion of engine and fuel system parts, likely reductions in engine wear and spark plug life, and longer oil-change intervals.

MULTIFUEL VEHICLES

A second strategy is to design vehicles with a multifuel capability. Multifuel engines would provide vehicle operators with flexibility and would mitigate the problem of initially sparse networks of outlets for the new fuels. Various efforts have been made to develop engines that burn on a range of fuels; these include the Texaco TCCS engine,[32] Ford PROCO ("programmed combustion") engine,[33] and the gas turbine engine. Development of the first two engines faltered in the late 1970s because of high production costs and high fuel consumption rates. Both were flexible with respect to petroleum products, but considerable development would have been needed to adapt them to alcohol. Development of the gas turbine has continued, but it is not likely to become commercially viable without major technical breakthroughs. More recently, the U.S. Department of Energy has funded research on the use of multiple fuels in the Stirling engine, but initial results suggest once again that these new engine technologies are long-term options at best.

The most promising approach for developing multifuel vehicles is refinement of modern spark ignition automobile engines that use microprocessor-controlled fuel intake systems. Most autos in the U.S. already have these systems; they have in the exhaust line oxygen sensors that send signals to the fuel intake system to adjust automatically the fuel flow to cylinders in order to attain optimal combustion as fuel composition and driving conditions change. Although the sensors on production vehicles normally have a narrow range, only enough to accommodate expected differences in gasoline composition, researchers at Ford Motor Company and elsewhere have developed more flexible sensors that can respond to any combination of alcohol and gasoline fuel. Ford developed what is called a flexible fuel vehicle (FFV) in which they reprogrammed the on-board computer, which controls fuel flow and the spark advance, to respond appropriately to a wide range of fuels. Fuel composition is monitored by an optical sensor. In a 1982 test all fuel combinations, including straight gasoline, were found to have equal fuel efficiency, drivability, and exhaust emissions.[34] In 1985 Ford Motor Co. successfully built and demonstrated a number of these experimental vehicles with favorable results.[35] As of 1987 the state of California was seriously considering buying a fleet of about 5,000 multifuel vehicles.

The advantage of a multifuel vehicle is the flexibility provided to the user; any fuel or fuel blend can be used. The driver is not necessarily inconvenienced by the scarcity of outlets for new fuels, and the problem of building a network of outlets for new fuels becomes less critical.

The disadvantage of a multifuel vehicle is that it cannot be designed to exploit fully the unique characteristics of any one fuel. For example, a multifuel vehicle could not be optimized to operate efficiently on alcohol because then it would operate inefficiently and with poor drivability on gasoline. One forgoes potential fuel efficiency increases to gain multifuel capability. Since multifuel vehicles must be able to operate on the lower octane fuel (which is gasoline) and because the engine would be a descendant of gasoline engines, the multifuel vehicle would be a somewhat compromised vehicle.

The production cost for a liquid fuel multifuel vehicle should be no greater than for a comparable gasoline vehicle—if large production runs are made. Costs are discussed more fully in the next section on fuel-specific vehicles.

The design of a multifuel vehicle that operates on both liquid *and* gaseous fuels is more complex and expensive than a multifuel vehicle that operates only on miscible liquid fuels, mostly because it would require redundant fuel systems. The CNG vehicles addressed under the category of "retrofitted" vehicles are, in fact, dual-fuel vehicles. Those vehicles require distinct fuel tanks and fuel lines as well as additional control valves and fuel induction systems for switching between the two fuels. The additional cost for mass-produced vehicles that operate on both liquid and gaseous fuels would be roughly $700—mostly the cost of providing the extra CNG tanks.

FUEL-SPECIFIC VEHICLES

A third strategy for introducing vehicles that operate on nonpetroleum fuels is to design vehicles for a specific fuel. These vehicles would provide the highest efficiency and best performance but would sacrifice flexibility. Fuel-specific engines are designed to exploit the properties of specific fuels, thereby improving power and fuel efficiency. Current motor vehicle engines and power trains are designed to utilize the properties of petroleum fuels. Alcohol-fueled vehicles would run most efficiently if they were completely redesigned to take advantage of alcohol's higher octane rating, faster burn rate, and greater tolerance of lean burning; similarly, a vehicle designed for CNG would be more efficient if redesigned to accommodate the unique properties of natural gas.

Vehicles designed for natural gas, hydrogen, or alcohol should be significantly more efficient on a miles-per-Btu basis than gasoline. One Japanese researcher asserts that a vehicle completely and exclusively designed for methanol could get 100 percent greater efficiency (on an energy basis) than current gasoline-powered vehicles.[36] That assertion assumes that the cooling system could be downgraded or eliminated because of methanol's low flame temperature, and that a number of other changes would be made. It also assumes that gasoline engine technology is not upgraded over time. While that estimate may be overstated, it nonetheless suggests the potential for efficiency improvements. A more realistic assessment, based on what automakers could be expected to produce at competitive costs and taking into consideration likely advances in engine technology, is that a specially designed fuel-specific alcohol production model would achieve about 10 to 20 percent efficiency gains over comparable gasoline vehicles.

Improvements in performance by fuel-specific CNG vehicles would be substantial, but slightly less than for alcohol cars. Based on mathematical simulation studies of CNG-optimized vehicles (the only true CNG vehicles that exist as of 1986 are twenty-seven hand-assembled Ford research prototypes), a research group found that a CNG vehicle would at most be 3 percent more efficient than a comparable gasoline vehicle, but that acceleration would be slightly slower.[37] If acceleration and range were set equal to a comparable gasoline vehicle, then the CNG vehicle was 10 percent less fuel efficient. Other tests have shown substantial gains for CNG vehicles.[38] Ford researchers state that their twenty-seven light-duty CNG "Ranger" trucks have equal performance but shorter range than comparable gasoline trucks. The Ranger pick-up trucks are only incrementally different from gasoline vehicles, using the same basic engine and vehicle design. Based on this sparse information, one would expect a true single-fuel CNG vehicle to be roughly comparable to a gasoline vehicle on a performance basis, while retaining its superior emission characteristics.

The greatest disadvantage of fuel-specific engines is their limited flexibility. A vehicle designed for a specific blend proportion (e.g., 50 percent methanol and 50 percent gasoline) or specific fuel (e.g., methanol, ethanol, CNG, or gasoline) could be sold only where that specific fuel or fuel blend proportion is available. These fuel availability requirements would severely hinder the initial market penetration of alcohol or CNG cars, since many vehicle owners are unlikely to purchase a vehicle that they might not be able to use outside their immediate locale.

The cost of alcohol and CNG vehicles is primarily a function of production volume. Two significant cost components are incurred in building vehicles: retooling costs, which are fixed costs and must be amortized over the production volume, and research and development costs. A third cost component, replacement of incompatible materials, is much less significant. Although the new materials used in alcohol cars are different, they are not necessarily more costly than the ones they replace. To make gasoline cars compatible with alcohol, typical modifications would be polyethylene tanks to replace terne-plated tanks and replacement of some plastic valves and gaskets and other materials in the fuel lines and fuel intake system. The cost of building material compatibility into vehicles would therefore be negligible as long as it was accomplished during initial vehicle assembly.

Research and development costs may be significant. Ford claimed in 1984 that it had spent $50 million in developing LPG, CNG, and alcohol vehicle technology.[39] When spread over 100,000 vehicles, that cost is $500 per vehicle, which is not an exceptionally large amount for a large company except that there is no guarantee that Ford will realize the benefits from its R&D investment.

Retooling is the most significant additional cost involved in introducing new vehicles; the major costs are with respect to the engine manufacturing plants, which are the most highly automated plants of the auto industry. Even small changes in engines require scrapping specialized production equipment.[40] The cost of building alcohol or CNG engines would entail very large initial capital costs in retooling engine plants, although if production runs are great enough, these costs would be comparable to those for any new vehicle model. A review of several studies suggests that the total fixed cost of developing a new engine and drive train and building or retooling a production plant would be about $800 million to $1 billion for each engine drive train combination.[41] Presumably the costs for an alcohol or CNG engine-drive train would be at least that high. While these costs are large, they are reasonable on a per unit basis if enough vehicles are sold.

According to an old (1971) but possibly still relevant study, it was determined that about 200,000 vehicles must be sold to justify the retooling expense of building a new model.[42] Another study in 1981 estimated that a 50,000 vehicle run would add about 5 percent to each vehicle's cost, and that 250,000 vehicles would be enough for an automaker to recoup all added costs.[43] Still another study agrees that the desirable annual production rate for an all-new vehicle would be on the order of 250,000 units.[44] One can surmise from these estimates that roughly 100,000 vehicles of each model must be sold to produce a new cost-competitive alcohol or CNG vehicle.

These start-up costs could be reduced in the future by the increased use of "flexible" manufacturing processes.[45] This new approach to manufacturing enables managers to use computer-controlled equipment to reprogram the manufacturing equipment to produce small batches. For example, one fuel injection system could be installed on only 1,000 vehicles after which "robot" equipment that does machining and assembly might be reprogrammed to build and install a different fuel injection system on the next 1,000 vehicles, and so on. If computerized robot-type equipment continues to be developed and deployed, the size of minimum production runs needed to recoup retooling costs for alcohol or CNG engines could be reduced significantly from those estimates cited above.

Another caveat regarding alcohol and natural gas vehicles is that the analysis in this section has dealt with near- and mid-term developments and prospects. In the longer term (over twenty years or so), other engine technologies might be developed that might be significantly different from the modified spark ignition gasoline engine. While the most likely engine developments are lean-burn technologies with high compression ratios and perhaps some hybrid of spark and compression ignition technologies, commercialization of the Brayton gas turbine and the Stirling engines is also possible, although neither is likely to dominate the motor vehicle market in the foreseeable future. One promising technology for alcohol is methanol dissociation, in which the methanol is decomposed into hydrogen and carbon monoxide in the vehicle. Initial research indicates potential efficiency gains of 30 to 40 percent over comparable gasoline engines,[46] although the gains relative to a liquid methanol engine would presumably be much less. Most likely, however, current engine technology will continue to evolve incrementally without radical changes.

VEHICLE AND FUEL REGULATION

The decision to modify vehicle technologies, whether through retrofit or redesign, encounters other start-up costs—those that result from government intervention to promote the public interest through technology regulation. Since the mid-1960s in the U.S., there has been a predilection for establishing regulatory procedures and standards based on the specific attributes of a piece of technology, particularly with respect to environmental regulation. This regulatory approach, which ostensibly seeks to protect the environment, may in reality hinder innovation and change and even have counterproductive results, as will be indicated below.

The most notable example of regulatory obstacles to the use of new fuels in motor vehicles is the emission certification process. Initial opposition by the auto industry to government regulation of vehicle emissions in the 1960s and early 1970s prompted Congress and the Environmental Protection Agency (EPA), which implements and enforces air quality laws and regulations, to define very strictly the procedures and rules that would be used to enforce emission reductions. Procedures and rules were made as inflexible as possible to ensure strict compliance. But that inflexibility, which EPA is reluctant to tamper with, may prove to be an obstacle to the use of new fuels.

Ironically, air quality regulations may be a barrier to alcohol and natural gas use, even though emissions from alcohol and natural gas combustion tend to be less threatening to the environment than those from gasoline and diesel fuel combustion. This apparent perversity is more a reflection on the regulatory process in general rather than on alcohol or natural gas emissions per se, but the effect is the same.

The Clean Air Act requires that commercially offered fuels and fuel additives be "substantially similar" to those used in certifying vehicles of model year 1975 and later.[47] A fuel producer proposing to market any fuel or additive substantially different from that in use in 1975 can apply to EPA for a waiver of this prohibition, but it must supply EPA with evidence that the fuel or additive will not cause violations of emission standards or failure of vehicle emission control devices. In 1978 the EPA allowed gasohol to be marketed on the basis of its very small share of the total highway vehicle fuel market and pending the development of further emission data.[48] The introduction of methanol additives was held up until November 1980 by the EPA's refusal to grant waivers for their use. At that time the EPA finally granted a temporary waiver to Sun Petroleum Products Company to use up to 5.5 percent Oxinol, a one-to-one blend of methanol and TBA (a higher order alcohol) in unleaded gasoline. Shortly thereafter the EPA granted another waiver for a gasoline blend containing a maximum of 12 percent methanol by volume, plus cosolvents, with the total alcohol content (methanol plus higher-order alcohol cosolvents) not to exceed 15 percent of the blend. This "Petrocoal" waiver was used to market small quantities of fuel, but was eventually revoked by EPA (in July 1985 and upheld in 1986 by refusal of the Supreme Court to hear the appeal). Both waivers are for blends containing unspecified cosolvents and corrosion inhibitors known only to the fuel producer seeking the waiver; the result is that only the fuel producer who sought the waiver can use it (unless another producer or refiner were to purchase the additive from the producer holding the waiver).

Fuel producers were therefore pleased when Du Pont's proposed waiver for an additive containing a generic cosolvent was approved by EPA in early 1985, although they were displeased with several conditions attached to the waiver. Those conditions are requirements that the blended fuel meet stringent evaporative emission and volatility restrictions and that extensive monitoring be established. Industry representatives claim the restrictions are more severe than those placed on (unblended) gasoline and that the monitoring requirements are excessively burdensome and expensive; there appears to be a consensus in the industry that the Du Pont waiver will not be utilized unless it is altered. The effect of these fuel certification procedures has been to limit alcohol fuel options; as of 1986 the only alcohols being used were ethanol in blends of less than 10 percent and methanol in branded Oxinol blends.

Related to this issue of "substantially similar" fuels and their approval by EPA is the cost of verifying that the fuel indeed will not cause emission standards to be violated. A representative of one fuel company that successfully obtained a waiver to blend methanol with unleaded gasoline stated that his company had spent around $1 million to provide the test information required by EPA.[49] While this official conceded that much of this required testing would have been performed by the company anyway, the existence of the stringent regulations and procedures creates uncertainty for prospective producers and marketers. It particularly affects small producers and marketers to whom the uncertainty creates an especially large risk. If these restrictions had been in place in the 1930s, the small firm that produced Agrol fuel would probably never have succeeded in bringing its fuel to market. The granting of generic waivers would greatly reduce the cost and uncertainty of introducing new fuels.

The emission certification process will also affect vehicle manufacturers. In 1987 the EPA was drafting emission rules for methanol cars. As of that date, only gasoline and diesel vehicles (and CNG and LPG vehicles in California) were regulated. But in the future manufacturers who want to sell nonpetroleum vehicles in the U.S. will undoubtedly be required to meet fuel-specific emission standards. These test procedures are extensive and costly; the testing costs may be $13 million or more for each engine line.[50] Again, this requirement would not necessarily be onerous for large manufacturers, but it would be for small ones.

Fuel regulation takes place on the state level also. States require fuels to meet various specifications; specifications most relevant to alcohol blends concern fuel volatility, which affects vapor lock and evaporative emissions. Since alcohol blends (but not straight alcohol) are more volatile than gasoline, strict state regulations have the effect of banning alcohol/gasoline blends. Over thirty states have some volatility restrictions.[51]

In California, for example, motor vehicle fuels are required to have a Reid Vapor Pressure (RVP) of less than nine pounds per square inch (psi) during summer months[52]—primarily to reduce evaporative emissions and reduce smog formation. (This standard is stricter in the summer-time because the presence of sunlight promotes the formation of photochemical oxidants.) Gasohol cannot meet the restriction and so it could not be marketed during summer months as long as that rule was in effect. Since it was not attractive to fuel retailers to market gasohol only during winter months, they generally did not market gasohol during any part of the year. In effect, alcohol fuels were banned from the California market. In 1980 a special waiver was enacted[53] which specifically exempted gasohol from the vapor pressure rule if the gasoline component in the gasohol had a RVP lower than nine psi. The gasohol business flourished in California as soon as the rule was changed.

Rules regarding labeling and regulation of fuel efficiency provide another illustration of the potential effect of highly specific regulatory rules. A 1975 law[54] established corporate average fuel economy (CAFE) rules that require automakers to attain specified fuel efficiencies, but these efficiencies are measured in miles per gallon regardless of the energy density of the fuel. As a result, diesel fuel is rewarded because it has about 12 percent greater energy per unit of volume than gasoline. Alcohols, on the other hand, are discriminated against because they have lower energy content per unit of volume, even though they have higher energy (thermal) efficiency. Also, energy efficiency labels on showroom vehicles are expressed in mile-per-gallon units, which, when viewed by a buyer, gives the impression that alcohol vehicles are highly inefficient.

These various rules need not be inhibiting. They could be modified in such a way as to promote alternative-fueled vehicles. For instance, the

CAFE rules, whose purpose was to conserve petroleum, could be modified so that alcohol vehicles do not count at all in the calculations (as was done with electric vehicles) or so that only the gasoline portion of the fuel counts. Then, by outfitting large cars with alcohol engines, automakers could sell more large cars (with high fuel consumption) and still meet the standards. (How the rules would be defined for multifuel vehicles is more problematic since it is not known what portion of the time the vehicle operates on each fuel.)

These are a sample of highly specific rules and regulations that were adopted to protect air quality but that in practice impose a large cost and considerable uncertainty on fuel producers and vehicle manufacturers who desire to introduce a new fuel. Two other examples of regulations that create uncertainty for nonpetroleum vehicles are antitampering restrictions, which prohibit alterations of emission control equipment (which in effect prohibits vehicles from being retrofitted for other fuels), and tailpipe emission standards. At present only the major gasoline pollutants are regulated; aldehyde, an alcohol pollutant, is not. Will aldehyde standards be promulgated? What will they be? And what effect will they have on alcohol fuel use?

In summary, an elaborate and highly specific regulatory system has been designed in the U.S. around the use of petroleum fuels. The effect of these regulations has been to create high costs and uncertainty for those seeking to introduce nonpetroleum fuels. Their net effect is inhibiting, especially to smaller firms that attempt to introduce unique products and fuels. The intent here has not been to criticize the existing regulatory process, for it has been effective in reducing air pollution, but rather to illustrate the array of conservative forces inhibiting innovation and change. Philosophically, the environmental regulation of fuels and vehicles represents a realization that unbounded technological innovation and economic growth is not unequivocally positive. The challenge is to be certain that regulation of one activity to achieve one purpose does not inhibit other activities that contribute to the achievement of that and other purposes.

CONCLUSION

The overriding conclusion from this chapter is that while major advances in engine and power-train development have occurred during this century, the benefits have accrued to engines that burn gasoline and diesel fuel. Alcohol fuels and natural gas may be used in these engines with relatively minor modifications, but the vehicles are optimized for either gasoline or diesel fuel. The transition to these alternative fuels will have to overcome uncertain regulations, pressures to be compatible with existing vehicle technology, and a limited market that does not support the development and production of optimized vehicle designs.

The obstacles in introducing alcohol fuels and CNG into vehicles are not technical problems or exceptional vehicle production costs—the obstacles are strictly of a start-up nature. Even if new fuels are less expensive than gasoline, as indeed natural gas has been, start-up barriers and regulatory and marketing uncertainties will severely restrict those fuels dissimilar to gasoline and diesel fuel.

THIRTEEN

Market Niches

One way to mitigate the large start-up barriers to new fuels is to target those market niches in which the new fuel has some advantage or attraction. The market niche approach is appealing because it fits neatly with conventional marketing strategies and with theories of technology development.

If a business aims to sell a new unique product, it generally identifies one or more initial market niches where it believes the product is most likely to succeed. Similarly, in a more theoretical vein, the process of technology development may be represented as follows. First, an innovation is introduced into market niches where the characteristics of that product or innovation are highly valued. For instance, the initial turbochargers developed for automobiles were unreliable and prone to mechanical breakdowns, but they provided the attraction of much greater power and acceleration. Certain consumers, especially those with diesel vehicles and sports cars, valued those attributes very highly, and therefore purchased turbochargers even though they were expensive and unreliable initially. Over time, as experience was gained in these market niches, the turbocharger was improved and made more reliable. As reliability increased, so did sales, and prices dropped because of economies of scale in production. The turbocharger soon spread to other market segments.[1] This turbocharger example is suggestive of a general case in which a new technology or product is introduced into a market niche where consumers tend to accept its relatively high cost and usually low reliability in return for some new positive attributes it offers. If a product is improved, and production costs are lowered, then it is accepted in other market niches until it gains widespread acceptance.

One or another of the alternative fuels is attractive in each of the following niches:

agriculture
gasoline additive
urban transit buses
fleet vehicles.

Large trucks are not included in this list for several reasons: Diesel engines are technically less suited to alcohol and natural gas fuels than spark ignition engines, truck operators highly value the long life and low maintenance requirements of diesel engines, and diesel fuel tends to be much less expensive per vehicle mile than gasoline. Stricter government regulation of particulate emissions may create an opportunity for methanol or natural gas in diesel engines, but that possibility is unlikely except in specialized circumstances such as transit buses (see table 40 for transportation fuel use by mode and fuel).

Other market opportunities exist in the large household vehicle market; the only real niche, however, is high-performance cars, and that is a very small niche. Second or third cars owned by households are not a likely target market because households are not likely to replace these usually older, lower-value cars with new and therefore more expensive nonpetroleum cars. The household vehicle market (including high-performance cars) is addressed in the following chapter.

Each of the market niches listed above will be analyzed in this chapter. One salient observation emerges from this and the following chapter —transportation market niches are relatively small and few in number. The transportation sector, unlike other energy-consuming activities, appears homogeneous to new fuels. In the industrial, commercial, and residential sectors, and in the electric utility industry, a veritable plethora of new energy sources have gained wide acceptance. Not so in highway transportation. Most fuel is consumed in just two different types of engines, and most automobiles and light trucks are called upon to serve a large number of different purposes and to have long driving ranges (per tankful of fuel), even though most vehicles are used mostly for short trips.

The emphasis of this and the following chapter is on market oppor-

TABLE 40 TRANSPORTATION ENERGY USE BY MODE AND FUEL, 1983, THOUSANDS OF OIL-EQUIVALENT BARRELS PER DAY

	Gasoline	Distillate Fuel Oil	Jet Fuel	Residual Fuel Oil	Other[a]
Automobiles	4191	49	0	0	negl.[b]
Urban transit buses	1	30	0	0	0
Intercity buses	0	13	0	0	0
School buses	22	0	0	0	0
Light trucks	1308	9	0	0	13
Other trucks	360	838	0	0	13
Airplanes	30	0	676[c]	0	0
Boats and ships	65	113	0	560	negl.
Pipelines[d]	0	negl.	0	negl.	425
Rail (pass and freight)	0	276	0	1	19
Military and other[e]	34	72	229	10	0
TOTAL	6011	1400	905	571	470

Total = 9,357,000 b/d

SOURCE: Oak Ridge National Laboratory, *Transportation Energy Data Book: Edition 8* (Springfield, Va.: NTIS, 1985, ORNL-6205), p. 1-9.

[a]Includes LPG, natural gas, electricity.

[b]Alcohol blended with gasoline is included under gasoline.

[c]Does not include fuel purchased abroad by commercial airlines.

[d]Breakdown of energy use in pipelines not available.

[e]Includes heavy duty off-highway vehicles.

tunities for fuels dissimilar to petroleum products, not the petroleum-like liquids from oil shale, oil sands, and directly hydrogenerated coal. These petroleum-like fuels would have little or no effect on end-use activities except in terms of price because they would be blended into the petroleum product stream at the process plant, petroleum refinery, or bulk storage terminals and would be indistinguishable to the vehicle operator and fuel purchaser. Their market penetration would be relatively insensitive to vehicle technology considerations.

Liquid transportation fuels are premium fuels in the sense that they have high energy density, and are clean burning and easily transported. In general, these fuels are valued more highly in transportation than other uses. New fuels that are attractive to transportation technologies and users are not likely to be diverted from transportation to other uses or to be valued as highly in those other activities. Thus alcohol fuels would generally be more highly valued as transportation fuels than as fuels for boilers or combustion turbines. The situation is different for gaseous fuels because they have a low energy density and thus less attraction as transportation fuels. Gaseous fuels such as CNG and, in the long term, hydrogen may or may not be more attractive for transportation applications, depending on how end users value their different attributes (relative to gasoline and alcohol).

In any case, the two most attractive nontransportation markets for natural gas and alcohol fuels are reviewed next, followed by more in-depth analysis of transportation market niches.

CHEMICAL MARKET

The chemical industry uses a number of key intermediate chemical materials, especially ethylene, propylene, methanol, benzene, toluene, and the xylenes, to produce a vast array of chemicals. Earlier in the century, biomass was the primary feedstock for these chemicals;[2] now these intermediate chemicals are derived almost entirely from petroleum and natural gas.

In the U.S. in 1985, about 75,000 b/d of methanol was used to manufacture resins, glues, solvents, plastics, and other chemical products;[3] worldwide chemical methanol consumption in the mid-1980s was about 250,000 b/d.[4] The U.S. and world chemical methanol market is expected to grow at a 4 to 5 percent annual rate into the foreseeable future (see chap. 22).[5]

Ethanol is also widely used as a chemical. It is usually produced from the intermediate chemical ethylene, which is a product of petroleum and natural gas, and is used as a solvent, germicide, antifreeze, and chemical intermediate in the production of other organic chemicals. Much less chemical ethanol is produced than chemical methanol. The U.S. produced over half of the world's total until about 1980; like the chemical methanol industry, petrochemical ethanol production has been shifting to countries with low-cost natural gas supplies.[6]

Chemical methanol and ethanol differ from fuel alcohol only in that they must be purged of impurities and water.

Petroleum-like liquids derived from coal and oil shale could also be used to produce intermediate chemicals, but the need to gain constancy of composition from the hundreds and thousands of chemical compounds in the liquids makes them less useful as chemicals than as fuels.

Liquids and gases produced from biomass, coal, oil shale, and oil sands would sometimes receive a higher price in the chemical market than in the fuel market, but even if those lucrative opportunities did exist, they would soon disappear as supply increased and prices dropped. Some specialized chemical market opportunities may appear for alternative fuels, but they are not likely to be significant. In the U.S. biomass products will initially be at a particularly large disadvantage in the chemicals market, not only because of costs but because of mismatches in location and size of process plants.[7] Virtually all chemical alcohol plants in the U.S. are located on the Gulf Coast[8] and are part of large integrated industrial complexes. The U.S. chemical industry is not likely to provide any significant early market niches in which to nurture alternative fuels.[9]

ELECTRICITY GENERATION

The electric utility industry shows somewhat more promise for nurturing the development of alternative fuels, but it is likely to play a modest role at most. One opportunity is to burn shale oil liquids in boilers (for either process heat for industry or steam heat for electric utilities), but the high cost of those fuels makes that option generally unattractive. Alcohol fuels and coal synthesis gases are technically and environmentally more attractive replacements for petroleum and natural gas in engines and boilers. Since they are expensive, they would only be used in selected applications; they would not be used alone in boilers (for generating heat in residential, industrial, and electric utility applications), since they would be much more expensive than the low-quality, inexpensive, heavy petroleum products typically used. The only unique attribute of alternative fuels valued by electric utilities is their clean combustion.

One possible market niche in the electricity-generating industry that is relevant to transportation fuels is the use of combined cycles to convert coal into a synthesis gas and then electricity. As indicated earlier, burning coal in a combustion turbine in a combined cycle facility generates substantially fewer air pollutant emissions than does direct coal combustion. Recent evidence indicates that combined cycle coal generation of electricity costs about the same as direct coal combustion with control of flue gases (see chap. 7). The successful Coolwater plant in California and Dow Syngas project in Louisiana are demonstrations of the technology. No firm estimates of the market potential of this technology have been made, but it should prove attractive in areas with high-sulfur coal, such as in Ohio, Illinois, and Indiana, and in other areas actively concerned with air pollution, such as southern California and Germany. While combined cycle coal gasification does not provide an auxiliary market for transportation fuels, it does provide for the development and manufacture of coal gasification technology, which would improve the performance and lower the costs for the coal-to-gas and coal-to-methanol plants. It also provides a lower-risk opportunity to coproduce synthesis gas and methanol, providing the flexibility to shift from one product to the other as market conditions change.

A second market niche in the electricity industry is in areas with air pollution problems where methanol may be used as a fuel for combustion turbines and for "overfiring" of petroleum fuels in boilers. Methanol burns efficiently in combustion turbines, which are generally used only in times of peak electricity demand and in rural areas. Combustion turbines now burn either natural gas or middle-distillate petroleum fuels (e.g., No. 2 fuel oil), but may be retrofitted or built specifically for methanol with only minor cost penalties.[10] Methanol generates as little as 25 percent of the nitrogen oxide (NO_x) emitted by No. 2 fuel oil, somewhat more hydrocarbons, and comparable quantities of carbon monoxide; compared with natural gas, methanol generates somewhat lower NO_x and similar quantities of other pollutants.[11] Methanol may therefore be attractive to electric utilities in areas with NO_x problems and unreliable natural gas supplies, although they would face considerable cost penalties in using methanol. The only area where ambient NO_x air quality standards are being violated in the U.S. is southern California.

Overfiring, as referred to above, is the simultaneous combustion of two different fuels—in this case methanol and a petroleum fuel—through different burners in a boiler. The attractiveness of this dual-fuel firing technique is that NO_x emissions are reduced far out of proportion to the fraction of methanol used. Initial tests indicate that substituting methanol for 20 percent of the oil has reduced NO_x emissions by up to 40 percent.[12] Boilers represent a much larger proportion of the electricity-generating market than combustion turbines (including combined cycle operations), but as with the combustion turbine option, overfiring in boilers is relevant only in areas with NO_x problems, which are few.

In conclusion, electricity generation does not provide important market niches for nonpetroleum transportation fuels, except perhaps in southern California and as a comarket for coal synthesis gases in areas with high sulfur coals.

GASOLINE ADDITIVES

The first market niche successfully penetrated by alcohol fuels has been the gasoline additive market. The petroleum refining industry is facing growing difficulties in producing large volumes of high-quality gasoline and is attracted to alcohol for two reasons. One reason is the ban of tetraethyl lead in gasoline. The U.S. began phasing out lead in 1974 because it destroys the catalysts in catalytic converters of engines; in 1985, because of lead's adverse health effects, the phaseout schedule was accelerated.[13] The European Common Market also decided to begin mandatory lead phaseouts, effective in 1989, although West Germany and some other countries will begin sooner.[14]

The second reason alcohol is becoming more attractive as an additive is the diminshing quality of petroleum supplies. As reserves of light, low-sulfur crude oil diminish around the world (but not yet in the Middle East), new production comes from heavy, dense oil, which is more difficult to refine into high-octane gasoline. Petroleum refiners slowed this transition to lower-quality oil in the mid-1980s by opting to purchase more high-quality stocks because of the low oil prices; the long-term effect of this purchasing pattern will be to accelerate the depletion of high-quality petroleum.

Mandated reductions in lead use and dwindling supplies of light, high-quality crude petroleum stocks threaten a century-long trend toward higher-octane and higher-quality gasoline.[15] The intent of automakers to use higher compression ratios in future vehicles (to achieve greater fuel efficiency) places still greater demands on refiners to produce high-octane gasoline. While the general perception in the oil industry is that there will be a serious and growing problem in meeting the demands for gasoline production, that situation will be very different for each petroleum refiner, depending mostly on the sophistication of refining equipment and access to high-quality crude petroleum. Generally, small and independent refiners will face the most difficult problems. They usually do not have assured access to high-quality crude stocks (unlike vertically integrated major oil companies), and they tend to have simpler, less sophisticated, and less flexible refineries. The problem of small, independent refineries in the U.S. was exacerbated by the withdrawal in 1983 of a special dispensation that allowed them to use lead additives. The new requirements force them to meet the same stringent phaseout schedule as large refiners.

Two principal approaches are used to increase octane. A refiner may refine the oil stock more intensively or may purchase octane-enhancing additives and blend those additives into the gasoline stock. Large refiners tend to upgrade gasoline stocks by subjecting them to more intensive processing in existing facilities (e.g., catalytic reforming and alkylation) or by building new processing facilities; in this way they produce high-octane aromatics such as toluene and xylene, which are blended back into the gasoline. The energy and capital cost for this additional processing is high, however. The alternative is to purchase or manufacture additives such as MTBE (methyl-t-butyl-ether), toluene, ethanol, and methanol. Small refiners with less sophisticated and flexible refinery equipment are more likely to resort to this second option.

The relative economics of these alternative options are difficult to determine, partly because they vary from one refiner to another, and partly because the information is proprietary and not publicly available. Generally speaking, though, industry analysts find MTBE to be less expensive than toluene and ethanol and an oxinol-type blend (see chap. 12) to be the least expensive,[16] but this varies among refiners and may change over time. MTBE is made from and is therefore more expensive than methanol, but MTBE has the advantage of being less sensitive to the presence of water. MTBE was first introduced in Europe in the mid-1970s and in the U.S. in 1979; by 1983 there were eight plants in the U.S., all in Texas, producing about 28,000 b/d, and by 1987 production capacity had reached 90,000 b/d.[17] MTBE is used in proportions of 4 to 7 percent of gasoline, with the upper limit fixed by EPA at 11 percent.[18] Economical production of MTBE is somewhat constrained by the availability of the refinery byproduct isobutylene, an ingredient used in MTBE production. The World Bank estimates that worldwide production of more than about 110,000 b/d of MTBE would exhaust inexpensive isobutylene supplies; above that level costs would increase substantially.[19]

Alcohols are highly attractive additives because of their high octane. Ethanol and methanol have octane ratings of about 101.5 ([RON + MON]/2), compared to ratings in the mid to high 80s for

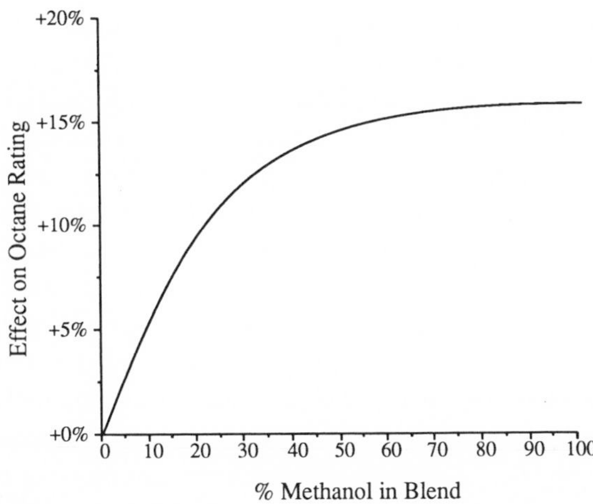

Figure 32. Effect on Octane Rating of Blending Methanol in Gasoline

Source: Roberta Nichols, "Technical Aspects of the Use of Methanol-Gasoline Blends as Transport Fuels", Presented to the Joint China-U.S. Seminar on Clean Coal Fuels, Taiyuan, China, 13–18 May 1985.

Note: Assumes constant temperature of 100°F at engine inlet. The octane boast is somewhat smaller at higher temperatures and with ethanol.

typical gasoline (in the U.S.). If used in small proportions with gasoline, alcohol will boost the octane of the fuel, while still being for the most part compatible with existing distribution and end-use systems (see chap. 11 and 12). As shown in figure 32, the addition of 5 percent methanol (or ethanol) to gasoline boosts the octane rating of the gasoline fuel by about 1.5 points. Adding larger proportions will raise the octane rating still more, but above 40 percent or so the gains begin to diminish.

Methanol is the least expensive and potentially the most abundant additive available, but because of problems of water sensitivity, materials compatibility, and volatility, it is also the most troublesome to use in the existing gasoline distribution system. Methanol's water sensitivity and volatility is reduced by adding ethanol or some other cosolvent, such as a higher-order alcohol; e.g., tertiary butyl alcohol (TBA). These cosolvents are more expensive than methanol (on a volume and energy basis) and therefore increase the cost per gallon.[20] Many refiners would not have cosolvents readily available and would have to purchase them; others would have a ready supply from petrochemical operations (e.g., ARCO has a ready supply of TBA as a coproduct of its propylene oxide manufacturing process).

Initial experiences with methanol additives in the early to mid 1980s were technically successful, but suffered public relations setbacks. In 1983–1984 in particular the media publicized stories of automobile operators who had purchased gasoline with methanol in it and had experienced severe drivability problems. The cause of these problems, as discussed in chapter 11, turned out to be local fuel distributors who did not thoroughly dehydrate their tanks and who illegally blended large proportions of methanol with gasoline. Although the actual number of illegal and sloppy blending incidents was small, the spate of bad publicity was another example, similar to the converted car phenomenon in Brazil, of how the initial introduction of a new fuel is subject to extraordinary scrutiny, often with devastating results.

The market for methanol additives is promising, despite those setbacks. The potential size of the methanol additive market is determined by economic and technical factors, but in the U.S. the ultimate arbiter will be the EPA, which regulates the emission characteristics of fuels and administers the "substantially similar" fuel requirement of the Clean Air Act. As mentioned in chapter 12, the first permission to use methanol was the 1980 waiver for a 5.5 percent Oxinol blend. Another waiver was subsequently granted (for Petracoal) which allowed a higher percentage, but that waiver also had brand-specific restrictions; it was used on a small scale but was revoked by the EPA in July 1985 (with final court appeals rejected in February 1986). A waiver was finally granted in 1985 ("Du Pont waiver") to allow the use of generic (not proprietary or brand-specific) alcohol cosolvents. However, it applies more restrictive conditions on the volatility and evaporative emission characteristics of blended gasoline than on straight gasoline and stringent monitoring requirements, which industry representatives claim are expensive and burdensome. The EPA responds that it intends eventually to apply those same standards to all gasoline. Meanwhile, the industry has been unwilling to use the "Du Pont" waiver, thereby restricting the use of methanol as an additive.

Ethanol is technically more attractive than methanol because it has less severe material compatibility problems and because ethanol/gasoline blends are less sensitive to the presence of water. Ethanol was initially used without cosolvents in blend proportions of up to 22 percent in Brazil, although mostly as a fuel extender and not an octane booster, and in the U.S. in proportions of up to 10 percent, also mostly as an extender. Ethanol was used as a fuel extender in the U.S. because of lucrative subsidies. In the mid-1980s, as production continued to expand, a few entrepreneurially minded firms began to take advantage of ethanol as an octane booster. Small independent refineries began to produce less costly "sub-octane" gasoline with octane ratings of about 84 (compared to 87 for regular unleaded gasoline). One major common carrier pipeline (Williams Bros.) altered operations and tariff structures to carry suboctane gasoline, and independent fuel marketers blended it with ethanol at local storage terminals. If the practice of producing and distributing suboctane gasoline becomes institutionalized, the market value of ethanol as an additive will be fully exploited.

In the near term the market potential for alcohols as gasoline additives is limited to less than 5 percent of the gasoline market for methanol and 10 percent of that for ethanol—partly because oil refineries would not have the flexibility to begin producing the low-octane gasoline stock that could take advantage of larger proportions of alcohol, but, more important, because motor vehicles are not designed to handle larger blend proportions. Based on these ceilings, the ultimate market penetration of alcohol additives in the U.S. is about 360,000 b/d of methanol or 720,000 b/d of ethanol. In practice, actual market penetration would be much lower, depending upon the following factors: the demand for high-octane gasoline; the cost and availability of competing additives and methanol cosolvents (e.g., TBA); the cost, availability, and quality of petroleum stocks; the flexibility and capital investment plans of individual refiners; and, of course, the relative price of gasoline and additive candidates. Market potential could expand beyond these respective 5 percent and 10 percent ceilings if larger blend proportions were used, but to do so would require major changes in vehicle technologies and fuel distribution systems. For analytical purposes, larger blend proportions are therefore treated here in the same way as straight alcohol.

In summary, the additive market is significant but not exceptionally large. It is attractive for initial alcohol producers because it is a fairly predictable and secure market and therefore conducive to long-term contracts. The ability to secure long-term supply contracts, or at least expressions of intent from marketers, may be the key factor in decisions by investors to proceed with construction of alcohol production facilities.

AGRICULTURE

The agriculture industry in the U.S. consumes over 400,000 b/d of gasoline and diesel fuel, about half of it in automobiles and trucks.[21] Thirty years ago the great majority of agriculture fuel was gasoline. Since then, farms have steadily shifted toward diesel equipment and vehicles; most new equipment now consumes diesel fuel. As late as 1974, farms used about 240,000 b/d of gasoline and 170,000 b/d of diesel fuel (as well as the equivalent of about 154,000 b/d of distillate fuel oil, liquefied petroleum gases, and natural gas).[22] By 1981 total gasoline consumption on farms had dropped about 40 percent;[23] data for diesel fuel use could not be found, but presumably it increased to replace the reduced gasoline consumption.

Fuel produced from biomass might be an attractive alternative fuel for agriculture if it were produced from locally available low-cost feedstocks and used on site or locally. In the far-off future, plant oils might be used as diesel fuel substitutes, and methanol or ethanol produced from crop residues and wood might be used as gasoline substitutes. The market potential for ethanol (and methanol in the future) in the agricultural market is small, however. In 1981 total agricultural gasoline consumption was only 63,000 b/d. The potential market for alcohols shrinks even further when one considers that some of that gasoline is used for farm equipment, irrigation pumps, and miscellaneous other engines that are unlikely to be available in an alcohol-fueled version, at least not during the initial transition period.[24]

URBAN TRANSIT BUSES

A more attractive but even smaller market niche is the urban transit bus industry. Virtually all urban transit buses (but not school buses) in all parts of the world use diesel engines. In the U.S. diesel transit buses date back to the late 1930s, when General Motors formed a holding company with an oil and rubber company and began a massive campaign to purchase economically distressed urban rail transit companies so that it could replace rail (streetcar) lines with GM diesel buses.[25] Diesel urban transit buses increased from only 62 in 1937 to 680 in 1940, and then continued to expand after the war, numbering over 60,000 in the 1980s.[26]

Urban transit buses are an attractive market niche because the use of methanol or natural gas significantly reduces their otherwise high levels of air pollution. Although diesel transit buses have been recognized as an environmental nuisance since their introduction in the late 1930s, they have not until recently received a high priority from environmental regulators. It was thought that since urban buses are such a small proportion of total urban vehicles, their emissions are insignificant. Moreover, there has been no ready alternative to the bus since trolley cars began to be phased out in the 1930s and '40s.

Several factors have led to a reexamination of that view.[27] First, recent evidence based on improved data and testing suggests that public exposure to diesel bus emissions is much higher than had been thought. The high exposure results from buses operating on the most populated streets and emitting pollution at street level. A second factor in the U.S. is a legislative mandate by the 1977 Amendments to the Clean Air Act to reduce nitrogen oxide and particulate emissions from diesel engines. In 1985, after years of delay caused by concerns that cost-effective control technologies could not be developed, the EPA promulgated stringent particulate and NO_x emission standards for diesel engines. Those standards take effect in 1991 for diesel buses and 1994 for diesel trucks and are expected to result in an increase in new engine prices and a slight decrease in fuel economy.[28] The third factor encouraging stricter diesel engine emission standards is the development of diesel engines that operate on methanol and natural gas, which presents a clean-burning technological alternative option that had not previously existed.

Even though the air quality benefits of using methanol and natural gas in diesel buses may be large, the quantity of fuel that would be used is small. Over 1,000 urban areas in the U.S. have urban bus transit operations (almost all owned by local governments) that operate over 60,000 buses, virtually all with diesel engines (96 percent); but these buses consume only about 30,000 b/d.[29] Moreover, the penetration of this relatively small market is hindered by the slow turnover of transit buses. About 3,000 to 5,000 new buses are purchased each year in the U.S.;[30] they are typically used for up to twenty years or more, and engines are typically rebuilt two to three times during the life of each bus.[31] If all new buses and all buses being overhauled were outfitted for methanol, methanol fuel consumption would increase from about 4,500 b/d in year one to about 30,000 b/d in year seven or eight.

Since the Urban Mass Transportation Administration (of the U.S. Dept. of Transportation), which subsidizes 80 percent of the cost of new buses, has indicated enthusiasm for methanol buses, there is a strong possibility that methanol buses will be deployed soon after the technology becomes commercially available. In 1986 two transit districts in the Los Angeles area ordered a total of thirty-three methanol transit buses.

Market penetration of the transit bus market is likely to proceed faster than in any other vehicle market segment, although the rate is sensitive to relative fuel prices, methanol bus prices, and diesel bus prices. In any case, the transit bus market is small and dispersed and is not a significant factor from a fuel perspective.

VEHICLE FLEETS

Vehicle fleets represent a much larger and more attractive early market. This attractiveness is greatest for those vehicle fleets that rely exclusively on their own centralized fuel-dispensing stations, obviating the difficult logistical and strategic deployment problems of establishing a network of fuel outlets. These centralized fleets also often have the advantages of having their own mechanics, who could be specially trained, and of having the capability to retrofit all or some of their vehicles before automakers begin to manufacture production-line vehicles. Vehicle fleets that retrofit their vehicles for natural gas (or alcohol) would generate part of the "critical mass" that would convince vehicle manufacturers to invest in the production of alternative fuel vehicles.

Virtually all light-duty fleet vehicles (in fleets of ten or more vehicles) are gasoline powered; less than 1 percent are powered by LPG and only about 2 to 4 percent are powered by diesel engines.[32] These estimates do not include fleets of fewer than ten vehicles, medium- and heavy-duty trucks, and off-road construction and mining equipment—all of which are likely to be less attracted to alcohol or CNG.[33] Altogether, fleet vehicles consume about 1.2 million b/d of gasoline in the U.S.[34] They represent a disproportionate share of new vehicle sales because of their rapid turnover—fleets consisting of ten or more vehicles account for 12 to 14 percent of new car sales.[35] Table 41 presents estimates of the number of vehicles in fleets by fleet category.

Penetration of the fleet market by nonpetroleum fuels will depend principally on the following factors:

cost of purchasing, operating, and maintaining nonpetroleum vehicles

cost and availability of fuels

TABLE 41 FLEET VEHICLES IN FLEETS OF TEN OR MORE

Fleet Category	Cars (1981)[a]	Light Trucks (1979)[b]
Business	3,187	1,509
Salesperson-owned	340	—
Leased	2,192	—
Company-owned	655	
Individually leased	1,700	NA[c]
Daily rental	491	75
Government (excluding police)	607	213
Federal	107	—
State and local	500	—
Police	253	21
State	40	—
Local	213	—
Utility	162	307
Taxi	155	17
Total (10 or more)	6,555	2,142
Total (4 or more)	(12,590)	NA

SOURCES: Fleet car estimates, except for individually leased cars, are from Aerospace Corp., Assessment of Methane-Related Fuels for Automotive Fleet Vehicles (Springfield, Va.: NTIS, 1982, DOE/CE/50179), tables 5-7 and 5-8. Individually-leased car estimates and light truck fleet estimates are derived from Joseph R. Wagner, Fleet Operator Data Book, Vol. 1: National Data (Springfield, Va.: NTIS, 1979, BNL 50904).

[a] More recent 1984 estimates are virtually identical except for the category of individually leased vehicles. See Oak Ridge National Laboratory, Transportation Energy Data Book: Edition 8 (Springfield, Va.: NTIS, 1985, ORNL-6205), p. 2-47.

[b] Light trucks include vans. Reliable estimates by fleet category do not exist. Estimates are made here by using ratio of fleet cars to fleet trucks from 1977 Bobit/BNL survey in Wagner, Fleet Operator Data Book, p. 12. See also Bureau of Census, Truck Inventory and Use Survey (1977 Census of Transportation) (Washington, D.C.: U.S. Dept. of Commerce, 1980, TC F.F-T-52). Fleet trucks are probably significantly underestimated here because of increased popularity of light trucks since the 1977 survey.

[c] No data available on light trucks leased to individuals; number is probably very small compared to leased cars.

TABLE 42 AVERAGE RATING FOR CRITERIA USED IN AUTOMOBILE PURCHASE DECISION IN FLEETS OF 10 OR MORE VEHICLES (0 TO 5 SCALE)

Fleet Type	Purchase Price	Resale Value	Maintenance Cost	Life-Cycle Cost	Reliability
Business	3.04	3.19	3.00	2.84	2.98
Rental	1.86	3.49	3.51	3.27	3.24
Government	4.51	2.55	3.26	3.11	3.22
Police	3.31	1.86	2.73	2.93	3.55
Utility	3.68	1.76	3.57	3.00	3.31
Taxi	3.43	2.13	3.53	3.48	3.45
All Fleets	3.29	2.66	3.19	2.96	3.18

SOURCE: Joseph R. Wagner, Fleet Operator Data Book (Summer 1977 Data), Vol. 1: National Data (Springfield, Va.: NTIS, 1979, BNL-50904), p. 17.

NOTE: 0 = least important; 5 = most important.
Ratings were weighted based on size of respondent's fleet.

reliability of vehicles

availability of parts and repair services

vehicle range and other performance attributes

organizational and individual resistance to change.

A precise analysis of vehicle purchase decisions by fleet operators is difficult and has not been attempted by myself or others. A difficulty is that, since operators have had little or no experience with non-petroleum fuels, they cannot be expected to give reliable responses to hypothetical questions. However, the following brief analysis of each segment of the fleet market provides some insight into likely responses of fleets to new fuels.

Certain fleet vehicle operators, including salespersons, operators of rental car companies, and companies that lease vehicles to individuals, would have little or no interest in those fuels that are dissimilar to gasoline and diesel fuel. These vehicles are refueled most or all the time at retail fuel outlets (not private depots) and are often used for long trips. These drivers would face the same fuel availability difficulties as drivers of household vehicles. Rental companies would also be uninterested because they sell their cars when they still have relatively low mileage and depend on the revenue gained from resale (see table 42). The resale value of unconventional (i.e., alternative-fueled) vehicles would be uncertain and probably depressed initially, making alternative fuels unattractive for rental companies.

Taxicabs are a somewhat more plausible early market for new fuels because they are operated within a limited geographical area. Refueling would be a relatively minor problem if a number of outlets were established in the cities in question. Drivers could be readily informed of the locations of those outlets. Taxis represent a potential market for both CNG and alcohol in those cities with air quality problems. Local governments regulate taxicabs and could mandate that taxicab operators use alcohol or CNG as an air pollution control strategy. Taxicab operators would be highly sensitive to vehicle and fuel prices, however, and would not be likely purchasers of alcohol cars unless subsidies pushed prices below those for gasoline cars.

Fleet vehicles for electric, gas, and telephone utilities are also a plausible early market for alternative fuels. Utility fleet vehicles accumulate an average of only forty-six miles per day[36] and usually have central fuel depots at which vehicles can be fueled overnight. Refueling and limited driving range are not problems. Utilities that sell natural gas are a logical market for CNG vehicles (and indeed have been the principal advocates of CNG fuel in Canada and New Zealand). Because utilities are regulated by state commissions, they are more sensitive to public concerns; their fleets could be targeted if state policies were adopted to promote those fuels.

Likewise, government fleets are subject to executive or legislative mandates. They are a plausible early market if a government promotes an alternative fuel. The relatively short trips of government vehicles (an average of fifty miles per day based on a five-day week for federal vehicles, and sixty-five miles per day for state and local vehicles)[37] renders them attractive candidates. A small pool of gasoline vehicles could be retained for drivers planning long trips that prevent them from refueling at local fuel depots. Police fleets are a more questionable market. On the one hand, they value high performance, such as provided by alcohol, but, on the other hand, they also travel relatively long distances. Police departments are also reluctant to adopt new vehicles if there is any possibility of reduced reliability or increased danger.[38]

Business vehicles constitute a very large proportion of fleet vehicles, but data on these vehicles are incomplete and often inconsistent. One survey, using probability sampling techniques, estimated that commercial fleets had 12.7 million light-duty vehicles, of which 5.6 million were automobiles and the remainder light-duty trucks and vans—but only about 40 percent were in fleets of ten or more vehicles.[39] Another study further disaggregated the analysis and estimated that for business fleets of ten or more vehicles only about 20 percent of the vehicles were company owned; the remainder were either leased (69 percent) or owned by salespeople (11 percent).[40] Most of the salesperson-owned vehicles would probably not represent an attractive early market for alternative-fueled cars because of erratic refueling patterns. The situation with leased vehicles is more uncertain. Businesses lease vehicles to reduce their capital expenditures, but it is unknown to what extent these vehicles are used differently from company-owned vehicles.

With data from the Bobit/BNL survey cited in table 42, the Jet Propulsion Laboratory (JPL)[41] used the following criteria to estimate the proportion of the fleet market that might be diverted to methanol fuel:

vehicles in fleets of 200 or more vehicles

vehicles that have required driving ranges of up to 150 miles per day

fleet operators who perceive that diesel fuel is available in enough retail fuel outlets to consider purchasing a diesel car

resale considerations.

The "large fleet" criterion of 200-plus vehicles was applied because larger fleets have greater resources, flexibility, mechanical expertise, and overall capability to handle different fuels and vehicles. Short driving ranges (shorter than 150 miles) are important because of the lower energy density and driving range of alcohol and CNG vehicles. The diesel fuel availability criterion was a proxy for sparse fuel outlets. Initially, alcohol and CNG refueling outlets would be at least as sparse as were diesel fuel outlets in 1977 when the survey was conducted. Survey data for these first three criteria are presented in table 43. The last criterion was based on the importance to the fleet operator of generating revenue from the sale of used vehicles; JPL estimated that based on resale considerations, one-half of all business fleets, and all police, government, and utility fleets would be willing to consider methanol vehicles, but that taxi and rental fleets would not.

Assuming that the restrictive effects of all these factors were indepen-

TABLE 43 INDICATORS OF ACCEPTABILITY OF NONPETROLEUM FUEL IN FLEET VEHICLES

	Fraction of Cars in Fleets of 200+ Vehicles[a]	Fraction of Trucks in Fleets of 200+ Vehicles[a]	Fraction of Cars for Which Diesel Fuel Availability was Considered Adequate	Fraction of Light Trucks for Which Diesel Fuel Availability Was Considered Adequate	Fraction of Cars Needing Range of 150 Miles or More	Fraction of Light Trucks Needing Range of 150 Miles or More
Business	0.85	0.87	0.34	0.81	0.82	0.77
Rental	0.94	0.92	0.65	0.91	0.60	0.77
Government	0.05	0.89	0.30	0.44	0.68	0.32
Police	0.93	0.76	0.40	0.45	0.97	0.49
Utility	0.92	0.91	0.43	0.66	0.43	0.28
Taxi	0.13	0.65	0.57	0.77	0.61	0.83
Weighted Averages[b]	NA	NA	0.39	0.69	0.74	0.48

SOURCE: Data for western states are available in Jet Propulsion Laboratory, _California Methanol Assessment_, Vol. II (Pasadena, Calif.: 1983), chap. 8. National data were published in Joseph R. Wagner, _Fleet Operator Data Book (Summer 1977 Data), Vol. 1: National Data_ (Springfield, Va.: NTIS, 1979, BNL 50904), pp. 12, 22, 26.

[a]These two columns were estimated based on data from western states only; they are probably similar to the rest of the country.

[b]Averages were weighted based on size of respondent's fleet.

dent (and additive), JPL estimated[42] that in the western U.S., 2.6 percent of annual fleet car sales and 25 percent of annual light-duty truck fleet sales could be diverted to methanol use. To estimate an upper bound of market potential, the four conditions were relaxed, and only the fleet size and resale criteria and the less restrictive of the diesel fuel and short trip criteria (but not both) were applied. This upper-bound estimate comes to 11 percent of fleet car sales and 50 to 55 percent of light-duty fleet truck sales. These two estimates of car and light truck sales translate into a range from 175,000 to 569,000 fleet vehicles sales per year, the majority of which are light-duty trucks. This analysis ignored cost considerations and is based on perceptions and fleet market characteristics from 1977–1978, but it is probably roughly accurate in estimating the upper limit of initial fleet market penetration by alcohol—if alcohol fuel and alcohol vehicles were to cost about the same as gasoline and gasoline vehicles.

Estimates of vehicles sales may be readily converted into fuel sales. If one assumes that fleets steadily increase their purchases of methanol vehicles from 40,000 in year one to 200,000 in year five (which accumulates to a total stock of 600,000 vehicles in year five), that each vehicle travels 15,000 miles per year (fleet cars generally travel more, light-duty fleet trucks less), and that each vehicle consumes one gallon of methanol every fifteen miles, then fleet use of methanol would increase from 40 million gallons in year one (2600 b/d) to 600 million gallons in year five (39,000 b/d).[43]

Several observations may be made about the potential of introducing alternative fuels into the fleet market. One is that the market penetration analysis above suggests that light-duty fleet trucks may be much more easily attracted to alcohol than fleet cars. Second, CNG is not attractive to any part of the fleet market, except on a cost basis; from a user perspective, the slower acceleration and reduced storage space of CNG vehicles are not compensated for by any superior attributes. Third, some of the barriers to alcohol and CNG may be mitigated; for instance, by redesigning the vehicles with larger fuel tanks, and by arrangements with the vehicle manufacturer to sell the used vehicle at a guaranteed minimum price.

Penetration of the fleet market may be even slower than suggested in the preceding analysis. One reason is that several nonpetroleum fuels and vehicles are competing for the same fleet vehicles niche. LPG, CNG, electric, and alcohol vehicles are all competing for the same centrally fueled, limited-range fleet vehicle market. Also, fleet managers, especially those of the larger fleets, which are presumably the more likely homes for nonpetroleum vehicles, tend to make their vehicle and fuel choices based on careful analysis of life-cycle costs—unlike individual purchasers, who may be more responsive to considerations of prestige, novelty, style, and other noneconomic criteria. Fleet managers will demand reliability, low maintenance, and low cost; many individuals will not be so demanding. The very low level of market penetration by light-duty diesel vehicles into fleets in the U.S.—about 2 to 4 percent at its peak—is evidence that fleet managers are conservative and unwilling to take cost and reliability risks, and that they will follow, not lead, the household market, even when fuel costs are lower than those for gasoline. In conclusion, fleet vehicles represent an important, but not a major, early market for alternative fuels. Even 39,000 b/d, the quantity estimated above for year five of a fairly aggressive fleet vehicle transition, is minor compared to the 6 to 7 million barrels per day of gasoline consumed in the U.S. While advocates of alternative fuels point to fleet vehicles as a major early opportunity, the reality may be very different.

METHANOL VEHICLES IN CALIFORNIA

The two largest fleets of methanol cars as of 1987 are in California. Beginning in 1980, the Bank of America began buying and converting cars to run on methanol; within several years, 292 out of its 2,400 vehicles were operating on methanol. An even larger fleet of over 500 methanol vehicles was purchased in the early 1980s by the California Energy Commission (a state agency) and placed in various state, county, and city government fleets around the state.

The Bank of America (B of A) was attracted to methanol in 1979 as a response to the gasoline shortages that spring. B of A concluded that whatever extra cost might be incurred to assure a reliable source of fuel would be worth the security it provided, especially since the cost of operating a vehicle fleet was a trivial part of their overall budget. B of A purchased 250 Ford cars, which were converted to methanol use by raising the compression ratio and making various material substitutions. An additional forty-two Chevrolet cars were converted to methanol but without changing the compression ratio.

B of A was very happy with the vehicles. It claimed that drivers came to prefer them over gasoline cars because of their power and smoother idling. They also claimed the vehicles provided substantial economic benefits and were much more energy efficient than gasoline vehicles—they reported that even the Chevrolet vehicles (with unchanged compression ratio) had 50 percent or more energy efficient improvements over comparable gasoline cars.[44] They acknowledged, however, at a meeting sponsored by the National Science Foundation in January 1985, that there were numerous irregularities in their record keeping. The B of A methanol experience was an impressive technical and operational success; it illustrates one motivation and market niche for alternative fuels, and demonstrates the ability of large fleets to support innovative endeavors. Their economic and fuel efficiency claims for methanol cannot be substantiated, however.

The state of California's purchase of methanol cars was the cornerstone of an effort by the state's Energy Commission to initiate a transition to methanol.[45] In 1980 the state converted four Ford Pinto cars to methanol and four to ethanol. In 1981 the state purchased forty Ford Escorts and nineteen Volkswagen Rabbits and light trucks that had been modified for methanol use (as well as twenty ethanol vehicles) by the manufacturers. In 1983 Ford made additional design changes and built 582 methanol Escorts on a special production run on an assembly line in one of its factories; 500 of these vehicles were sold to California. They had slightly increased compression ratios and carbureted engines. All of the California methanol vehicles were integrated into government fleet operations.

To support these 500 methanol vehicles, the state also subsidized the establishment of eighteen methanol stations across the state. In 1987 Atlantic Richfield (ARCO) and Chevron signed agreements to establish twenty-five to fifty additional stations. The intent was to continue expanding this network over time and to make the stations accessible to the public.

The methanol vehicles provided an important learning experience for Ford and the State of California. Various small technical and fleet management problems were encountered and resolved. The vehicles generally performed well, but reliable statistical results are not available because of erratic record keeping and lack of a control group. In any case, the vehicles used carburetors (not electronic fuel injection) and only slight modifications of the basic gasoline technology; they were far from being optimized for methanol.

The salient conclusion by the Energy Commission from the methanol fleet experience was that methanol vehicles were ready for the marketplace, but that the scarcity of methanol fuel stations was an overwhelming obstacle. Drivers were so displeased and nervous about the scarcity of stations that they sometimes abandoned the vehicles when the fuel gauge reached one-quarter full, even when they were as few as ten miles from the garage destination. As a result of those reactions, the State of California decided to shift from dedicated methanol vehicles to multifuel vehicles that operate on either methanol or gasoline or any blend of the two. As of 1987 it was preparing to purchase about 5,000 such multifuel vehicles.

CONCLUSION

It is impossible to accurately forecast market penetration of each niche identified in this chapter. However, to highlight points made earlier and to provide an order of magnitude review of the sizes of different market

TABLE 44 MARKET POTENTIAL FOR ALCOHOL FUELS, GASOLINE-EQUIVALENT BARRELS PER DAY

	Total Size of Fuel Market, 1984	Actual Market Penetration, 1984	Low-level Penetration in 15 Year Time Frame	Maximum Plausible Penetration in 30 Year Time Frame
Gasoline additive[a]	300,000	80,000	75,000	150,000
Agriculture[b]	63,000	negl.	5,000	60,000
Urban transit buses	30,000	0	5,000	20,000
Vehicle fleets	1,200,000	negl.	50,000	400,000
Household vehicles[c]	6,500,000	0	50,000	3,000,000

See text for elaboration.

[a]Included in 1984 figures are MTBE and biomass ethanol. In practice, most of the ethanol is used because of tax incentives; in most cases its octane-enhancing qualities are not exploited. In that sense ethanol is not truly valued as an additive. Estimates of market penetration are based only on alcohols valued for their octane-enhancing qualities.

[b]Agriculture used 63,000 b/d of gasoline and over 200,000 b/d of diesel fuel in 1981 with the trend toward less gasoline and more diesel fuel use. Plausible market penetration in agriculture includes a large share of gasoline market plus a portion of diesel engine applications that over time might be switched to alcohol use.

[c]See chapter 14 for elaboration.

segments, table 44 is presented. The numbers in the table should be treated as indicative, not definitive. They suggest that all market niches are small, even taken as a sum, and that a transition to nonpetroleum fuels must focus on the household vehicle market.

FOURTEEN

Household Vehicle Fuels Market

Household vehicles provide by far the greatest market opportunity for new transportation fuels, an opportunity much greater than those market niches identified in the previous chapter. But penetrating the household vehicle market will be difficult and slow even when the fuel is less expensive than gasoline. The following three factors are the principal deterrents to penetration of the large household transportation fuels market: the lack of growth potential in the vehicle fuels market, the absence of a network of retail outlets for the new fuels, and the conservative nature of vehicle purchase decisions. These three factors and their implications for alcohol and natural gas fuels are elaborated upon in this chapter.

STAGNANT FUEL MARKET

The consumption of fuel in personal vehicles is not expected to grow into the foreseeable future. Aggregate gasoline sales will be the indicator used to measure the size of the household vehicle fuels market, since almost all personal vehicles use gasoline and since about 80 percent of all gasoline is sold to the household market (the rest goes to agriculture and fleets). Gasoline sales were increasing until 1978, when they peaked at about 7.55 million barrels per day (b/d); sales then quickly decreased to around 6.7 million b/d and stabilized at that level through 1986. According to almost all forecasts, U.S. gasoline sales will decrease slightly further into the foreseeable future.[1] These forecasts of flat or slowly decreasing gasoline demand are the product of two opposing trends: slowly increasing population and per capita vehicle ownership, which tend to increase vehicle usage, countered by improving vehicular fuel efficiency. The fuel efficiency of new cars increased from about thirteen miles per gallon in 1974 to about twenty-six mpg in 1986. Even if fuel efficiency does not increase beyond twenty-six mpg (which will depend mostly on petroleum prices), fuel efficiency of the overall vehicle population will continue to improve for another ten to fifteen years as old inefficient cars are replaced by new more efficient cars. In contrast, during this same time period, diesel fuel sales for commercial trucks are expected to increase significantly—by about one-third to one-half between 1980 and 2010.[2] Diesel fuel sales were about one-third those of gasoline sales in 1986; by 2010 the diesel fuel truck market may be as much as one-half the size of the gasoline market. But diesel fuels are much less susceptible to being replaced by alternative fuels than are gasoline fuels.

The stagnant gasoline market discourages energy companies from taking the risk of investing in dissimilar fuels. It results in a "not-me-first" phenomenon. Unlike investors in growing markets such as computers in the 1980s or petroleum until the 1970s, investors in energy companies have few opportunities to create new markets and new demand. It is more difficult to gain market share in a stagnant competitive market than in an expanding market. The risk for the first company that enters the market is huge. The risk for later companies is lower, since they can wait to see if a market develops. Due to the homogeneity of the product, it is not likely that the first company will be able to gain and retain a large share of the alternative fuels market solely because it was first into the market. Thus the risk of being first is large, while the potential benefits are modest at best.

This reluctance to be first into the market also extends to vehicle manufacturers. This reluctance is again the result of great risk that is not matched with a potential for large rewards. The absence of large rewards to vehicle manufacturers is due to two factors: first, as will be demonstrated below, market penetration of nonpetroleum vehicles, especially in the U.S., will be slow; and second, it is widely known how to build alcohol and CNG vehicles. Indeed, Ford engineers stated in 1986 that even with their considerable investment in designing and building a group of 550 methanol cars, they felt that they would have only about a six- to nine-month lead over GM (and other manufacturers) in putting a vehicle in the marketplace.

MORE FUEL OUTLETS NEEDED

The second reason for slow market penetration is the lack of incentive to fuel marketers to establish a network of retail fuel outlets and the corresponding reluctance of consumers to purchase a vehicle for which fuel is not readily available. The lack of incentive to fuel marketers is part of the same problem cited above: high initial cost with little promise of substantial benefits. As indicated in chapters 10 and 11, the cost of setting up a methanol or CNG (or LNG or hydrogen) fuel outlet can be substantial; the revenue that is generated at those first few stations will be negligible for several years or more. A rule of thumb in the fuel business is that a particular fuel must account for at least 10 percent or so of total fuel sales at a station to justify devoting a tank and pump to that fuel; for fuels that incur special costs (such as CNG and alcohol), the minimum threshold would be even higher. Those 10 percent sales levels will not be realized for initial fuel outlets.

This chicken-or-egg stasis that keeps fuel marketers from setting up a fuel outlet and motorists from purchasing a vehicle to operate on that fuel is widely acknowledged (see chap. 17), but it is difficult to determine how important it really is. Only fragmentary evidence is available for determining the relationship between the size of a fuel network and the willingness of consumers (and fleet owners) to purchase a vehicle that operates on that fuel.

Evidence comes from three experiences: CNG in New Zealand and Canada, and diesel cars in the U.S. Brazil is *not* a relevant example for several reasons: ethanol was introduced at a large number of fuel outlets by the state-owned petroleum monopoly before the first ethanol cars were marketed, only ethanol and not gasoline was sold on weekends, and the government sent strong signals (initially) that gasoline would be phased out and ethanol would be phased in.

The Canadian and New Zealand experiences indicate that limited fuel outlets are a major deterrent to CNG market penetration. In New Zealand the number of vehicles converted annually to CNG increased from 1,600 in 1979 to 25,000 in 1986; the cumulative total was 130,000 in 1986, which represented about 10 percent of all light-duty gasoline vehicles on the North Island (see chap. 5). During that same time period, the number of CNG fuel stations increased from zero to 300. Surveys conducted in 1980[3] and 1984[4] found that fuel availability was one of the most important factors discouraging people from converting to CNG, and, for CNG vehicle owners, fuel availability concerns were about equal in importance with vehicle performance as the most important negative factor associated with ownership. The most telling factor, however, is that despite the price of CNG being about half that of gasoline (and the cost recovery period for conversion being less than two years for almost all owners), only 10 percent of the vehicles had been converted after seven years.

A similar situation prevailed in Canada, although on a smaller scale. The number of fuel stations was even more limited (see chap. 5), the CNG fuel price was again about half that of gasoline, and market penetration was less than 1 percent, even in the area (Vancouver) with the greatest concentration of fuel stations.

In both these cases it is clear that fuel availability was a barrier, but it is not clear how large the barrier was. Studies of the diesel car phenomenon in the U.S. provide a more quantitative indication of how many stations are needed. Until the mid-1970s, the only diesel cars in the U.S. were those sold by Mercedes-Benz and, for a short while, Peugeot. They represented less than 0.1 percent of the car population. In 1976, just before Volkswagen and General Motors began mass marketing diesel cars, a study in California found that diesel fuel was being sold at about 9 percent of the retail fuel outlets.[5] That same study found that as diesel cars increased their market penetration of new car sales from 1 percent in 1977 to a peak of 9 percent in 1981, the proportion of retail fuel outlets serving diesel fuel increased to 22 percent (see table 45).

Even with this fairly large number of fuel outlets, both diesel and gasoline car drivers expressed substantial concern about diesel fuel availability. In a 1986 random sample survey of 535 diesel car drivers in California, 39 percent of drivers said that at the time they had purchased their vehicle they had been somewhat (27 percent) or very (12 percent) concerned about diesel fuel availability.[6] This level of concern was found to be relatively unchanged between those who purchased their vehicles in the early years (1977–1981) when fewer outlets existed, and those who had purchased their vehicles in later years when a much larger network of outlets existed. It was also found that those with pre-purchase concern about fuel availability in fact encountered about the amount of difficulty they had expected; it was also found that this relationship between pre-purchase concern and post-purchase difficulty was independent of when the vehicle was purchased. These findings indicate that concern for fuel availability did not decrease as the network of fuel outlets expanded. The explanation for this apparently

TABLE 45 DIESEL FUEL STATIONS AND NONCOMMERCIAL DIESEL VEHICLE SALES IN CALIFORNIA

	Diesel Fuel Stations		Noncommercial Diesel Vehicle Sales	
Year	Authors' estimate	% of retail stations serving diesel (based on authors' estimate)[a]	No. of vehicles	% of total sales[b]
1976	1,200	9	-	-
1977	1,300	10	9,000	1
1978	1,440	12	20,500	2
1979	1,650	15	33,100	3
1980	2,000	19	44,400	6
1981	2,250	22	60,000	9
1982	2,500	24	48,000	7
1983	2,500	23	34,400	4
1984	2,500	25	18,000	2
1985	2,500	-	15,100	1

SOURCES: Daniel Sperling and K. Kurani, "Refueling and the Vehicle Purchase Decision: The Diesel Car Case," SAE 870644 (1987).

[a]The total number of retail fuel outlets in California decreased from 13,066 in 1976 to 10,771 in 1980, and remained at about that level through 1985. Retail fuel outlets are defined as those establishments that derive at least half of their retail sales from motor fuel sales.

[b]Estimated total passenger vehicle sales for California were 1.13 million in 1978, declined sharply in 1980, reached a low of approximately 650,000 in 1982, and gradually increased to 1.04 million in 1985.

counterintuitive finding is that the initial group of diesel car buyers accepted the fact that few fuel outlets existed and that they would have to accommodate themselves to this reality. But as diesel car ownership expanded beyond a core group into the general car-buying public, limited fuel availability was not so readily tolerated.

Consistent with this explanation that a tolerant core group exists is a finding that the general gasoline car–owning public would be even less reluctant to accept the inconvenience of limited fuel availability. In response to survey questions asking whether they would consider purchasing a diesel car, 33 percent of 1,530 gasoline car drivers said fuel availability would be a "very important" concern and another 28 percent said they would be "somewhat concerned."[7] This total of 61 percent concerned drivers was much higher than the 39 percent response by diesel owners, reinforcing the suggestion that as one moves beyond the self-selected early adopter group, tolerance for limited fuel availability diminishes sharply.

Because diesel fuel had been available at retail outlets for many years (to serve long-haul trucks) and because diesel cars get about 50 percent more driving range than gasoline cars (and 100 percent more than alcohol and gaseous fuel vehicles), diesel cars are not necessarily a good test case for nonpetroleum vehicles. Even so, the following conclusions can be drawn. First, a small number of individuals are willing to put up with the inconvenience of a limited number of fuel outlets if the vehicles offer some positive attributes, such as lower cost or higher performance. But this group of "early adopters" is very small, probably less than 1 percent of the vehicle-buying population.

Second, to gain significant market share (more than about a 1 percent share of new car sales) in a targeted region, *at least 15 percent* (and probably much more) of the fuel outlets must supply that fuel. Having dual-fuel capability in principle solves the fuel availability problem, but initial evidence indicates that this assumption may not be accurate. The New Zealand surveys cited above and a Canadian survey[8] found that people who converted to CNG relied very heavily on CNG and very little on gasoline; the Canadian survey found that despite limited fuel networks, 83 percent of convertees did 70 percent or more of their driving on CNG, and the 1984 New Zealand survey found that 73 percent of the vehicles ran at least 90 percent of the time on CNG and that one-third ran exclusively on CNG. Apparently, when owners make a commitment to a new fuel, it is more than just a financial outlay; it represents a personal commitment. The Canadian survey (based on focus group interviews) found that drivers of CNG vehicles were "carrying a minimum (often a very bare minimum) of gasoline in their tanks, purely for unforeseen (and highly undesirable) emergency situations, and are intent on using nothing but natural gas as their fuel" and goes on to conclude that "every time they are forced to resort to the gasoline switch, then, they are forced to admit a small defeat, and to the extent that this happens they are unhappy with natural gas vehicles."[9]

In conclusion, unless the sales of nonpetroleum vehicles are somehow mandated, a sizable network of fuel outlets must be put in place before nonpetroleum vehicles can be mass marketed, even if the fuels have a dual-fuel or multifuel capability. To the extent that the network of fuel outlets is limited, the sales of vehicles will be sharply constrained.

VEHICLE PURCHASE BEHAVIOR IS CONSERVATIVE

In a more general sense, most consumers are reluctant to make a major purchase, such as for a motor vehicle, unless the risk is minimal. In the case of an alcohol or gaseous fuel vehicle, this risk may involve the possibility of fuel shortages and a high fuel price in the future, low resale value of the vehicle, and technical failings that result in high repair bills, safety problems, inferior performance, poor fuel efficiency, starting problems, high noise levels, and so on.

It is very difficult to predict under what conditions consumers would purchase a nonpetroleum vehicle. Still, enough evidence exists to indicate that in general the most important factors in the decision to purchase a nonpetroleum vehicle are the following: fuel price, vehicle (or retrofit) price, vehicle power and acceleration, startability in cold weather, luggage space, driving range (per tank of fuel), and fuel availability (see table 46). Other generally less important factors include noise, safety, and air pollution. Maintenance is an important factor, but it is uncertain how it will be perceived and how it will in fact be different for each fuel-engine combination; maintenance is therefore left out of table 46.

More precisely, the principal factors determining the market penetration potential of each nonpetroleum vehicle are as follows (see chap. 12 for details). Since hydrogen vehicles would have a much shorter range than any other liquid or gaseous fuel vehicle and require much more space and/or weight, as well as costing significantly more, hydrogen vehicles would be competitive only if on-board fuel storage were significantly improved and fuel costs were significantly reduced.

Battery vehicles, which are not included in table 46, are hindered by the very low energy density of batteries. Barring some unexpected breakthrough, which is extremely unlikely, the only price-competitive batteries will be those that are not performance-competitive; they will be handicapped by energy densities one to two orders of magnitude less than those of liquid fuels.[10] Thus battery vehicles will continue to have very low performance, a very short driving range, and/or a very bulky and heavy battery pack. As a result, the only market potential for

TABLE 46 IMPORTANT ATTRIBUTES OF NONPETROLEUM VEHICLES RELATIVE TO GASOLINE VEHICLES FROM A CONSUMER PERSPECTIVE

	CNG Dual Fuel	CNG Dedicated	LNG Dedicated	Alcohol Multifuel	Alcohol	Hydrogen	Electric Vehicles
el price	Much lower	Much lower	Much lower	Slightly more	About same	Much more	Varies
nicle price[a]	More	Slightly more	Slightly more	About same	About same	More	More[b]
formance	Worse	Same	Slightly better	Same	Better	Same[c]	Much worse
artability	Same	Better	Better	Worse	Worse	About Same[c]	Better
ggage space	Much less	Less	Slightly less	Same	Same	Less	Much less
iving range	Much less on CNG	Less	Less	Much less on alcohol	Less	Less	Much less

sumes large production volumes.

cluding batteries.

r liquid hydrogen. If hydrogen is stored in hydrides, then performance would be slightly worse and ld start somewhat more difficult.

battery vehicles is for dedicated short-distance trip making; for instance, as a lightly used second car, as a commuter car, or in fleet use.

The only near- and medium-term options for introducing clean transportation fuels are natural gas and alcohol (usually methanol). Dual-fuel CNG vehicles have been, and will continue to be, attractive to consumers because of their continuing low fuel costs. Their disadvantages are high vehicle costs (for retrofitting), short range (on CNG), and reduced luggage space. Dedicated CNG vehicles would be more attractive than dual-fuel vehicles in all ways except that they have a shorter range (because of no gasoline backup) and are totally dependent on CNG fuel outlets.

LNG vehicles would be even more attractive than dedicated CNG vehicles principally because they would have a much longer driving range (although there may be some unfounded safety concerns). Other than dependence on a limited network of fuel outlets, the major barrier to consumer acceptance of LNG vehicles will be the boil-off problem.

The principal attraction of alcohol vehicles relative to gasoline and natural gas vehicles is their power and acceleration. Relative to natural gas, they also have the advantage of somewhat lower initial vehicle cost (because tanks are much cheaper) and somewhat greater vehicle range. Alcohol also has the advantage of being physically more similar to gasoline and therefore possibly more acceptable (initially) to consumers; the physical similarity also makes it easier to store and dispense alcohol at existing gasoline stations. The disadvantages relative to natural gas are difficult starting in cold weather and, most important, a higher fuel price. Methanol's price will remain higher than the price of natural gas fuels because methanol will be made principally from natural gas well into the next century and because methanol (and to a lesser extent natural gas) prices will probably be closely linked in the world market to gasoline prices; as gasoline prices rise, so will methanol prices. The introduction of methanol fuel therefore hinges on the willingness of consumers to accept a fuel that has little or no price advantage over gasoline, or upon the willingness of government to subsidize alcohol fuel prices.

The pivotal question in assessing the market potential of methanol and CNG is the importance of fuel prices. Under what conditions if any would an individual purchase a methanol car if methanol prices were equal to or higher (on an energy basis) than gasoline prices? Would consumers be more willing to buy a CNG vehicle with its lower energy cost and various disadvantages than a methanol vehicle?

All evidence indicates that individuals (and fleets) will *not* purchase a nonpetroleum vehicle unless there is a very strong economic incentive.[11] They need that economic incentive to offset the various uncertainties and negative attributes associated with those vehicles. For instance, even though CNG was available at half the price of gasoline in New Zealand, Canada, and Italy, market penetration was slow. Apparently, consumers are reluctant to veer away from the mainstream of gasoline car buying even with significant economic incentive.

Additional evidence of this conservativeness is provided by the U.S. diesel car experience discussed earlier. Diesel cars did indeed gain a significant market share in the U.S. in a fairly short period of time, with sales increasing from practically zero in 1976 to more than 6 percent of new car sales in 1981. Diesel cars became popular almost solely because of their low fuel costs. This cost advantage was about 25 percent through the early 1980s; it was due to a fuel efficiency advantage of roughly 15 percent and a cost per gallon advantage of roughly 10 per-

TABLE 47 U.S. DIESEL CAR SALES AND DIESEL AND GASOLINE FUEL PRICES

Year	Diesel Car Sales	Sales as % of Total U.S. New Car Sales	Avg. Retail Diesel Fuel Price (cents/gallon)	Avg. Retail Unleaded Regular Price (cents/gallon)	Price Difference Between Diesel and Gasoline (cents/gallon)
1970	~5,000	negl.	---	---	---
1976	22,735	0.2	52.8	61.4	-8.6
1977	37,498	0.4	57.9	65.6	-7.7
1978	114,880	1.1	59.9	67.0	-7.1
1979	271,052	2.6	84.2	90.3	-6.1
1980	387,049	4.3	112.4	124.5	-12.1
1981	520,788	6.1	131.0	137.8	- 6.8
1982	354,690	4.4	127.4	129.6	- 2.2
1983	197,710	2.1	125.4	124.1	+ 1.3
1984	150,548	1.5	131.2	121.2	+12.0

SOURCE: Oak Ridge National Laboratory, Transportation Energy Conservation Data Book: Editions 7 and 8 (Springfield, Va.: NTIS, 1984 and 1985, ORNL-6050 and ORNL-6205).

Negl. = negligible (< 0.05%)

cent. In 1978 the cost of fuel for a typical diesel car was about 2.5 cents per mile versus 3.4 cents for a comparable gasoline car (assuming 20 mpg for the gasoline car). In 1980 the cost advantage was still about 25 percent: 4.8 cents versus 6.2 cents. If the vehicles traveled 15,000 miles in 1980, the cost advantage for the year would have been $210 (excluding the somewhat higher purchase price of diesel cars).

When the differential between diesel and gasoline prices began to shrink, diesel car sales began to drop (see table 47). But even when diesel fuel prices began to exceed unleaded regular gasoline prices in 1983, diesel cars still had a fuel cost advantage because of their superior fuel efficiency. Although other factors played a role in the collapse of the diesel car market in the U.S.—in particular, technical problems with General Motor's diesel cars and diminished concern for energy—consumer surveys indicate that fuel prices indeed were the most important factor both in the diesel car surge and in the subsequent collapse.[12]

The diesel car and CNG experiences illustrate the conservative nature of consumers. Those experiences suggest that a new-fuel vehicle must have a significant cost advantage over gasoline cars to succeed in the marketplace. Those vehicles most similar to gasoline vehicles—in terms of performance, physical characteristics of the fuel, refueling, and fuel availability—will require the smallest cost advantage; those more dissimilar will require larger cost advantages. Since natural gas and methanol prices are and will be linked to petroleum prices and will therefore increase more or less in concert with gasoline prices, one must conclude that without strong government support no new fuel-engine combination will be able to penetrate the U.S. market or possibly any market in the foreseeable future, even with significant increases in petroleum prices.

GOVERNMENT AND UNCERTAINTY

Obviously, government will have to play a major role if new fuels are to be introduced in a timely manner. It will have to reduce the cost of buying and operating a new-fuel vehicle to that of gasoline vehicles and will have to reduce uncertainty and risk for the buyer. The Brazilian experience provides some insight into what this role could or should be (see chap. 4); it demonstrates that governmental action can be extremely effective, both positively and negatively.

Most impressive, given the experiences elsewhere with CNG, is that the Brazilian government was able to reduce consumer uncertainty to such an extent that gasoline cars were perceived to be riskier investments than ethanol! People were convinced in 1980 and 1981 that gasoline would be a scarcer and more expensive commodity than ethanol. Substantial subsidies and incentives were needed to make ethanol attractive and to convince people of the government's determination to phase out gasoline, but another major factor was public persuasion. In summary, continuing public pronouncements by officials, backed up by the steady stream of incentives and subsidies, were effective in reducing people's concern about the future price and availability of ethanol.

The effectiveness of government (and industry) in reducing uncertainty is critical. Consumers need assurance that the new fuel will not disappear or become very costly in the future.

Brazil demonstrated that an active commitment to a new fuel can be effective in overcoming the conservativeness of consumers. In this case the fuel was ethanol, which is more similar to gasoline than is gaseous fuel. With only a small economic incentive, practically everyone was willing to switch to the new fuel.

CNG OR METHANOL—A CONSUMER PERSPECTIVE

If one were to compare dedicated CNG and methanol vehicles, and assume that methanol prices were about 25 percent higher than CNG prices and that fuel availability was adequate for both, then I would expect most consumers to select CNG. That choice would also be preferred by society because air pollution and total economic costs would be lower with CNG. The problem is how to get from here to there. This is a case where the "getting there" part of the process is so difficult that the socially optimal choice might not be made.

The reason methanol might be preferred to CNG is that a transitional methanol vehicle—i.e., a multifuel vehicle—would be more attractive to most consumers than a transitional (dual-fuel) CNG vehicle. The multifuel methanol vehicle is much less inferior to a dedicated methanol vehicle than is a dual-fuel CNG vehicle relative to a dedicated CNG vehicle. This is because a multifuel methanol vehicle has only one fuel system, will cost about the same (at high production volumes) as a gasoline vehicle, and will be only about 5 percent to 10 percent less efficient than a dedicated methanol vehicle. A dual-fuel CNG vehicle, however, requires a dual fuel system and would therefore cost at least several hundred dollars more. It would also be roughly 15 to 30 percent less efficient and considerably less powerful than a dedicated CNG (or LNG) vehicle.

Most consumers in affluent countries such as in U.S., Europe, and Japan would probably be reluctant to accept a dual-fuel CNG vehicle under most plausible energy price conditions. They have already shown their willingness to pay very high gasoline prices—as much as $1.60 per gallon in the U.S. (in 1981) and over $3.00 in Europe. Those motorists who wish to reduce energy costs have the option of purchasing a smaller or more efficient car. There is no need to convert their vehicles to CNG. In most cases, though, consumers are affluent enough to be willing to pay more to have the better performance, greater convenience, and lower risk associated with gasoline vehicles. To initiate a CNG program, government would have to make a significant commitment in terms of political initiative to develop a unified policy and in terms of financial resources to subsidize the establishment of CNG fuel outlets.

The situation may be different in the many developing countries that have large natural gas reserves (a total of as many as fifty and perhaps more). Most of those countries have not yet developed their gas reserves because there is no export market for it (except as LNG) and because it has been difficult to justify the substantial cost of developing a domestic pipeline system. They generally rely on petroleum for transportation and most industrial and commercial applications. There are two reasons why CNG would be more attractive in these countries. First, the population is far less affluent and, compared to the population in affluent countries, probably far more interested in reducing operating costs even if it means lower performance and greater inconvenience. Second, they have a far smaller institutional and economic commitment to petroleum fuels and vehicles than does the U.S. These differences suggest that dual-fuel CNG would be much easier to introduce in less developed countries than in the more affluent countries.

The World Bank has come to the same conclusion (although with a more limited scope in mind).[13] It has determined that when oil is priced under $25 per barrel, it is economically attractive for developing countries to convert all highway vehicles to (dual-fuel) CNG. If dedicated CNG vehicles were available, then it would be economically attractive to switch to CNG when oil prices were between $15 and $20 per barrel. For some transportation applications, switching to CNG was attractive at prices as low as $10 per barrel. This analysis did not include the cost of building a pipeline infrastructure, but if other sectors also converted to natural gas, then the cost to the transportation sector would be modest. The point is that CNG might be very attractive in many developing countries and that it would face far fewer barriers than it would in more affluent countries.

THE PREMIUM GASOLINE MARKET

The opportunities for targeting certain vehicles and households in the household transportation fuels market are very limited. One oft-mentioned market opportunity for new vehicle technologies is secondary household vehicles—those vehicles for which reliability is less important and which are used mostly for local trips. But it is for those reasons that new-fuel vehicles would not be highly valued. Few households would pay a premium for a car that is used intermittently or only for short trips. Most secondary household vehicles are older vehicles with low value.

For alcohol vehicles, a more likely market—in fact, the *only* attractive opportunity initially—is the premium fuel market. About 20 percent of all gasoline sold in the U.S. in 1986 was gasoline with an octane rating of 91 or more.[14] This premium (unleaded) gasoline sells for ten to fifteen cents per gallon more than regular (unleaded) gasoline. Alcohol is likely to be more attractive to premium gasoline users because of premium gasoline's higher price and the higher octane. Although no published analysis of this premium fuel market is available, it is probably safe to say that the users of premium gasoline belong to one of the following groups: owners of high-performance sports cars, owners of pre-1973 cars with high-compression eight-cylinder engines, affluent people who are relatively insensitive to high fuel price and conjecture that their expensive cars will probably run better on high octane fuel, and people who imagine or truly experience pinging (knocking) in their vehicles even though the vehicles were designed to run on regular gasoline. The relative size of each of these groups in the premium fuels market is unknown, but some rough estimates can be made. First, a small number of sports cars actually are designed to run on premium fuel. All other light-duty vehicles are designed to run on regular gasoline. The number of pre-1973 high compression vehicles was tiny in 1986 and will be negligible by the early 1990s. The only other vehicles that must run on premium fuel are those few vehicles that through the vagaries of quality control do not perform as designed and indeed do "knock" and need premium gasoline. But this number is probably very small. The bulk of the premium fuel market is therefore made up of people who for one reason or another think their vehicles run better on high-octane gasoline. One explanation given by automotive engineers for the inclination to buy premium gasoline is the tendency for some cars to ping slightly when climbing hills. They emphasize that this pinging does not hurt the engine. Nevertheless, drivers apparently believe the contrary and opt to use premium fuel in the belief that it is better for the engine (the oil industry is obviously not inclined to dissuade them).

If the foregoing analysis is correct, it has mixed implications for alcohol. On the one hand, the only part of the premium fuel market that might be strongly attracted to alcohol is that of high-performance sports cars, which represent a small percentage of the market at most. Marketers of alcohol cars could, and probably should, target the overall sports car market because it is the market that is most attuned to performance vehicles, but it will be an uphill battle, since most of those vehicles are designed to run on regular, not premium, gasoline.

On the other hand, anyone who is willing to pay ten to fifteen cents per gallon extra for premium gasoline alternatively may be willing to pay that same amount for methanol. Further research is needed to investigate potential demand for alcohol fuels.

The high-performance market is not penetrable by CNG because CNG vehicles, especially dual-fuel vehicles, have less power than comparable gasoline cars. The only initial market where CNG is attractive is vehicles that are heavily used in targeted local areas. Heavily used vehicles will benefit the most from the low cost of CNG, but these vehicles will have greater difficulty finding CNG fuel on long trips outside the region because CNG fuel is not likely to be as available elsewhere. The number of vehicles that meet these criteria for CNG is small.

TIMING OF MARKET PENETRATION

The process of introducing new fuels is deceptively slow for one very simple reason: vehicles last a long time. Automobiles have a median life of almost eleven years; light trucks, about fourteen years.[15] Since the introduction of new-fuel vehicles will undoubtedly begin with a very small percentage of new car sales and increase gradually from there, the replacement of gasoline (and diesel) will proceed slowly. To give some sense of just how slow that process will be, a market-penetration scenario has been created (table 48). It is based on the actual market penetration rate of diesel cars in the U.S. during the years when sales were increasing (from 0.2 percent in year one to 6.1 percent in year six), and then on postulated increases that reach 20 percent of new car sales in year ten.

This penetration scenario represents a moderate schedule. More extreme schedules are those of unleaded gasoline in the U.S. and ethanol in Brazil. In both cases it took about eight years from the time the "new-fuel" vehicles were first mass marketed to replace 50 percent of the gasoline. The more moderate schedule presented here is slower than it would be if the fuel were mandated or heavily subsidized, but it is faster than what a free market would accomplish on its own (except under crisis conditions). In other words, this is the type of deliberate controlled growth that would not be too disruptive. As shown in table 48, even though vehicle market penetration reaches 20 percent of new car sales in ten years, the corresponding replacement of gasoline is only 8.6 percent.

This observation regarding the inherently slow process of gasoline

TABLE 48 POSTULATED MARKET PENETRATION SCHEDULE FOR U.S. BASED ON U.S. DIESEL CAR EXPERIENCE

	(a)	(b)	(c)	(d)
Year	Nonpetroleum Car Sales as Fraction of Total Car Sales (N_i)	Approximate Nonpetroleum Car Sales	Nonpetroleum Fuel Consumption as Fraction of Gasoline Sales	Nonpetroleum Fuel Sales (b/d)
1	0.002	20,000	0.0003	2,000
2	0.004	40,000	0.0009	6,000
3	0.011	120,000	0.0026	17,000
4	0.026	260,000	0.0063	41,000
5	0.043	440,000	0.0123	80,000
6	0.061	610,000	0.0203	132,000
7	0.080	800,000	0.0304	198,000
8	0.110	1,100,000	0.0439	285,000
9	0.150	1,500,000	0.0621	404,000
10	0.200	2,000,000	0.0859	558,000

Column (a): Postulated scenario. Sales rates for years 1 to 6 are identical to those of diesel cars in the U.S. from 1976 to 1981; rates for years 7 to 10 are postulated. Actual diesel car sales dropped after 1981.

Column (b): Column (a) x 10 million cars/year (approximate car sales volume in U.S.).

Column (c): $F_i = N_i \sum_{j=1}^{10} C_j$ for i = 1 to 10

for $C_j = 0.15$; $C_{j-1} = 0.14$; $C_{j-2} = 0.12$; $C_{j-3} = 0.11$; $C_{j-4} = 0.10$;

$C_{j-5} = 0.09$; $C_{j-6} = 0.07$; $C_{j-7} = 0.06$; $C_{j-8} = 0.05$; $C_{j-9} = 0.03$.

$C_j = 0$ for $j<1$

where F_i = Nonpetroleum consumption in year i as fraction of gasoline sales.

N_i = Nonpetroleum cars in year i as fraction of total car sales in year i (see Column a).

C_j = Age-specific fuel consumption of cars of age j in current year. i (represented as fraction of total fuel sales in year i by vehicles of age j). These fuel consumption factors are based on 1978 data; they have changed somewhat since then, but not significantly so.

Column (d): Column (c) x 6.5 million b/d

replacement is important. It implies that if a nation truly desires to have an orderly transition to methanol or natural gas (or hydrogen or batteries) and would like to replace a significant, though not necessarily large, proportion of petroleum fuels within ten to fifteen years, then it had better get started immediately. Another implication of this slow market-penetration process is that governmental initiatives in support of nonpetroleum fuels will not unduly harm the oil refinery business by reducing demand for oil or by disrupting the balance between petroleum products. Knowing that demand for gasoline will decrease by less than 10 percent in ten years gives the oil refinery industry sufficient time to make adjustments without extreme hardship.

CONCLUSION

In conclusion, fuels dissimilar to petroleum products face major end-use and marketing barriers. Even though alcohols and natural gas fuels can be used in conventional spark ignition engines with little modification, market penetration will be slow. The homogeneity of the motor vehicle fuels market discourages an incremental, market niche approach, while the conservative purchase behavior of vehicle owners frustrates marketing initiatives by vehicle manufacturers and fuel marketers. Unless a new fuel is identical to petroleum fuels (or superior in every way), it will be resisted.

The best hope for new fuels, other than the tiny transit bus market and certain small segments of the fleet market, is the large household vehicle market. Although at first glance vehicle fleets appear to be an attractive niche, the reality is that fleet buyers tend to be even more conservative than individuals. In almost all cases, notwithstanding the Bank of America, fleet buyers conduct a careful economic analysis before making a vehicle purchase. They will not buy alcohol vehicles because they will be more expensive than other choices. As indicated in chapter 13, they already have indicated an unwillingness to buy light-duty diesel and CNG vehicles, even though both choices are economically superior to gasoline. Apparently, fleet buyers are not willing to purchase a vehicle technology, even if it is less expensive, until it is well proven in the marketplace. While individual consumers are also conservative, many tend to be less careful about analyzing the economics of the choices, many are more willing to value noneconomic attributes such as greater acceleration (and perhaps lower emissions), and at least some are willing to accept inconveniences such as limited fuel availability.

Penetration by alcohol, natural gas, or hydrogen into the transportation fuels market will not occur in a timely manner—and possibly not at all—without substantial government involvement. Without active involvement by government to support these fuels, energy production investments will inevitably be attracted toward fuels that are similar to petroleum. That would be very unfortunate, because those fuels tend to be more expensive and, as will be shown in the next two chapters, much more disruptive and destructive of the environment.

PART VI

Environmental Degradation

A theme of this book is that technological choices and the evolution of human-built systems respond to goals and purposes of societies, not to immutable laws. This point is nowhere better illustrated than with environmental quality. The environmental characteristics of technologies are manipulated by perceptions, goals, and values that are codified into rules and laws. The environmental effects of a particular technology or project may vary over a very wide range, depending upon the level of control required by the society or jurisdiction and by the enforcement of those requirements. It will be shown with one notable exception that environmental effects are less a function of inherent pollution-generating features of technologies than of the will of a society to curtail degradation of the environment.

During the initial growth of fuel activities (or any new industrial activity), perceptions of environmental effects are fuzzy. Environmental goals and values are poorly focused and public awareness is low. Knowledge of likely environmental effects is lacking. As a result, modification and codification of rules and laws lag and environmental factors tend to play a minimal role in initial investment and deployment decisions, except to the extent that the inevitability of future environmental rule making creates uncertainty and risk.

The notable exception is hydrogen produced from solar energy. As will be shown, photovoltaically produced hydrogen stands out as the single major transportation energy option that is inherently superior to all others environmentally, in terms of both production and end use.

Electric vehicles that use electricity generated from solar (or wind or water) energy are also environmentally attractive but are unlikely to be important. Natural gas and, to a lesser extent, methanol produced from natural gas are the other options that are attractive environmentally. All other options have major negative effects. It is not possible to rank those remaining options because it depends on which environmental effects are ultimately judged to be more or less acceptable, and upon the political will to control those different effects.

FIFTEEN

Air Quality

(WITH MARK DELUCHI)

One of the most prominent environmental concerns with new fuels has been air quality. The principal sources of air pollution are fuel production and fuel combustion. This chapter will first address fuel production sources, then end-use sources.

AIR POLLUTION FROM FUEL PRODUCTION

Air pollution effects from fuel production are less certain than those from fuel combustion. The uncertainty is due to the diversity of feedstock-production activities and the lack of reliable data. The use of diverse feedstocks and process technologies results in very different emission characteristics, while diversity in the scale, geographical location, and industrial experience of plant owners and managers suggests the likelihood of nonuniform enforcement of rules by governments. Data for most of the processes considered here are derived from pilot and demonstration plants; extrapolation of those results may or may not be representative of commercial-scale plants. The analysis that follows covers ethanol production by fermentation, the thermochemical conversion of coal and biomass to make methanol, the production of methanol from natural gas, the direct liquefaction of coal into synthetic gasoline, oil pyrolysis to make synthetic gasoline and diesel fuels, and the production of natural gas and hydrogen.

TABLE 49 AIR POLUTANT EMISSIONS FROM ENERGY CONVERSION PROCESSES WITH CONTROLS, GRAMS PER MILLION BTU OF OUTPUT

Conversion Process	Feedstock	Particulates	SO_x	HC (organic compounds)	NO_x	CO	CO_2
Pyrolysis (surface) retort)	Oil shale	10-35	3-16	3-15	50-150	3-16	55,000
Direct Liquefaction[a]	Bituminous coal, 4% sulfur	10-25	18-60	0.3-3	4-210	3-5	50,000
Methanol (thermochemical conversion: gasification)[b]	Subbituminous and lignite coal 0.4-0.6% sulfur	1-25	30-200	100-500[c]	15-150[c]	NA	65,000-90,000
Great Plains SNG Plant[d]	Lignite coal	11	108[g]	no limit[d]	63	no limit[d]	no limit[d]
Methanol (thermochemical conversion: gasification)	Wood	0- 30[c]	negl.	NA	10-200[c]	NA	NA
Ethanol (bioconversion: fermentation)	Corn	45-370	37-1500	5-140	100-830	10-170	38,800-47,400
Ethanol (acid cellulose hydrolysis)	Wheat straw and corn residue	100-200	800-1100	NA	500-600	NA	NA
Electricity-generating, coal-fired plant[e]	Bituminous coal, 2% sulfur	20	200-1000	very low	very low	100-500	NA
Petroleum refinery[f]	crude oil	2	11	10	9	NA	NA

SOURCES: A. U. Hira, J. A. Mulloney, and G. P. D'Allessio, "Alcohol Fuel From Biomass," Envir. Sci. Tech. 17, no. 5 (1983): 202A-213A; Pace Co., "Comparative Analysis of Coal Gasification and Liquefaction" (Prepared for Acurex Corp. and California Energy Commission) (Denver, Colorado, 1981), p. 78; M. A. Chartock et al., Environmental Issues of Synthetic Transportation Fuels from Coal (Springfield, Va.: NTIS, 1982); OTA, U.S. Congress, Energy from Biological Processes, vol. 2 (Washington, D.C.: GPO, 1980); OTA, U.S. Congress, Increased Automobile Fuel Efficiency and Synthetic Fuels (Washington, D.C.: GPO, 1982); Argonne National Laboratory, Environmental and Economic Evaluation of Energy Recovery from Agricultural and Forestry Residues (Argonne, Ill.: 1979, ANL/EES-TM-58); U.S. DOE, Synthetic Fuels and the Environment: An Environmental and Regulatory Impacts Analysis (Washington, D.C.: DOE, June 1980, DOE/EU-0087) appendix III-G; and SRI International, Alcohol Fuels Production Technologies and Economics (Prepared for U.S. DOE, Menlo Park, Calif., November 1978, EJ-78-CO-01-6665). One other source of emissions data is E. J. Bentz and E. J. Salmon, Synthetic Fuels Technology Overview with Health and Environmental Impacts (Ann Arbor, Mich.: 1981). The emission rates in this last source were consistently much higher for coal and oil shale processes than reported in any other source, and were therefore not incorporated into the table.

NA = not available

[a]Based on Exxon Donor Solvent and Solvent Refined Coal II processes.

[b]Based on various gasifier technologies. The upper values refer to low-temperature Lurgi gasifiers.

[c]The upper value is highly suspect.

[d]Emission rates for this plant are those established in the plant's air quality permit. All emissions limits were met except those for SO_X. See text for discussion of SO_X emissions.

[e]Based on new source performance standards (NSPS) for new powerplants.

[f]Based on California data for 1981 provided by Air Resources Board; emission rates are somewhat higher elsewhere in U.S.

[g]Actual SO_X emissions in 1986 were 360 grams per million Btu.

1. BIOCONVERSION: FERMENTATION

Bioconversion plants generate more air pollution per unit of energy output than any other fuel production process except possibly cellulose hydrolysis processes. In bioconversion, unlike other processes, the production rate of air pollutants is due more to the energy material used for process heat than to the feedstock or the process technology used. If coal is used to fire the boilers to create heat, then large quantities of nitrogen oxides (NO_x), sulfur oxides (SO_x), particulates, and volatile organic compounds are emitted. If petroleum is used for process heat, similar pollutants, but in smaller quantities, are generated. If cellulosic biomass (e.g., bagasse or crop residues) is used, then particulates are the most significant pollution generated. The range in emission rates for bioconversion (fermentation and distillation) is illustrated in table 49. Those estimates are based on engineering studies conducted by four different research laboratories and design firms (Argonne National Laboratory, Stone and Webster Engineering, Mueller, Oak Ridge National Laboratory) for eight hypothetical plants ranging in size from 1 to 50 million gallons per year (65 to 3,260 b/d).[1]

2. THERMOCHEMICAL CONVERSION: METHANOL FROM COAL AND BIOMASS

In some respects, the thermochemical conversion of coal would generate air pollutants similar to those generated from the thermochemical conversion of cellulosic biomass: the major difference between the two, as discussed below, is that biomass contains much less nitrogen and sulfur than coal and thus would generate fewer NO_x and SO_x emissions, and that coal contains significant quantities of many trace metals such as arsenic, boron, beryllium, and uranium.

In thermochemical methanol production, air pollutant emissions are generated from the following activities and waste streams: feedstock pretreatment, gasification and quenching, tar separation and stripper flash gases, acid gas and sulfur removal, decommissioning and regeneration of catalysts, incinerator and boiler flue gases, and cooling tower evaporation. In addition to these stack gases, other pollutants are released as gases leaking from valves, compressors, fittings, and other miscellaneous sources. Most of these fugitive emissions could be eliminated with diligent detection efforts and various technical modifications and controls, but the cost would be relatively high. The vast number of these potential leaks—in the thousands for a typical coal plant—suggests that in practice, fugitive emissions would continue. These fugitive emissions would contain not only conventional combustion products but also toxic trace elements and gases.

In general, methanol produced from biomass is less threatening to air quality than methanol produced from coal—because of lower NO_x and SO_x emissions, absence of trace metals, the small size of the plants, and also because it produces few polycyclic hydrocarbons (which are likely to be mutagenic and carcinogenic). The problem of trace elements is potentially severe. One estimate indicates that a large coal gasification plant would generate the following quantity of trace metals in a single year:[2]

Arsenic	14,000 tons
Cadmium	1,000 tons
Lead	10,000 tons
Manganese	54,000 tons
Mercury	200 tons

While the estimate may be overstated, the presence of these trace elements nevertheless poses a major threat because these elements resist detoxification and are nondegradable. They persist in the environment and have the potential to accumulate in organisms until they reach harmful levels. Some of these metals, such as mercury, arsenic, and cadmium, volatilize into a gas relatively easily and pose air quality threats, especially on site and near the plant. The more serious threat, however, will likely be trace elements that enter the environment through water discharges and solid waste disposal.

Polycyclic hydrocarbons are molecular compounds with multiple rings of carbon atoms. Many of these compounds are proven carcinogens, but little is known about them and many have not yet even been identified.[3] These hazardous organic compounds may enter the environment as particles in gaseous fugitive emissions and in cooling tower mists, via evaporation of waste water streams and in sludges and solid waste.

3. METHANOL FROM NATURAL GAS

The bulk of emissions generated in thermochemical production of methanol occurs during the gasification process rather than during the methanol synthesis step. It is thus clear, although quantitative estimates are not available, that production of methanol from natural gas is much cleaner than other methanol production processes.

4. LIQUEFACTION OF COAL INTO SYNTHETIC GASOLINE

Direct liquefaction of coal may pose a greater threat than other alternative fuel processes, not because of the conventional pollutants emitted, but because of toxic and carcinogenic emissions. As with coal gasification, there is no reason to doubt that advanced control technologies will be highly effective in keeping emission rates to moderate levels. But again, just as with coal gasification, fugitive emissions are a major threat. An environmental impact statement (EIS) prepared for the proposed SRC-II direct liquefaction plant identified 5,000 possible fugitive emission sources and estimated that the relatively small demonstration plant might release up to 750 tons of hydrocarbons per year as fugitive emission.[4]

Two other risks from direct liquefaction emissions, similar to those of coal gasification, are trace metals and polycyclic hydrocarbons. Although pollution data for direct liquefaction plants are sparse and based strictly on pilot plant tests, there is substantial evidence that these plants will generate considerably more carcinogenic and mutagenic material than coal gasification.[5] The cancer-causing effects of coal liquids are further addressed in the next chapter under the topic of occupational safety.

Once again, a definitive determination of pollution effects is not possible. To a very large extent the quantity of emissions will depend on the efficiency and effectiveness of control techniques and plant design. The greatest concern most likely will prove to be the frequency and severity of accidental releases of conventional pollutants and toxic and carcinogenic gases.

5. OIL PYROLYSIS TO MAKE SYNTHETIC GASOLINE AND DIESEL FUEL

Oil shale pyrolysis plants also generate significant volumes of air pollutants, but again, on an energy output basis, when controlled with available modern techniques, the emission rates are relatively low (table 49). The emissions, especially particulates, would be higher from surface retorting as a result of the exposed preparation and retorting of the shale.[6] Oil shale retorting tends to produce relatively few trace metals and polycyclic organic matter.[7]

6. THE PRODUCTION OF HYDROGEN AND NATURAL GAS

The production of hydrogen from water is emission-free if the electrolysis process uses electricity from solar energy. Natural gas fuels are also very attractive, since no pollutants are generated other than the minor quantities of gas that escape during handling and through valves.

COMPARISON OF FEEDSTOCK-PRODUCTION OPTIONS

Clearly photovoltaic hydrogen and natural gas are far superior to all other options. Methanol produced from natural gas ranks next on the list. Other production processes generate much larger quantities of air pollutants, continuing downward on the scale from thermochemical conversion of biomass to coal and oil shale processing and to the very worst, the hydrolysis and fermentation-distillation of biomass.

While emission rates of acid and enzymatic hydrolysis processes (that convert biomass to ethanol) have not been specifically addressed, sparse evidence indicates that on a unit energy basis they would tend to generate higher levels of sulfur oxides than any other fuel option, and NO_x and particulate levels similar to those for fermentation (table 49). The emission rates of processes that directly liquefy coal (e.g., EDS, SRC II, H Coal) are less certain than those for coal gasification. Very roughly, on a unit energy basis, direct liquefaction processes may be considered equivalent to the coal gasification process in terms of emissions of conventional, regulated pollutants. Direct liquefaction processes, however, generate more toxic and carcinogenic materials than any other process.

In any case, under normal conditions, assuming fugitive emissions are not excessive, air pollution from the production of alternative fuels in general does not pose a major threat to environmental quality, although as we shall see later it can affect plant siting and regulatory decisions. As a comparison, note that the production of alternative fuels is likely to generate much less air pollution than coal-fired electrical generation plants, although many assumptions underlie this generalization. The most important assumption underlying these estimates and generalizations is that modern air pollution control technology is applied. Numerous documented examples are available of relatively uncontrolled oil shale and coal gasification facilities outside the U.S. —in particular, a gasification plant in Yugoslavia—that have an extremely hazardous impact on air quality.[8]

Another caveat to the above generalization is that emission rates may vary widely even within each generic process technology. For instance, the Lurgi gasifier (used in the Great Plains SNG plant) operates at lower temperatures than other gasifiers and produces much higher levels of heavy oils, tars, and other hydrocarbons that contain toxic and potentially carcinogenic compounds.

One other irregularity illustrates the impreciseness of emission estimates: operations under "upset" conditions. During upset or emergency conditions relatively large amounts of some pollutants (especially SO_2 and particulates in the case of thermochemical processes) will be spewed out; in a worse case as much SO_2 might be released in two hours as would normally occur in ten days of operation.[9] How frequently such upsets might occur, however, is unknown.

INSTITUTIONAL CONTEXT

To some extent, the institutional setting determines the severity of environmental degradation from any particular project or set of options. To be complete, an assessment in the U.S. of expected air quality impacts should be conducted on the basis of the following institutional considerations: promulgation of national new source performance standards (NSPS), local institutional response to air quality impacts, and future regulatory treatment of currently unregulated pollutants.

The U.S. Environmental Protection Agency (EPA) established NSPS for all industrial facilities. However, most processes (except methanol from natural gas and ethanol derived via bioconversion) are not commercially established, and the data base from which to devise new emission standards is insufficient. In the past the EPA has written "pollution control guidance documents" for each type of pollution-generating industrial facility (e.g., copper smelters, steel plants) to guide the development of standards and issuance of permits. These very detailed and comprehensive documents describe the control systems available for each waste stream in a plant and the level of control judged to be attainable. A series of these documents were drafted for coal- and shale-based fuels but final reports were never completed, apparently because the Reagan Administration cut off funding. Although the EPA has promulgated new source performance standards for boilers that generate heat for processing and other component units that comprise existing industrial processes, neither EPA guidance documents nor NSPS have been approved for new component technologies such as biomass and coal gasifiers. As a result, process plants to this date have been designed and built with no regularized emission targets or approved control devices.

Due to the absence of NSPS, the siting and design of alcohol fuel process plants are subject to greater local review. According to the elaborate regulatory structure created by the 1970 and 1977 amendments to the Clean Air Act, local areas must attain ambient air quality standards established by EPA. If trends for an area indicate that an area will not attain the ambient standards (by a legally established date that has been periodically postponed from 1975 to 1977 to 1982 to 1987 and to an undetermined future date), then they must prepare a "State Implementation Plan" detailing what changes and controls they intend to pursue to reach those ambient levels. Areas not expected to achieve the ambient standards (called "nonattainment" regions) must determine whether to allow the siting of a new industrial plant and if so, what controls to require. These determinations are made on the basis of the EPA-approved State Implementation Plans. Usually, new plants are allowed only if they meet NSPS and the project sponsors can demonstrate that they, the proponents, will offset the new emissions by reducing emissions from other sources elsewhere in the region.

In practice, considerable discretion is exercised in this approval process, despite the statutory requirement to attain ambient standards. The EPA has generally not exercised the considerable power granted to it by Congress to punish nonattainment areas. If local areas show some sort of effort to improve air quality, then EPA tends to accept it as meeting the intent of the law.

For areas already meeting ambient air quality standards or expecting to by 1987, a different set of rules applies: those that call for prevention of significant deterioration (PSD). Areas that are, or shortly will be, in attainment—most rural areas and small cities—are categorized in one of three PSD classes: Class I, no emission increment; Class II, some increment; and Class III, the most. Each proposed project must show on the basis of mathematical air quality modeling that the allowable increment will not be exceeded.

For illusrative purposes, consider a situation in which a number of synthetic fuel process plants are clustered in one area—a situation that

TABLE 50 EFFECT OF COAL LIQUIDS PLANTS ON EMISSIONS IN THE EVANSVILLE-OWENSBORO-HENDERSON AIR QUALITY CONTROL REGION

	Thousands of tons per year				
Emission Source	SO_X	Particulates	NO_2	CO_2	HC
1976: total	266	50	57	100	29
Proposed powerplants (1976-1992)	267	20	180	11	3
Four proposed synfuel plants	50	4	74	15	7
Total without synfuel plants	533	70	237	111	32
Total with synfuel plants	583	74	311	126	39

SOURCE: James E. Jones and Harry Enoch, "The Kentucky Synfuel Industry: A Basis for Assessment and Planning," Kentucky Dept. of Energy, 1981, printed in U.S. Congress, The Socioeconomic Impacts of Synthetic Fuels, hearing before Subcomm. on Energy Development and Applications and Comm. on Science and Tech., House of Rep., 97th Cong, 2d Sess., 28 April 1982, (Washington, D.C.: GPO, 1982), p. 261.

was actually encountered in Kentucky when four independent and serious proposals to build coal liquefaction plants were put forward in the late 1970s. The four plants, located in a single air pollution control district, were to produce 260,000 barrels of oil-equivalent fuel. In table 50 the estimated emissions from the synthetic fuel plants are compared to existing and projected levels. Parts of the region were in "nonattainment" only for particulate and ozone standards. While operation of the four plants would have caused a projected increase over the base case of only 8 percent in particulate concentrations, they would have had a very severe effect on attainment of ozone standards. The four plants would have increased hydrocarbon emissions and NO_x emissions, which react in the presence of sunlight to form ozone (O_3), by 25 percent and 130 percent respectively. Moreover, the increase in SO_x and NO_x emissions (though not leading to violation of ambient standards in this Kentucky case) contributes to acid rain and deposition.[10]

The U.S. Department of Energy conducted a county-by-county analysis to determine the importance of air quality rules and other environmental factors on the siting of coal gasification and liquefaction plants.[11] The report identified 159 counties in fifteen states that had large enough coal reserves to support at least a 100,000 b/d plant for twenty-five years. It found that enforcement of nonattainment air quality rules would eliminate ninety-six counties from consideration and that PSD rules would eliminate another twenty-two counties. A large number of those eliminated were in the eastern U.S., where many other emission sources already exist. While new coal plants will rarely be as large as 100,000 b/d and although air quality has been slowly improving since the 1970s in most areas, this siting analysis nonetheless demonstrates that emissions from fuel production plants can greatly restrict the location of these plants.

In the future the location and licensing of fuel conversion plants may be further restricted by new regulations for currently unregulated pollutants. At present just the following ambient pollutants are regulated: particulates, NO_x, SO_x, ozone, hydrocarbons, carbon monoxide, and lead. No ambient standards (as distinguished from emission standards) have been adopted for any of the other toxic and carcinogenic particles and gases. Many unregulated pollutants have not yet been identified as hazardous. In the future these currently unregulated pollutants may be regulated at the state and federal levels, thus affecting future siting decisions.

Another gas that is not considered a pollutant and is completely unregulated but threatens major long-term environmental damage is carbon dioxide (CO_2). CO_2 is effective at absorbing infrared radiation that is reflected from the earth's surface and reradiating it back to earth. Increased accumulations of CO_2 in the atmosphere create a "greenhouse effect" in which an increased amount of heat is retained within the earth's atmosphere, causing an increase in the temperature of the earth's surface. This global warming would shift global precipitation patterns, disrupt agriculture, and eventually melt portions of the polar ice caps and threaten coastal cities worldwide with flooding.

The major human-caused (anthropogenic) source of CO_2 is fossil fuel combustion, accounting for 56 to 71 percent of the total CO_2 released in the world in 1980.[12] Although there are many unknowns and uncertainties, which leaves much room for differences of opinion about the causes and consequences of a greenhouse warming, researchers are in almost unanimous agreement that substantial increases in CO_2 concentrations will cause substantial warming. Various studies have concluded that if current trends continue, global temperatures will increase 1 to 5.5° C (2 to 10° F) in the next fifty to one hundred years.[13]

While CO_2 is by far the most important gas contributing to the

TABLE 51 EMISSIONS OF GREENHOUSE GASES

Fuel/Feedstock	CO_2-equivalent Emissions[a]	
	GT/Yr[b]	% change relative to petroleum
EVs from solar power	0	-100
Hydrogen from solar power	negligible	~ -100
CNG, LNG, or methanol from biomass	0	~ -100
EVs from current power mix	0.74	- 43
CNG from natural gas[c]	1.09	- 16
Methanol from natural gas	1.20	- 8
LNG from natural gas[c]	1.21	- 7
Gasoline from crude oil	1.30	---
Hydrogen (hydride) from coal	1.55	+ 19
Liquid hydrogen from coal	1.76	+ 35
Methanol from coal	2.19	+ 69

SOURCE: Mark DeLuchi, Robert Johnston, and Daniel Sperling, "Transportation Fuels and the Greenhouse Effect," Transportation Research Record (forthcoming).

[a]Does not include N_2O emissions from motor vehicles. Emissions were calculated for production, distribution, and end-use activities.

[b]GT = gigatons (10^9 tons).

[c]Includes CH_4 emissions adjusted to CO_2-equivalent basis.

greenhouse effect, note that other trace gases, including methane (CH_4), N_2O, and ozone (O_3) also have some effect. Only methane has much significance for transportation fuels.

The fuel choices that generate the fewest CO_2-equivalent emissions, as shown in table 51, are hydrogen produced with solar energy, electric vehicles with electricity produced from solar energy, and CNG, LNG, and methanol produced from biomass. These three sets of options would eliminate CO_2 emissions by the transportation sector. In the case of biomass fuels, the net effect is roughly zero emissions because the biomass used as feedstock absorbs roughly the same quantity of CO_2 from the atmosphere before harvesting as is released when the manufactured fuels are combusted. The fuel options with the most severe threats, worse even than gasoline and diesel from petroleum, are those that use coal as feedstock. Natural gas–based fuels are somewhat better than petroleum fuels, even taking into consideration the methane gas that is released during combustion.

Acid rain or, more generally, acid deposition is another type of air pollution that is not directly regulated. When the link between NO_x, SO_x, and acidity is more definitively specified in the future, and when more stringent controls and standards are imposed, then the cost of process plants and their siting will be affected. The imposition of acid deposition controls will discourage investments in Lurgi gasifiers and those fermentation and hydrolysis plants that generate large amounts of SO_x and/or NO_x. Most alternative fuel plants would, however, be far less of a threat than coal-fired power plants. Indeed, as argued by two EPA officials in their 1985 book,[14] gasification-based processes may prove highly attractive in areas with high-sulfur coal, since they generate few SO_x emissions.

END USE

Gasoline and diesel fuel are inherently "dirty" fuels. The combustion of gasoline generates large amounts of nitrogen oxides (NO_x), carbon monoxide (CO), and hydrocarbons (HC). The combustion of diesel fuel in compression ignition engines also generates large amounts of these pollutants, as well as of particulates. Because petroleum fuels contain very little sulfur, sulfur oxide (SO_x) emissions from both engine types are negligible. Regulations have been promulgated to control the three primary pollutants from motor vehicles (see table 52), as well as particulates for diesel automobiles (and for diesel buses to take effect in

TABLE 52 U.S. EXHAUST EMISSION STANDARDS FOR NEW AUTOMOBILES (GRAMS PER MILE)

	Precontrol	Model Year 1968	Model Year 1981+
Hydrocarbons	6-10	5.9	0.41
Carbon Monoxide	60-90	51	3.4
Nitrogen Oxides	4-8	no standard	1.0

1991 and heavy diesel trucks in 1994). Similar emission standards are in place in Japan and, in weakened form, in Australia and Western Europe. Note in table 52 that U.S. vehicle emission standards for 1981 and later are dramatically lower than typical emission levels in 1968—CO and HC by over 90 percent and NO_x by about 75 to 85 percent. But air pollution remains a health threat in most cities, and motor vehicles remain the primary source of that pollution, accounting for almost all the carbon monoxide and half or more of the hydrocarbons and nitrogen oxides emitted in urban areas. Alcohols and natural gas hold out promise for additional reductions in motor vehicle emissions; and hydrogen and electric vehicles for practical elimination of vehicular emissions.

ALCOHOL IN SPARK IGNITION ENGINES

Alcohols are relatively clean fuels, in the sense that their combustion tends to degrade air quality less than gasoline or diesel combustion.

Alcohol combustion releases about the same quantity of carbon monoxide as gasoline combustion. Carbon monoxide emissions may be reduced by increasing the ratio of air to fuel in the engine to make combustion "leaner." This happens when small amounts of alcohol are added to gasoline (and there is no oxygen sensor to make adjustments) or when the engine is redesigned using "lean burn" technology to operate at lean air-fuel ratios. Generally, though, CO emissions are similar for both alcohols and gasoline.

Alcohol is somewhat more effective at reducing hydrocarbons or, more accurately, reactive organic matter. The combustion of gasoline results in relatively high emissions of several types of hydrocarbons, such as alkanes, alkenes, and aromatics. Some of these organic compounds are considered toxic, but the primary justification for their regulation has been their central role in the formation of photochemical oxidants (ozone). Alcohol combustion produces very low levels of these types of hydrocarbons but relatively high levels of other organic matter—principally unburned alcohol and aldehydes (including formaldehyde). Testing has shown that the emissions of organic matter from methanol vehicles—mostly unburned methanol and aldehydes—are roughly equivalent on a mass basis to the hydrocarbon (i.e., organic) emissions from gasoline vehicles. But different organic compounds have varying levels of reactivity and therefore have varying impacts on ozone formation: in the case of alcohol emissions, unburned alcohol has a low reactivity, while formaldehyde is highly reactive. On balance, because of the smaller proportion of formaldehyde, methanol vehicles are expected to contribute somewhat less to ozone formation than gasoline vehicles with equivalent emission controls, although there is much uncertainty, as discussed below.

Methanol engines are most effective at reducing NO_x emissions, potentially generating about 25 to 50 percent less NO_x than gasoline engines (although, as suggested earlier, the lack of incentive for automakers to reduce emissions below the standards will mean that these potential benefits may be traded off for other benefits).[15] These NO_x levels are low because methanol burns at a lower temperature than gasoline and because NO_x formation decreases as temperature decreases. Since in the U.S. only the Los Angeles area is in violation of ambient NO_x standards, there are two possible responses to alcohol's inherently lower NO_x emissions: (1) continue to deploy the sophisticated (and expensive) three-way catalytic converters now used on cars and reduce NO_x emissions lower than current levels, thus enabling Los Angeles to meet the NO_x standards; or (2) simplify the catalytic converters by eliminating the NO_x reduction catalyst on alcohol vehicles, thereby reducing vehicle production costs by an estimated $200 or so per vehicle, but resulting in no reduction in NO_x from current levels. This latter strategy also eliminates the need to import from South Africa, the U.S.S.R., and elsewhere the expensive catalyst metals used for NO_x reduction.

Although most areas of the U.S. are not violating ambient NO_x standards, there is still a strong motivation to reduce NO_x levels because NO_x, along with hydrocarbons, causes the formation of ozone, more popularly known as smog (actually, ambient standards are expressed in terms of NO_2, while emission standards are in terms of NO_x). Ozone is a major pollution problem in almost all large cities. Another reason for paying attention to NO_x even when ambient standards are not being violated is that as indicated earlier, NO_x is a cause of acid deposition, especially in the western part of the U.S.

Some indication of relative emission benefits of methanol is presented in table 53. The data are based on a group of forty methanol cars manufactured by Ford (1981 Escorts) and operated in California during the early 1980s (see chap. 13); they provide the best source of empirical data for comparing emissions from gasoline and methanol vehicles. (The larger 500-plus fleet of 1983 methanol Ford Escorts was not systematically tested using standard procedures and therefore comparable

TABLE 53 EMISSION FACTORS AT 50,000 MILES FOR 1981 METHANOL-FUELED AND GASOLINE-FUELED FORD ESCORTS, GRAMS PER MILE, BASED ON FEDERAL TEST PROCEDURE DRIVING CYCLE

Pollutant	Methanol-fueled Vehicles	Gasoline-fueled Vehicles
NO_X	0.41	0.63
CO	6.80	8.80
RHC*	0.48	0.50

SOURCES: R. J. Nichols and J. M. Norbeck, "Assessment of Emissions from Methanol-Fueled Vehicles: Implications for Ozone Air Quality," Paper presented at Annual Meeting of Air Pollution Control Association, Detroit, Michigan, June 1985; see also F. J. Wiens et al., "California's Alcohol Fleet Test Program--Final Results," Proceedings of the Sixth International Symposium on Alcohol Fuel Technology, Ottawa, Canada, May 1984.

*Reactive hydrocarbons (calculated as $CH_{1.85}$)

emissions data are not available, and Brazilian ethanol cars were not controlled for pollution or reliably tested). The 1981 Escorts had carbureted engines and were converted production-line gasoline vehicles (i.e., not designed specifically for methanol). Emission data from the forty methanol cars and ten gasoline control vehicles were standardized and average rates calculated for vehicles after 50,000 miles of use. While the estimates show those methanol cars to be superior to their gasoline counterparts, the findings should not be generalized to actual production-line methanol vehicles.

The air quality issue of overriding importance for alcohol vehicles is ozone formation. It is still fairly uncertain, however, how large a benefit would result from alcohol use. Several major modeling studies have found that if all gasoline cars were replaced by methanol cars, then significant though not necessarily large reductions in ozone levels would result. A study by the Jet Propulsion Laboratory found that the reduction would be 14 to 18 percent in Los Angeles,[16] and a study by Ford Motor Company (using the EPA EKMA-II photochemical model) found reductions from zero to 36 percent for a group of twenty major metropolitan areas.[17]

It is the promise of these ozone reductions which has attracted so much support for methanol in the U.S. While a 16 percent reduction may seem modest, this is not the case. Most urban areas in the U.S. with ozone problems have already implemented those controls that are politically acceptable, but even with those controls they still are not expected to attain the primary health-based standards for ozone—perhaps ever in some cases.[18] In total about 80 million people live in ("nonattainment") areas where ozone air quality standards are being violated. An air pollution official in the Los Angeles area has stated that "the transition to neat methanol fuels for all motor vehicles represents the most significant opportunity for air quality progress which exists between now and the end of the 20th century."[19]

These promises of ozone benefits may be overstated. The two studies cited above were based on single-day episodes; that is, the effect of methanol emissions on same-day air quality. Ozone episodes often persist for several days, however, when temperature inversions keep pollutants trapped near the ground. Preliminary (but unpublished) results have indicated that ozone levels from methanol emissions may not be reduced as much on the second and third day of ozone episodes as on the first day. Because of the inherent lack of precision in air quality modeling and the relative sparsity of research on methanol emissions, none of these modeling results can be considered definitive. It is difficult, then, to make blanket statements about the impact of methanol on ozone. Modest gains could be expected in some areas, but in others, including Los Angeles, the gains might be disappointingly small.

The emission effects of blends of alcohol and gasoline are a different story; emissions from blends may be much different than those from either alcohol or gasoline. In fact, the use of blends tends to cause slightly worse air pollution than either alcohol or gasoline alone. The overall effect of 5 percent methanol blends on exhaust emissions is minimal, but evaporative emissions may increase.[20] The blending of alcohol and gasoline leads to the formation of new azeotrope compounds (light hydrocarbon molecules) that evaporate easily. This evaporation may take place during storage at terminals and service stations, during transfer to the motor vehicle (vapor losses), directly from the engine and fuel tank especially on hot days (diurnal evaporation), and immediately after vehicle operation (hot soak emissions). Evaporative emissions from motor vehicles are reported to increase by 10 to 90 percent when alcohol-gasoline blends are used in place of gasoline.[21] The formation of azeotropes and their evaporation can be reduced by removing certain volatile petroleum components from the gasoline, but this raises the overall cost. The use of cosolvents may reduce volatility.

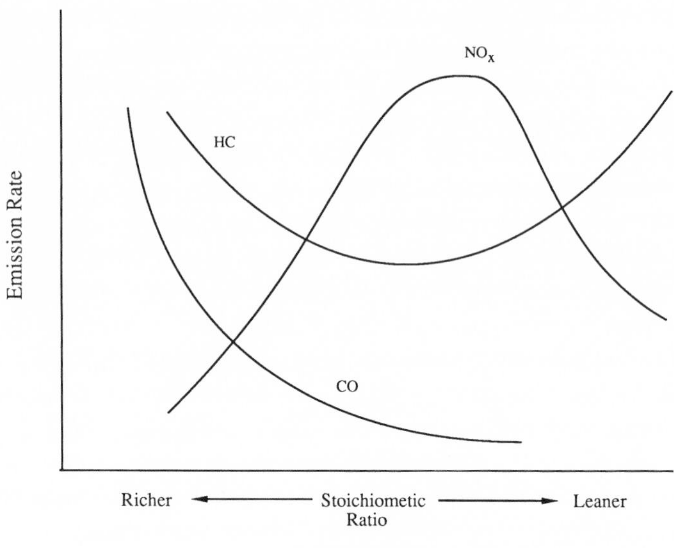

Figure 33. Emissions as Function of Air-Fuel Ratio

Assuming the gasoline base is not altered, taking into account increased evaporative emissions in the fuel distribution system (storage tanks, pipelines, etc.) and from the vehicle, the net effect of blends on total regionwide emissions of organic compounds would be a modest increase of up to about 3 percent.[22]

NATURAL GAS IN SPARK IGNITION ENGINES

In general, natural gas combustion in spark ignition engines is far superior from an air quality perspective to gasoline combustion and possibly superior to methanol—especially so in engines designed for and dedicated to natural gas fuels. NO_x emissions are about the same as for gasoline, but CO emissions tend to be almost totally eliminated and most of the hydrocarbon emissions are nonreactive.

A major factor explaining these differences is the ability of natural gas, like alcohol, to burn very efficiently at lean air-fuel ratios. Gasoline in contrast needs to be burned at or near stoichiometric ratios to be efficiently combusted. As shown in figure 33, CO and NO_x emissions are in theory much lower at lean air-fuel ratios, while hydrocarbon emissions are higher at lean air-fuel ratios, although this tends not to be a problem for natural gas because 80 to 90 percent of the hydrocarbon emissions are methane,[23] which is not photochemically reactive and therefore does not react with NO_x to form ozone.[24] For a dual fuel vehicle, the reactive fraction of HC emissions is about 10 to 20 percent with CNG use and 55 to 90 percent with gasoline.[25] In practice, NO_x emissions from natural gas combustion may be higher than indicated in figure 33 because the spark timing will be advanced to start combustion earlier than with gasoline and because adjustments will be made to increase power (except perhaps for LNG vehicles).

In dual-fuel vehicles, various tests indicate that when the vehicle is operating on CNG, CO emissions are nearly eliminated, NO_x emissions are comparable or slightly less than gasoline combustion in a comparable engine, and nonmethane HC emissions are about half that of gasoline.[26] Thus dual-fuel engines when operating on CNG are far superior to gasoline engines and, on balance, similar to alcohol engines.

Dedicated CNG engines are substantially cleaner burning than dual-fuel engines. Tests run on a 1983 Ford V-6 engine more or less optimized for natural gas and with all pollution control equipment removed showed that the engine could meet the California CO and nonmethane HC emission standards but not NO_x standards, while achieving better thermal efficiency than the equivalent stock gasoline engine.[27] Other tests have similar results, showing that CO and nonmethane HC emissions are greatly reduced relative to gasoline but that NO_x levels are roughly comparable.

HYDROGEN IN SPARK IGNITION ENGINES

Hydrogen is much cleaner burning than gasoline, alcohol, or natural gas. The main combustion product of hydrogen and air is H_2O, although at high combustion temperatures significant amounts of NO_x may be formed. In general, though, hydrogen engines can be expected to generate almost no pollution if some form of exhaust gas treatment is used to reduce NO_x emissions. It is possible that a fully optimized hydrogen engine could meet U.S. emission standards for NO_x without any controls whatsoever.[28]

ALCOHOL IN DIESEL ENGINES

Natural gas and alcohol fuels offer an even greater air quality benefit in compression ignition (diesel) engines than in spark ignition engines, although, as shown earlier, the potential penetration of the diesel market is limited. Diesel engines emit large amounts of particulate matter and nitrogen oxides (NO_x) as well as other pollutants, most of which could be sharply reduced by using natural gas or alcohol in place of diesel fuel.

Experience with compression ignition engines that operate on alcohol is limited (see chap. 13). There is more experience using natural gas, but only in stationary and marine diesel engines. Nonetheless, based on theoretical analysis, one may conclude that diesel engines operating on natural gas or alcohol would emit low levels of particulate matter (smoke) because of the absence of carbon-carbon bonds and of fuel impurities such as lead and sulfur. In addition, low particulate emissions permit the use of a catalytic converter to reduce CO and (in the case of alcohol) organic emissions. Moreover, alcohol emits less NO_x, just as in spark ignition engines because of its low flame temperature, while natural gas diesels emit less NO_x at part-load.[29]

Data from prototype methanol-fuel diesel bus engines and tests with dual-fuel natural gas compression-ignition engines support these theory-based determinations (see table 54). Particulate emissions from the General Motors bus in that table are higher than expected and would presumably be reduced in future designs. The MAN NO_x value is also higher than one would expect, but the low GM value indicates the potential for reduced NO_x emissions. The expectation of lower CO emissions with a catalytic converter is supported by results from the MAN engine. A careful EPA review of air pollutant emission from methanol-fueled compression ignition bus engines developed by GM and MAN conclude that

> in summary, the use of methanol in urban buses would result in significant reductions in particulate matter and oxides of nitrogen in urban areas. Emissions of other pollutants, such as smoke and sulfur oxides, would also be reduced. While methanol buses will emit higher levels of methanol and formaldehyde, catalytic converters will ameliorate those problems and because of lower hydrocarbon levels the total photochemical reactivity impact of methanol buses will likely be less than that of diesel buses.[30]

Similarly, other investigators have shown that methane-fueled diesel engines, using a small amount of diesel fuel, produce very little smoke and reduce the emissions of reactive (nonmethane) hydrocarbons.[31]

The positive air quality effects of natural gas and methanol transit buses are important for several reasons. First of all, despite the 1977 Amendments to the Clean Air Act requiring significant reductions of

TABLE 54 EMISSIONS FOR NEW DIESEL AND METHANOL BUS ENGINES (GRAMS PER HORSEPOWER-HOUR)

Pollutant	Typical New Diesel Bus Engine[a]	New M.A.N. Methanol Engine With Catalyst[a]	New General Motors Methanol Engine With Catalyst[b]
Particulate	0.57	0.04	0.17
Oxides of Nitrogen	6.25	6.60	2.20
Carbon Monoxide	3.22	0.31	--
Total Organics	1.61	0.68	1.28
Hydrocarbons	1.51	0.001	--
Methanol	--	0.68	1.13
Aldehydes	0.10	0.001	0.15

SOURCE: U.S. Environmental Protection Agency. See C. Gray and J. Alson, Moving America to Methanol (Ann Arbor, Mich.: University of Michigan Press, 1985), p. 122.

[a]The first two columns contain directly comparable data that were generated over the EPA transient engine test, which involves operating an engine over a simulated cycle that consists of constantly changing engine speed and load conditions.

[b]GM data collected from steady-state engine testing.

NO_x and particulate matter from heavy-duty diesel engines—75 percent for NO_x and "greatest emission reduction achievable" for particulate matter—the EPA had found that the cost of modifying and controlling those engines to attain those standards was prohibitive. Promulgating those emission standards would in effect have resulted in a banning of heavy-duty diesel engines.[32] The use of methanol, however, allows attainment of those standards. Second, the use of methanol or natural gas probably would result in reduced ozone formation. Third, the suspected carcinogens emitted by diesel engines (as particulates) would be eliminated with the use of natural gas or methanol. Fourth, the use of methane or alcohol in urban buses would provide benefits of modestly improved aesthetics and air quality in downtown areas. While past studies have dismissed the pollution effects of diesel buses because of their minor contribution to regional emissions, recent unpublished evidence from General Motors and local environmental analyses indicate that up to half the particulate matter along downtown streets may be due to transit buses. The elimination of noxious smoke from diesel buses would enhance the quality of downtown environments.

The technical difficulties of reducing diesel emissions motivated the agency that regulates air quality in the Los Angeles basin to take the desperate and extreme position of proposing (but never actually implementing) a ban on the use of diesel vehicles.[33]

SUMMARY

The transition to hydrogen fuels produced with solar energy would result in major improvements in air quality. Natural gas–based fuels would also provide air quality benefits, although the actual benefits depend on the use and maintenance of emission control technologies on vehicles. All other fuel options lead to some degradation of air quality, again depending on the nature of emission controls. Most production technologies, for instance, can be designed and built to assure a minimal impact on air quality—in any case, far less impact than coal-fired power plants—while motor vehicle engines can likewise be designed and built to control emissions. The 95 percent or so reduction in emissions from new automobiles in the U.S. over a period of about fifteen years is indicative of this ability to control emissions. In countries without an interest in or commitment to clean air, however, the transition to new fuels, other than hydrogen, and perhaps natural gas and natural gas-based methanol, could be environmentally disastrous.

In addition, one other potential air quality problem threatens disaster—not only for environmentally unconcerned countries but for the entire world. That is the greenhouse effect. At present, the relationship between global warming and emissions of carbon dioxide, methane and other gases is still uncertain. The use of any fossil fuel, including methanol made from coal or natural gas, contributes to the greenhouse effect. Hydrogen and electricity produced from solar energy do not release carbon-containing compounds, and thus do not contribute to the greenhouse effect. Biofuels (methanol and natural gas made from crops or waste) do not contribute net CO_2—and thus do not contribute to global warming either.

In conclusion, even apart from questions of the greenhouse effect, and despite the likely absence of major impacts, air pollution will be a major issue for alternative fuels, especially in the U.S. Controversy will reign as a result of the pervasive institutional structure that has been created. The design and siting of production facilities will be hindered by the numerous standards, permits, and plans they must comply with. At the center of the debate over new fuels will be questions about the magnitude of the air quality benefits and costs and the extent to which air quality benefits should be rewarded.

SIXTEEN

Impacts on Water, Land, and Human Settlements

Air pollution may receive more attention, but other environmental effects are likely to pose at least as great a threat to the environment. These other environmental effects are more difficult to discern, however, because in the U.S. the rules and institutions for controlling them are less developed and less pervasive than those for air quality.

In this chapter the effects of new fuels on water quality and water use, soil erosion, "boomtown" communities, and human safety are analyzed. The greatest potential environmental dangers appear to be soil erosion from intensive cultivation of wood and agricultural plants, and the generation of toxic materials.

WATER QUALITY AND SUPPLY

Water pollution and water supply needs pose significant environmental concerns for all alternative fuel production plants except hydrogen, but from an institutional perspective neither consideration is likely to be a major factor in the cost feasibility or siting of new plants. First, consider bioconversion plants.

Wastewater from bioconversion plants is generated principally from the following four sources:

1. Stillage from the distillation process;
2. Washwater from fermenters and other plant equipment;

3. Condensate return from evaporators, coolers, condensers, and other heat exchangers; and

4. Boilers and cooling tower blowdown.

The waste streams generated by ethanol plants do not contain toxic materials but are acidic, deplete the dissolved oxygen levels in receiving waters, and cause odors and discoloration. A wide variation in effluent levels is found from one ethanol plant to another depending on process configuration (e.g., continuous vs. batch fermentation), extent of effluent recycling and treatment, and plant size.[1] By far the greatest environmental problem with ethanol plants is stillage—the watery residue resulting from distillation. Roughly ten to eighteen gallons of stillage is produced for each gallon of alcohol. Stillage severely depletes dissolved oxygen in receiving waters, having a biochemical oxygen demand count roughly 200 times that of raw municipal sewage (per unit of volume).[2] Although the liquid volume generated by an ethanol plant is not exceptionally large in comparison with typical volumes of industrial effluents or with the hydraulic capacity of conventional waste water plants, the pollutant loading impact is enormous.[3] An ethanol plant producing 25 million gallons per year, for instance, would generate roughly one million gallons of stillage per day, which is equivalent on a BOD basis to about 200 million gallons of municipal sewage. Assuming a U.S. average of about 100 gallons of sewage per capita, the wastes from a single 25-million-gallon-per-year fermentation plant are roughly equivalent to the wastes from a city of two million people.

This pollution can be handled in several ways. Wastewater may be treated by placing it in evaporation ponds, spreading it on agricultural fields, or treating it in conventional sewage treatment facilities. Stillage in some cases may also be dehydrated for use as animal feed. There are many options, some with substantial benefits. The treatment and disposal procedures are fairly routine for industrial facilities in economically advanced countries, and even though the costs are large, they do not have major effects on overall costs.

The most severe effects on water quality would be in countries with less stringent wastewater treatment requirements. Reports from Brazil indicate that major environmental catastrophes have occurred in numerous rivers from the discharge of untreated stillage (see chap. 4). In countries with strong pollution control programs the problem of pollutant loads is readily resolved. From a water quality perspective the decision to support an ethanol bioconversion program should be questioned only in those countries where there is doubt whether wastewaters will be properly disposed.

Water supply presents a similar situation for ethanol bioconversion plants. The plants consume large quantities of water—about forty gallons for cooling and ten gallons for processing per gallon of ethanol.[4] This ration of fifty gallons of water per gallon of ethanol is substantially greater on an energy output basis than the water required per gallon of fuel produced from coal or shale, or even to generate an equivalent amount of electrical energy from coal—but two important differences render water consumption by ethanol plants a minor issue. First, because ethanol plants are much smaller, a single plant will generally not require a significant proportion of local water supply. Second, and more important, ethanol plants would be located where biomass grows in abundance and water is plentiful. Thus water supply would tend not to be an issue for any biomass conversion process, including thermochemical conversion.

Water supply will be a more important issue than water quality for coal and oil shale plants. Coal-based synthetic fuel plants will be designed (in the U.S.) to have little or no discharge of pollutants into receiving waters.[5] The treatment and elimination of pollutant discharges will be facilitated by recycling water through the process plant and the cooling system and thereby concentrating the various toxic and nontoxic pollutants. However, the risk still remains that something may go wrong in the production or clean-up process and that highly toxic pollutants will escape untreated.

In the western U.S., where low rainfall and high evaporation rates are common, process plants will likely use evaporation ponds to achieve zero discharge of water effluents.[6] In the eastern part of the U.S., zero discharge may be impractical, but effluents would be highly treated. The effect of zero discharge and high levels of treatment is concentration of effluents, which, in effect, shifts pollution from a water quality to a solid waste problem.

In practice, even with zero discharge, some pollutants would be released into surface water and groundwater through holes in liners of evaporation ponds, from fugitive sources (such as wastewater drains and pumps, and various seals and valves), and through leaching from solid waste and coal storage sites. Water runoff from coal storage piles and solid waste sites can become acidic (e.g., by forming sulfuric acid) and may contain trace amounts of numerous toxic metals. These risks were realized when two workers died in 1984 after falling into an on-site disposal pond at the Union oil shale plant. But generally these secondary forms of water pollution constitute a minor threat to the environment.[7]

A more substantial issue is water supply. For coal-based plants, water consumption may be an issue not because of insufficient water, but because of institutional obstacles and uneven distribution of water resources. Some perspective may be provided to this sometimes controversial issue by noting that plants producing coal liquids and gases would use only about one-third to one-half the amount of water that a coal-fired powerplant would consume.[8] Note also that the cost of procuring water will be a small fraction of total fuel production cost, typically less than 1 percent on an annualized basis, even in the most extreme situations.[9]

A gasification facility will typically use about 1.9 to 3.4 barrels of water for every barrel of fuel produced depending on process technology, plant design, and site conditions.[10] Lurgi gasifiers, for instance, may use twice as much water as other gasifiers, while "dry" cooling techniques (which use air and are energy intensive and expensive) use much less water than "wet" cooling (which uses either cooling towers or evaporation ponds). A direct liquefaction coal plant would typically use somewhat less water, although that depends on various factors such as type of coal and cooling facilities used (see table 55).[11] In any case, a typical 50,000-barrel-per-day coal fuels plant would use about the same amount of water as a city with a population of 30,000. Note, however, that only about 5 to 10 percent of all fresh water in the U.S. goes to household use. Nationally, about 70 percent is used in agriculture; in the western part of the country where irrigation is common, the percentage is even higher. Thus, relative to other consumers, coal-based synthetic fuel plants are not large water users.

Water supply problems will be a greater issue in arid areas of the western U.S. than elsewhere, but even there the problems would not be due to physical availability or cost. For instance, a U.S. government study cited earlier found that of forty-one counties with large coal reserves that would not be constrained by air quality problems (i.e., by PSD and nonattainment rules), *all* had sufficient excess water resources to support a 100,000 b/d production facility. Even if water were somewhat scarce and these large plants could afford to pay much more for water, water costs still would be a tiny proportion of total cost[12] (or, alternatively, water efficiency could be readily improved).

Water problems will be much more severe for western oil shale

TABLE 55 WATER REQUIREMENTS FOR FUEL PRODUCTION, BASED ON 50,000 OIL-EQUIVALENT BARRELS PER DAY

Resource and Conversion Process	Water Consumption (Acre-Feet Per Year)
Oil Shale	
Surface processes	3,200-12,000
In-situ process	4,200-12,000
Coal	
Gasification	4,500- 5,800
Gasification	6,000-19,000
Gasification	7,200- 9,000
Indirect Liquefaction	5,350-16,150
Indirect Liquefaction	9,900-10,800
Direct Liquefaction	3,500-11,000
Direct Liquefaction	2,200- 7,000
Direct Liquefaction	5,000- 7,000
Oil Sand/Heavy Oil	
Mining-Extraction	17,500
In-situ	6,100- 8,500

SOURCE: U.S. Synthetic Fuel Corporation, Comprehensive Strategy Report, Appendices (Washington, D.C., 1985), p. D-13. Estimates are based on reports by TRW, U.S. Geological Service, and University of Denver Research Institute and Department of Civil Engineering.

plants. Oil shale production generally requires somewhat more water than the production of coal fuels. A more critical factor, however, is the location of rich oil shale reserves in the arid Colorado River Basin where most of the water is already legally allocated.[13] Rights to that water have been more actively litigated than have water rights anywhere else in the world. Claims have been made by Indian tribes, the federal government, Mexico (where the water eventually empties) and others, which, when combined, exceed the total flow in the basin. Currently about 80 percent of the existing rights are for agricultural use.[14] Most studies agree that water supplies are physically available for a sizable oil shale industry, perhaps as much as 2 million b/d or more.[15] The problem for project sponsors will be to gain access to that water.

To gain this access, oil shale plants would purchase rights from others, withdraw water from underground aquifers, or transport water from other river basins. In the Colorado River basin all of these options would be problematic. Transporting water from the nearest water-rich basin, the Columbia River, would be expensive, requiring a massive network of pipelines, dams, and storage areas. Even if this were economically attractive there would be strong political opposition. Concerned about such a possibility, residents in the Columbia River area successfully lobbied for federal laws that ban until 1988 even the study of water diversions from the Columbia River Basin. Underground aquifers are also not a viable supplier since major aquifers are already being depleted.[16] The purchase of existing irrigation rights is feasible, but the effect would be to raise the price and shrink the supply of water for farmers. This prospect would elicit strong opposition from powerful agricultural interests in those states.

In summary, the water supply problem is a site-specific problem, mostly based on institutional and legal conflicts. The severity of these conflicts would be influenced by the level of public support being given to energy production. Where water supply is limited, the question is who gets to use it and who pays the economic, social, and environmental costs.

SOLID WASTE

Solid waste from fuel production would probably be a greater threat to the environment than water or air pollution. These wastes degrade the environment by occupying large areas of land and by introducing toxic materials that leach into groundwater and surface water supplies. Generally, conversion of biomass and coal into transportation fuels generates large amounts of solid waste, but much less than oil shale production.

In ethanol bioconversion, solid waste is generated as sludge from wastewater, ash remaining after combustion of coal or other process heat sources, residue from air pollution treatment (i.e., from scrubbers), and dust from feedstock preparation. These quantities are relatively small and contain no toxic materials (see table 56). Hydrolysis pretreatment generates much larger quantities of soil waste since the lignin constituents of lignicellulose remain as a residue.

In thermochemical conversion of biomass and coal, the main source of solid waste is the ash that remains after the organic portion of the feedstock is converted to gaseous products. Other sources of solid waste from coal fuels production are sludges from air and wastewater treatment (e.g., from scrubbers, evaporation ponds), spent catalysts, and

TABLE 56 SOLID WASTE FROM ENERGY PRODUCTION, KILOGRAMS PER MILLION BTU

Conversion process	Ash	Other[a]	Total (excluding coal prep. refuse)	Coal Preparation Refuse
Corn fermentation (ethanol)	7.4-22	NA	14-38	--
Wood gasification (methanol)	2.5-6	1.5	4-9	--
Lignite gasification (methanol)	8.0-9.6	NA	9-13	0-86[b]
Coal gasification (methanol)	5.8-6.7	NA	7-11	0-86[b]
Cellulose hydrolysis (ethanol)	--	25-257	25-257	--
Coal-fired powerplant (electricity)[c]	4.8-24	5.0-15.4	10-40	0-86[b]
Direct coal liquefaction (liquids)	8-15	NA	9-25	0-86[b]
Oil shale pyrolysis (liquids)[d]	0	145-390	145-390	--

SOURCES: U.S. Department of Energy, Synthetic Fuels and the Environment (Washington, D.C.: DOE, 1980, DOE/EV-0087); Argonne National Laboratory, Environmental and Economic Evaluation of Energy Recovery from Agricultural and Forestry Residues (Argonne, Ill.: 1979, ANL/EES-TM-58); SRI International, Alcohol Fuels Production Technologies and Economics (Prepared for U.S. DOE) (SRI, Menlo Park, Calif., 1978, EJ-78-C-01-6665); SRI, Mission Analysis for the Federal Fuels from Biomass Program (Springfield, Va.: NTIS, 1978).

[a]Other = scrubber waste, wastewater sludge, dust, spent shale. Sulfur and other marketable products not included.

[b]Lower value refers to case of thick coal seams. Not included in total column.

[c]Range in values represents two cases based on equivalent quantity of coal and equivalent quantity of output energy, as well as sensitivity to other factors, such as ash content of coal.

[d]Lower value is for modified in-situ process and upper value for surface retort processes.

dust from feedstock handling. Lower-rank coals such as lignite will generate larger amounts of solid waste in the form of ash. Sulfur is also produced in large quantities from coal gasification, but is sold as a coproduct.

The retorting of oil shale generates much larger amounts of solid waste than either coal or biomass fuel production. The volume of spent shale is 20 to 40 percent greater than that of the raw shale;[17] even if tightly compacted, spent shale still occupies more space than it did before processing. Oil shale production processes generate about five to ten times as much solid waste per unit of energy output (in the form of spent shale) as do coal conversion processes. In-situ processes, if they are developed, will enable much of the spent shale to be returned underground. The huge quantities of spent shale create an aesthetic problem. Since the physical and chemical properties of processed shale make it poorly suited to supporting vegetation, it would probably be necessary to provide a soil cover in order to improve the aesthetics of the large mounds of waste.[18] An even more important problem is that shale (and coal) wastes contain many trace metals that come from the feedstock and spent catalysts. These toxic substances can leach or percolate into water supplies. Various efforts are being made to design a technique for creating an impermeable layer of spent shale to prevent toxic substances from entering groundwater and surface water, but none is yet considered fully effective.[19]

The large quantities of waste from all these energy production activities pose a significant handling and disposal problem, especially for plants located in or near populated areas. Smaller biomass plants have the advantage of generating small quantities of solid waste at any one site and that little or none of the solid waste would be toxic. In contrast, oil shale wastes will be voluminous, concentrated at a few plants, and contain toxic substances.

As with other environmental factors, the influence of solid waste factors on decisions to introduce nonpetroleum fuels depends on the perception of those effects, and not on some readily quantifiable technical analysis. In support of this contention, note that solid wastes from some processes, including Lurgi gasifiers and direct liquefaction, could be declared "hazardous" (under the 1976 Resource Conservation and Recovery Act). If this determination were to be made, those process plants would be subject to very stringent disposal and monitoring requirements. The effect would be to limit feasible plant locations and to increase costs. If the "hazardous" material designation is applied, it is estimated that solid waste disposal and treatment costs for a coal-to-methanol plant using Lurgi gasifiers could increase from as little as $1 per barrel of methanol to as much as $20 per barrel.[20] Environmental perceptions and values are influenced but certainly not determined by scientific analysis, since "toxic" and "hazardous" are ultimately subjective terms. These environmental perceptions and values as manifested in government regulations and restrictions strongly influence the future of fuel production investments.

TERRESTRIAL IMPACTS

The most important terrestrial concerns are surface mining, the quantity of land used, and soil erosion.

MINING IMPACTS

The extraction of mineral resources could have major environmental impacts in some areas. The most severe effects are from surface mining in arid regions because so much land is disturbed and because it is so difficult to reclaim the stripped soil and spent shale. In the U.S. these would occur in the arid West where oil shale and low-sulfur coal resources are located. Other important mining impacts are acid water drainage from coal mines, alkaline runoff from oil shale mines, and black lung disease and a high injury and fatality rate for workers in underground coal mines.

The most prominent and difficult mining problem associated with expanded coal and oil shale production is likely to be land reclamation in the arid areas of the western U.S. Oil shale presents a special problem because of the vast amount of crushed rock that must be disposed of. A one million barrel per day mining operation using shale with an oil content of 25 gallons per ton will leave almost two million tons per day of crushed shale.[21] Coal operations generate about one-seventh as much refuse. The cost of restoring the land is not necessarily great—the cost of reclaiming western surface mines is $1,000 to $5,000 per acre or only about $0.02 per million Btu and somewhat more for Appalachian coal mines[22]—but the question is how long it will take for the land to recover. In the case of oil shale, the volume of material is so great that some proposals have even suggested filling canyons with the spent shale. The disposal issue alone could render a large oil shale industry unacceptable.

LAND USE

The area of land diverted to energy use can be very large for some of the options. As shown in table 57, biomass and photovaltaic hydrogen both require vast areas of land.

TABLE 57 LAND AREAS USED IN ONE YEAR TO PRODUCE ENERGY EQUIVALENT TO 50,000 B/D OF PETROLEUM[a]

Feedstock and Output	Acres
Biomass	
Wood-to-methanol[b]	1,277,000
Corn-to-ethanol[c]	3,066,000
Coal[d]	3,000
Oil Shale[e]	13,000
Hydrogen[f]	
Photovoltaics-water electrolysis	99,000

SOURCE: S. Schurr et al., Energy in America's Future: The Choice Before Us (Baltimore and London: Johns Hopkins University Press, 1979), p. 371; chapter 6 of this book; and M. A. Deluchi, D. Sperling, and, R. A. Johnston, Comparative Analysis of Transportation Fuels (Berkeley, Calif.: Institute of Transportation Studies, 1987).

[a]Land sites for coal and oil shale are normalized to a 30 year lifetime and then prorated annually. Thus the total land consumed over the life of a 50,000 b/d plant is 30 times the area indicated.

[b]5 tons/acre, 120 gallons/ton. Estimates of land areas in Schurr, Energy in America's Future, are more than an order of magnitude less than those estimated here.

[c]100 bushels/acre, 2.5 gallons/bushel.

[d]Average for surface and underground mining. About 80 to 90 percent of land is for mines. Land areas are similar for gasification and liquefaction processes.

[e]Above-ground retorts. In situ mining would use only a fraction as much land area.

[f]Based on estimates that a hydrogen plant located in North Africa would require 1055 square kilometers of land to produce 2.94×10^8 gigajoules of hydrogen energy. See C. Voight, "Material and Energy Requirements of Solar Hydrogen Plants," International Journal of Hydrogen Energy 9, 6 (1984): 491-500.

Diverting this amount of land to energy production is not necessarily a problem as long as the full costs of using the land are included in land use decisions. Using large amounts of land for biomass energy production in the U.S. would not be particularly undesirable except, as indicated in the next section, when it causes net soil losses from erosion. In other countries with limited agricultural land and deforestation problems, diverting land to biomass energy production may not be desirable.

The diversion of land to photovoltaic solar is not likely to have undesirable consequences. The best sites are in arid, sunny areas, where most of the land is undeveloped and unsuitable for farming. In any case, land devoted to photovoltaic electricity production could be easily converted back to other uses. As discussed in the previous section, the diversion of land to coal and oil shale mines is more problematic because of the long-term impacts on land quality and vegetation.

SOIL EROSION

Perhaps the most important attribute of biomass fuel options is their long-term sustainability. But this sustainability is threatened by soil erosion. Unless carefully monitored and controlled, more intensive use of lands, in this case for energy production, will accelerate the loss of topsoil, thereby reducing long-term productivity. Soil erosion also leads to sediment loading and pollution of rivers and lakes. Erosion of both farm and forest lands is the most serious environmental impact posed by increased biomass energy production.

Soil erosion is already a serious problem in the U.S. The U.S. Soil Conservation Service (SCS) estimates that national sheet and rill erosion (erosion due to rainfall but not wind) from all crop land is 4.77 tons per acre-year (about eleven metric tons per hectare).[23] Others estimate that total erosion, including wind erosion, may be as great as nine tons per acre per year.[24] The 4.77 figure above clearly understates the soil loss problem not only because it ignores wind erosion, which is the dominant factor in some areas, but because it camouflages wide variations. For example, Missouri and Iowa, major agricultural areas, have average sheet and rill erosion rates of 11.38 and 9.91 tons per acre-year, respectively, while Texas crop land has a total erosion rate (including wind) of about eighteen tons per acre-year. The U.S. Environmental Protection Agency estimates that each year eight tons or more of sediments per acre are lost on 20 percent of the nation's crop land, three to eight tons per acre on 50 percent, and less than three tons on the remaining 30 percent.[25]

The SCS claims that good soils could sustain long-term productivity with erosion rates of five tons per acre-year.[26] This estimate is not universally accepted, however; some claim that soil formation rates are much lower, as little as 1.5 tons per acre-year.[27] In any case, much of U.S. crop land is clearly experiencing net soil loss.

The effect of soil erosion on biomass production is not easily generalized. The effect on productivity depends on plant characteristics, soil nutrients, soil structure, topsoil depth, drainage, temperature, and moisture. SCS estimates that for each inch of topsoil lost from a base of thirteen inches or less, corn yields are reduced by an average of about 4 percent, and wheat yields by 5 percent.[28]

It takes about 100 to 1,000 years to generate one inch of topsoil from bedrock. Although this process of rebuilding topsoil could be hastened by plowing crops and crop residues into the soil, the process is still very slow. While typical soil losses will not lead to imminent ruin of crop lands—soil loss of five tons per acre is equivalent to about 0.03 inches—the ultimate effect of continued soil loss is devastating. Since the advent of agriculture in the U.S., it is estimated that about one-half the original topsoil has been lost from one-third of the nation's crop land.[29]

Opinions differ about the environmental effects of bringing additional acreage into production to increase biomass supply. From its 1977 inventory, the SCS concludes that the potential for expanding cropland production is so great that expanded biomass production would not create serious soil erosion problems.[30] An analysis of SCS inventory data indicates that existing prime crop land could be expanded by about 20 percent if new high-potential land (e.g., pasture, range, and forest) were brought into production, and by an additional 45 percent if medium-potential land were also cultivated for agricultural use. Others counter that expansion of farm lands will lead to the utilization of marginal lands that have steep slopes, less fertile soils, and less rainfall, which in turn would lead to higher erosion rates and greater use of fertilizers and pesticides, which would then wash into rivers and lakes and further contaminate them.[31]

There is no guarantee that those lands less prone to erosion, no matter how abundant, will be those selected for increased agricultural production. Convincing evidence indicates, however, that the additional land that is likely to be brought into production "is not radically different in its erosion qualities from land currently being utilized for intensive agricultural production."[32] The Office of Technology Assessment (OTA) of the U.S. Congress estimates the average erosion rate on new land put into intensive production will be only about 1.25 tons per acre-year greater (7.5 vs. 6.26) than land currently intensively cultivated.[33] But this analysis is misleading.

The important point is that soil loss is a major and continuing threat to land productivity. Regardless of whether the new land brought into production loses less soil than existing crop land, if the effect is a net soil loss, then the soil loss problem is being exacerbated. Continuing a bad practice is not an acceptable rationale for increased biomass production. The fact is that U.S. farms are following a dangerous path in allowing a steady loss of topsoil. Most farmers will not suffer the consequences of inadequate soil conservation during their lifetime, but the long-term consequences will be truly devastating. It would take hundreds of years to rebuild the fertile layer of topsoil.

Another soil erosion problem could be created if residues were removed from crop land and forest land for energy production. Removal of crop and forest residues poses environmental risks that are not great if carefully monitored, but may be disastrous in extreme cases.

Removal of crop residues from land has the negative effects of exposing soil to water and wind erosion and removing organic matter that otherwise would replace nutrients and eroded soil. On the other hand, *little or no advantage is gained by leaving all residues on crop lands*; the quantity that may be safely removed depends on plant physiology and root systems, site-specific considerations of soil and climatic conditions, and soil management practices. An analysis of residues of nine leading crops indicates that those residues contain about 10 percent of the phosphorus, 80 percent of the potassium, and 40 percent of the nitrogen currently applied in synthetic fertilizers to those nine crops.[34] If residues are removed, then these nutrients would most likely have to be replaced by additional quantities of synthetic fertilizers. Some portion of residues should therefore be left on the ground. Also, some residues should be left to provide sufficient protection against wind and water erosion. One group of researchers claims that about 58 percent of residues in the fertile corn belt region, 40 percent in Virginia, Georgia, and the Carolinas, and 10 percent in Mississippi and Alabama could be removed without adverse environmental impacts.[35] In order to protect against wind erosion, few or no residues should be removed from arid and windy areas such as the Great Plains.

Erosion of forest lands is also a potential problem. Forests typically lose less than one-tenth of a ton of soil per acre-year.[36] This slow rate may be deceiving, however, in that erosion during and after logging is much higher. As much as hundreds of tons per acre per year are reportedly lost on steep slopes in high rainfall areas because of poorly managed clearcuts (i.e., forest land that is completely cleared).[37] Soil loss of about seven to eight tons per acre-year has been estimated as the average value for recently harvested forests, although variation around this mean is large.[38] These losses generally occur at distant time intervals (when timber is harvested), but they represent serious hazards because forest land typically has much thinner layers of topsoil than crop land and is therefore especially vulnerable to increased biomass production.

The major cause of erosion in forests, however, is not the actual cutting of trees.[39] Vegetation usually regenerates quickly and reestablishes a protective cover on the land, thereby preventing surface erosion, except in areas where other aspects of the logging operation have damaged the soil. Erosion problems in forests are created primarily by the construction and use of roads in which the operation of equipment compacts or exposes the soil. This compaction caused by the operation of heavy machinery reduces the porosity and water-holding capacity of soil, which encourages erosion and restricts the growth of vegetation that eventually would help reduce erosion. Trails and roads typically cover up to 20 percent of the harvest area. The greatest erosion problems are created when roads are cut into erosive soils or unstable slopes on steep terrain.

Clearly, accelerated soil loss is threatened by expanded forest production, whether by gathering currently unused woody biomass from forests or by using short rotation techniques. More intensive management of forests would accelerate soil erosion as a result of more extensive and frequent use of heavy equipment on forest soils, greater disturbance of soil, and greater use of roads and trails. The practice of removing whole trees would provide more biomass but weaken the soil structure. The gathering of currently unused logging residues, surplus growth, and dead trees, and the use of short rotation practices would increase the use of heavy equipment and leave less time for recovery between harvests. The magnitude of soil erosion from these activities has not been quantified, in large part because it would vary greatly from one situation to another.

The special case of residue removal from forests, however, is not as environmentally negative as the general case of energy production in forests. In fact, it may not be environmentally harmful. More intensive logging efforts to remove residues will further disturb and compact the forest floor and will deplete nutrients from forest soils (though these effects are not well understood). On the positive side, though, residue removal serves several ecological and land management functions: it improves the forest aesthetics, allows revegetation, and removes a habitat for disease and pest organisms. The net environmental effect of residue removal from forests is unclear; it certainly is not highly negative.

TABLE 58 SOIL EROSION IN THE U.S.

Land Use	Typical Erosion Rates (tons/acre-year)	Land Area (million acres)
Cropland	4-6	412
Rangeland	1-5	408
Pastureland	0-4	367
Forestland	0-1	133

SOURCES: U.S. Congress, OTA, Energy from Biological Processes, Vol. II (Washington, D.C.: 1980); and Council on Environmental Quality, "The President's Environmental Program," 1979, cited in Aerospace Corp, Environmental Control Perspective for Ethanol Production from Biomass (Germantown, Md.: 1980), p. 20.

As shown in table 58, much less erosion occurs on pasture lands and grasslands than on crop lands; the soil is covered and disturbed less frequently and less severely than soil on crop land. The effect of increased biomass production on these lands depends, as suggested earlier, on the characteristics of the land. Some areas have deep layers of topsoil, good drainage, and gentle slopes and are considered highly attractive and well suited to intensive farming. Other land is less attractive—"medium or low potential" in the language of the SCS—and its utilization would result in lower yields and higher erosion rates.

The general problem of soil erosion encompasses problems of sedimentation and pollution. Water runoff carries soil and deposits it into bodies of water. The sedimentation problem is one of eroded soil filling up reservoirs, silting harbors and navigation channels, obstructing drainage and irrigation ditches, and raising the cost of water treatment, water distribution, and hydroelectric power generation. Because soil carries phosphorus, pesticides, heavy metals, and bacteria, the deposition of soil is also a pollution problem. The soil sediments cause algal blooms, reduce the recreational value of streams and lakes, and introduce heavy metals and poisonous pesticides into the water system. Sediments washed into the U.S. waterways cause damage estimated at $1 billion annually.[40] Although the sedimentation and water quality effects of increased biomass energy production are important, they are likely to be difficult to monitor and control because they are site-specific, highly variable, and not easily quantified.

The question of soil loss is crucial. It would be unwise to make a commitment to biomass fuels if that were in any way to accelerate the net loss of soil. It does not necessarily follow, however, that increased biomass production would automatically accelerate soil erosion, sedimentation, and water pollution. Soil erosion on farm and forest land could be prevented by judiciously restricting intensive land utilization to more resilient areas, using appropriate soil conservation practices and avoiding steep slopes and fragile soils. Conventional soil conservation practices such as contour farming, terracing, reduced fall plowing (to leave a soil cover), and the use of cover crops, combined with innovative practices such as reduced or no tillage, would greatly reduce soil erosion—to levels that enable permanent sustained production on those lands. The problem is that land managers have insufficient incentive to pursue these soil conservation practices. The cost to farmers in time and money associated with implementing an active conservation effort requires deviation from practices that proved successful in the past. And the benefits generally will accrue in later decades, probably to future generations. Moreover, in the case of reduced water contamination and sedimentation, the benefits accrue to others. Meanwhile, most farmers are caught in a precarious cycle in which they borrow each spring huge amounts of capital to be repaid after a hoped-for successful harvest in the fall. Most are unwilling to borrow more capital to finance soil conservation activities or to hurt their borrowing ability by undertaking new and possibly risky practices. Not surprisingly, the benefits of soil conservation tend to receive short shrift.

A serious plan to follow a biomass fuel path must be accompanied by a serious plan to control soil erosion. The risks of not doing so are too great. The current reality in the U.S. is that existing economic and regulatory incentives for biomass suppliers and users to protect the environment are weak.[41] The opportunity exists to install incentives and regulatory controls, but it requires a concerted commitment that currently does not exist.

SOCIOECONOMIC IMPACTS

A controversial consideration in the siting of large fuel production plants has been the ability and willingness of local communities to provide public facilities, services, and housing for the labor force that builds, operates, and provides support services for the new facilities. This impact is not on the physical environment, but on the human

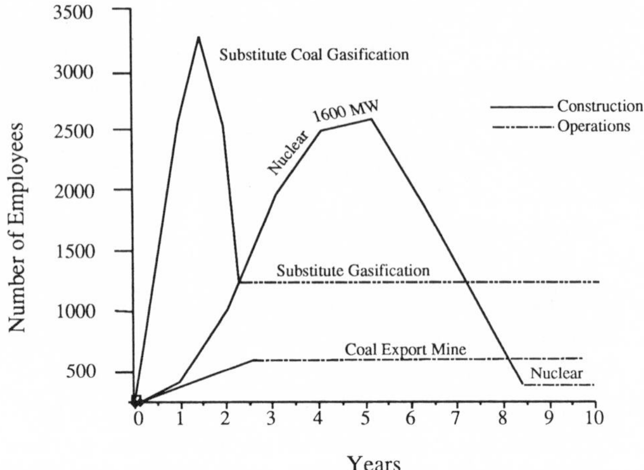

Figure 34. Employment Patterns for Selected Energy Projects
Source: U.S. Department of Housing and Urban Development, "Rapid Growth from Energy Projects," reproduced in statement of R.H. Ihara (Kentucky Department of Energy) before U.S. Congress, House of Rep., Subcom. on Energy Development and Applications, 28 April 1982, pp. 51–61.

environment. The impact would be greatest for large projects; thus impacts from biomass, hydrogen, and natural gas plants would tend to be less severe than those from coal and oil shale plants.

A highly impacted community would typically be located in rural, isolated areas of western U.S., and would lack the labor pool required to fill the new jobs (see fig. 34). This community would experience rapid and substantial immigration followed by characteristic boomtown effects: social disruptions, inflation, and shortfalls in provision of public facilities, services, and housing. In a previously cited analysis by the U.S. Department of Energy, it was determined that of forty-one counties with sufficient coal to support a large fuel production plant and not constrained by air quality considerations, thirty-one would be subject to boomtown effects.[42] In that study a county was considered sensitive to boomtown problems if local urban settlements housed fewer than 50,000 people.

The following factors distinguish growth caused by large fuel production facilities in rural areas from more easily handled "normal" growth and urbanization.

Uncertainty: The unpredictable nature of future development patterns increases the risk of initial investment in community facilities. Raising public funding necessary to construct new communities—including the provision of water, sewer, road improvements, schools, and other services—for a still hypothetical and contingent new population is difficult for small existing communities.

Pace of development: Many of these small rural communities have often experienced stable or declining populations for many years preceding development and are thus ill-prepared for the sudden surge of growth.

Temporary and fluctuating nature of the growth: Significant peaks and valleys of population influx generally occur during construction; the permanent operating force is often much smaller than the construction force. Communities must deal in a flexible manner with a temporary bulge in construction work force in order to avoid substantial excess growth in services during the construction phase (see fig. 34).

Front-end financing: Significant expansion and new construction of services and facilities generally must occur during the first few years of project construction, before a sufficient tax base has been established.

Jurisdictional problems: The new energy facility, with its increased tax base, may be located in different political jurisdictions than where the population growth occurs.

These factors make local funding (for instance by issuing bonds) an impractical solution; they create the need for federal, state, local, and industry cooperation.[43]

The ill effects and therefore opposition to new projects caused by rapid growth are not likely to be as critical as once thought. Since the energy boomtown debacles of the early 1970s, states and local governments have required corporations to participate in the planning, financing, and construction of urban facilities and housing. For instance, Arco Coal spent $17 million to build a small town near one of its projects,[44] while Union Oil Company spent $63 million through 1983 for housing, roads, schools, parks, and fire, police, health, water, and sewer services in an area surrounding their Colorado oil shale project.[45] State governments, especially those in the western U.S., are also imposing large severance taxes (taxes for severing a finite resource) of up to 30 percent on coal, oil, and gas for the purposes of financing new public facilities; Montana had a 30 percent tax on coal production through 1985. As communities and corporations become more experienced with boomtown growth, the negative effects will be mitigated, although never eliminated.

TABLE 59 OCCUPATIONAL HAZARDS OF ENERGY PRODUCTION

Production Activity	Fatalities: Deaths per 10^{15} Btu	Fatalities: Deaths per Thousand Worker Years	Lost Work Days: Thousand Worker Days Lost per 10^{15} Btu	Lost Work Days: Worker Days Lost per Thousand Worker Years
Conversion of biomass wastes to fuel[a]	1.3-15	0.23-0.25	3.2-41	570-670
Wood production and conversion[b]	10-28	0.28-0.29	36-91	910-1100
Liquid fuels from coal[c]	17-25	0.83-0.85	28-42	1400
Oil extraction and refining	1.1-1.2	0.31-0.32	1.8-2.1	510-550

SOURCE: Gregg Morris, "Integrated Assessment Issues Raised by the Environmental Effects of Biomass Energy Systems," Energy and Resources Group, University of California, Berkeley, 1980, ERG-WP-80-6.

[a]Based on assumed facility lifetimes of thirty years; includes both bioconversion and thermochemical conversion facilities.

[b]Based on intensive silviculture cultivation.

[c]Includes coal mining.

OCCUPATIONAL AND SAFETY HAZARDS

Fuel production and handling pose health hazards to workers and nearby residents,[46] and fuel transport poses hazards to the general public and to animal and plant life. These hazards and risks tend to be greatest for mineral fuels and smallest for renewable fuels. Liquid and gaseous fuels produced from coal, oil shale, and biomass present greater hazards to workers than petroleum production; hazards from natural gas production are similar to those of petroleum, and solar hydrogen presents the least hazard. Pipeline transport is less hazardous than truck transport, and therefore gaseous fuels (natural gas and hydrogen) are safer to transport than liquid fuels.

The directly measurable hazards facing workers at the energy-production end of fuel activities are quantified in table 59. While the figures are subject to considerable uncertainty and variation due to questionable reliability of data and to methodological difficulties, they suggest that the greatest dangers are associated with coal-related production and processing; most of the risk attributed to coal-based fuels is associated with coal mining. The safest activity of those listed is petroleum (and natural gas) production because it is highly automated. There are no data on the hazards of large solar-hydrogen plants, but it is reasonable to assume that they will not be much more hazardous than solar-photovoltaic facilities, which are considered relatively safe.

ENERGY PRODUCTION ACTIVITIES

The technologically simple processes of fermentation and distillation and solar-hydrogen production pose little health risk to workers. Although conclusive evidence is not available, workers who are exposed to process streams and fugitive emissions at gasification, direct liquefaction, and pyrolysis plants will tend to be subject to much greater risks. While workers are protected in the U.S. by federal (and sometimes state) occupational health and safety standards, the risk of accidental exposure to the many dangerous substances and routine exposure to fugitive emissions cannot be eliminated. The actual danger to workers posed by carcinogens and chemicals is difficult to determine because of limited experience with synthetic gases and liquids and, in the case of cancer, because of the long latency period of fifteen to twenty years or so. Although data are not available for the Sasol plants in South Africa because health records of workers were not kept, epidemiological studies of 350 workers conducted at the Union Carbide direct liquefaction pilot plant in West Virginia during the 1950s found that workers suffered from sixteen to thirty-seven times the expected rate of skin cancer. The sample was too small, however, for the results to be statistically significant.[47]

Although there is little data on worker safety, it is well known that the production of liquids and gases from coal and oil shale creates a wide range of known carcinogens and mutagens (which alter chromosomes). The carcinogenic threat posed by coal gases has been known since the 1800s and was medically detected in the case of oil shale liquids in 1876 in Scotland.[48] Tests on animals have identified many carcinogens produced during coal and oil shale conversion processes, especially polycyclic hydrocarbons and trace metals. Workers also face the acute physical hazards of working with chemicals. The relatively short history of large modern coal and oil shale conversion plants, and uncertainty over the effectiveness of safety standards to curtail accidents and fugitive emissions, make definitive determinations of worker safety impossible. It is clear, however, that the risk is great.

Nearby residents are also at risk. Skin contamination may result from airborne dusts, vapors, fumes, and mists that are released during the manufacturing process. Again, there is no definitive evidence of the magnitude of this danger.

TRANSPORTATION AND DISTRIBUTION ACTIVITIES

The transportation of liquids poses safety and health risks to the terrestrial, marine, and aquatic environments and to the general public. These risks are due to accidents and spills. The safest and least threatening mode of transportion is pipelines; pipelines have the fewest spills, and the spills that do occur tend to be relatively small, since pipelines shut down automatically when decreases in pressure are detected in the line. Trucks have the most accidents, but spills are small because tank capacity is small. Barges and marine tankers have accident rates that lie between those of trucks and pipelines, but because of their huge size (some carry over a million barrels), a single barge or tanker spill can have a devastating effect.[49] As an indicator of relative safety, fatalities per ton-mile can be used. Relatively old but still relevant data indicate that bulk movement of hazardous materials by pipeline caused about 3 percent as many fatalities as marine shipping on a ton-mile basis, 0.5 percent as many as rail transport, and only 0.1 percent as many as truck transport.[50]

The relative accident rates of these modes suggest that biomass fuel transportation might pose the greatest safety hazard to people, since trucks and railways, the dominant means of transporting biomass fuels, are in greater and more regular contact with humans than other modes. However, the difference would not be great, since all liquid transportation fuels would be delivered to retail fuel outlets by truck.

The transport of hydrogen and natural gas is and will be relatively safe because it relies almost exclusively on pipelines, the safest and most reliable mode of transportation. Most CNG and compressed hydrogen fuel outlets would receive fuel by pipeline, as would most retailers of liquefied natural gas or hydrogen; liquefaction would generally take place on site. In general, then, transport of methane and hydrogen would be relatively safe because of their reliance on pipelines.

There has been some public concern in the U.S., however, that LNG terminals and ships are a serious threat to public safety. The danger is that if an LNG container is punctured (or leaks), the very cold gas will not disperse as quickly as CNG vapors (at ambient temperatures) and,

TABLE 60 COMPARATIVE TOXICITY RATINGS

	Eye Contact	Inhalation	Skin Penetration	Skin Irritation	Ingestion
Gasoline	(2)	(3)	(3)	(1)	(2)
Methanol	2	2	2	1	1
Ethanol	2	1	1	1	1
Methane (CNG)	0	3*	0	0	NA

SOURCE: J. V. Steare, ed., CRC Handbook of Laboratory Safety, 2d ed. (Boca Raton, Fla.: 1971).

1 = mild, 5 = extreme toxicity ( ) = varies depending on composition.

*Dangerous only to the extent that it displaces air and acts as an asphyxiant.

unlike CNG vapors, vapor clouds will persist near ground level. For cases in which huge amounts of gas are present (e.g., in LNG ships), these LNG clouds if ignited would have tremendous explosive power, capable of devastating large areas by explosion and fire. However, the likelihood of such large leaks and the possibility that a LNG vapor cloud would be ignited and cause damage are minimal. Large LNG terminals have been operating around the world for several decades without major mishaps. Published assessments indicate that LNG risks are negligible by almost any standard.[51]

END-USE ACTIVITIES

All fuels pose significant health and safety hazards. Gasoline and other petroleum products and petroleum-like fuels are considered more toxic than ethanol, methanol, and methane (see table 60),[52] and also contain proven and suspected carcinogens and mutagens that can be inhaled or readily absorbed by the skin.[53] Alcohols, on the other hand, pose safety hazards because they are more easily ignited than gasoline in closed containers (due to broader flammability limits), although because of their higher flash point (the minimum temperature at which the vapor/air mixture ignites), they are less easily ignited in spills and open storage containers.[54] Alcohol combustion also is a safety concern because of the invisibility of flames from alcohol combustion; blending small amounts of gasoline with the alcohol mitigates that problem, however. Ford Motor Co. recommends that methanol fuels contain 15 percent gasoline (by volume) to assure that flames will be visible in case of accidents.

Natural gas fuels would have fewer adverse health impacts than other fuels.[55] Natural gas poses relatively little risk to health because, as indicated in table 60, methane is not toxic; the only significant health risk is that if present in high concentrations, methane displaces air, can be ignited, and can asphyxiate a person. The overall safety risks of natural gas fuels are similar to or less than those of gasoline.

The greatest safety risk is that in a confined space leaking gas could concentrate sufficiently to be flammable. Methane diffuses more quickly, burns colder, and ignites at a higher temperature than gasoline or methanol. Therefore, if methane leaks in an unconfined space such as at an outdoor refueling pump, then the gas will disperse quickly to a density below its lower flammability limit and therefore not create any safety problem. Gasoline vapors, in contrast, are a much greater fire hazard in an unconfined area because they are heavier than air and ignite at a much lower temperature than methane.

In a confined space such as a garage, methane leaks may pose more of a hazard than gasoline leaks because the gas spreads throughout the space and may become concentrated enough to be flammable. When methane displaces between 5 and 15 percent of the air it can be ignited. A 5 percent level of concentration would be achieved if, for instance, about one-third of all the gas in the two tanks of a CNG vehicle (i.e., thirty-two pounds of fuel) leaked into a closed garage measuring twenty feet in length and width and ten feet in height. If all the gas leaked out, then the gas would be too dense (above 15 percent concentration) to ignite.

The use of LNG presents two additional hazards. First, contact with the ultracold fuel can cause severe frostbite. Second, the fuel in the tank vaporizes as it warms, and the accumulating vapors, called "boil-off" gases, must be vented or burned periodically. An LNG vehicle would either have to be parked in a well-ventilated or outdoor area, driven periodically (every two weeks or so), or the gas would have to be recaptured or rendered harmless. The most practical solution would probably be the installation of a pilot light that automatically burns vented boil-off gases.

The probability of gas leaks and ruptured tanks is very low. To guard against leaks, the New Zealand and Canadian governments require periodic testing of all tanks (every few years). Because tanks are required to survive severe test crashes, tank rupturing even in collisions and intense fires is highly unlikely. In short, natural gas presents modest risks to vehicle users.

Many of the health and safety risks associated with hydrogen are similar to those for natural gas.[56] The most important difference between natural gas and hydrogen, whether in compressed or liquefied form, is that hydrogen is much lighter. Thus hydrogen disperses much more quickly, creating less danger of fire or explosion in unconfined areas. However, hydrogen ignites within a wider range of concentrations than does natural gas.

In summary, the least threatening fuels to occupational and public health and safety, taking into consideration fuel production, distribution, and end-use activities, tend to be hydrogen and compressed natural gas, especially when transported by pipeline and not in liquid form. Biomass fuels pose somewhat more risk. The processing of coal and oil shale and the handling of coal and oil shale fuels pose the greatest overall occupational, health, and safety hazards.

CONCLUSION

The overwhelming conclusion of this chapter is that substantial negative environmental impacts are associated with fuels produced from coal, oil shale, and biomass. The most environmentally attractive options are hydrogen produced with solar energy and fuels using natural gas as the feedstock.

Sorting and ranking the effects of biomass, coal, and oil shale fuels is difficult because the effects of biomass fuel options are dispersed and tend to be individually small, while the effects of coal and oil shale plants are concentrated and more prominent. One could argue convincingly that some kinds of concentrated effects are more easily handled than dispersed impacts, that concentrated pollution sources are easily identified and monitored and easily controlled with advanced and highly efficient pollution control technology, while the small, dispersed sources associated with biomass fuel are not easily monitored and not suited to advanced control technologies.[57] According to this view, biomass sources are the greater threat to environmental degradation. One might argue, however, that the dispersed nature of environmental stress allows the resilient environment to absorb the insults without serious degradation. Furthermore, it could be argued that large polluters have sufficient economic power to stymie government efforts to promulgate and enforce environmental control rules.

A rigorous analysis does not resolve these conflicting views. Decisions regarding required controls are to a large extent made in the political arena, where technical analyses are only one consideration. Other factors that influence what controls will be required are popular perceptions and sentiments (which are determined in part by the treatment of environmental issues in the mass media), the power and savvy of interest groups, the organization, resources, and traditions of governmental agencies, and party politics. Nevertheless, the analysis of these two chapters on environmental impacts supports several observations.

First, oil shale and coal liquefaction and gasification plants are likely to generate the most heated environmental debate. The facilities are physically large and expensive and many of the environmental impacts of these projects are large, immediate, and clearly undesirable: the more prominent impacts include the destruction of land; the creation and release of toxic substances; intrusive and health-threatening air pollutants; the generation of huge amounts of solid waste and, in the case of oil shale, waste rock; aesthetic degradation of western wilderness areas; and hazards of coal mining. Also, the plants are built and run by large energy companies, which are in the public eye and are often viewed unfavorably. Environmental groups will closely scrutinize proposals for large energy plants and will likely oppose most if not all of them.

In contrast, biomass fuel production plants are likely to generate less environmental controversy, and thus are less likely to be stopped, stalled, or modified for environmental purposes. Biomass benefits from a perception that it is "clean" and compatible with the natural environment (see chap. 20 for examination of soft energy paths) and from its association with farmers and farming. Perhaps most important, the primary negative impact of biomass systems, soil erosion, is not an immediate or prominent impact, and is not generally recognized as a severe problem.

Technical analysis, as presented in these two chapters, will play an important role in guiding the environmental debates. It will demonstrate that several options are clearly superior—solar hydrogen, followed by CNG, LNG, and methanol from natural gas—and that all others create the potential for environmental degradation. The decision to adopt less environmentally damaging options will depend on the willingness of a society to weigh environmental damage and risk heavily in the evaluation process.

PART VII

Getting from Here to There

Energy planning and policy analysis that focuses on far-off end-state scenarios is much less useful than planning and analysis that addresses changes needed to redirect energy decisions. As indicated in chapter 2, end-state thinking does not help define societal goals and decision criteria, does not nurture and facilitate consensus formation, and is usually highly inaccurate. In later chapters directions of change are explored in terms of five alternative energy pathways. For now, transition strategies are identified that could break the powerful inertia that is overwhelming efforts to introduce those transportation fuels different from petroleum.

First, chapter 17 examines the proposition that the introduction of new fuels is impeded by the structure of fuel supply and demand activities and by the nature of the technologies used for producing and transporting fuels and manufacturing motor vehicles. This proposition is expressed in terms of a "chicken-or-egg" metaphor in which it is hypothesized that neither fuel producers nor motor vehicle manufacturers are willing to make an initial commitment to new fuels.

In chapter 18 specific strategies are explored for overcoming the "chicken-or-egg" and "not-me-first" syndromes that stymie investments in new fuels. It is shown that several widely promoted strategies lead to dead ends or follow inefficient detours. The most efficient strategy is one that targets spatially segmented end-use markets.

SEVENTEEN

The Chicken-or-Egg Stasis

An expectation of market economies is that through the ideal workings of a competitive marketplace, superior products will emerge. However, there are many cases and situations in which the superior product will not emerge unless it is assisted by government intervention. When the successful introduction of that product is dependent upon a number of independent investments, the process becomes more complex and the start-up costs even more overwhelming. This problem of multiple independent investments may be characterized in terms of the "chicken-or-egg" metaphor. This metaphor may be generalized to any situation in which it is presumed that first, in order for a new product to be widely accepted, two or more major investment activities must proceed concurrently, and second, that unless an independent but complementary investment is also made, no investment will be made. Obviously there would be a paralysis of investments because the initial commitment for one set of investments would be contingent on a commitment for the other; neither commitment would be forthcoming because of the great risk associated with making the initial commitment.

In the case of alcohol fuels, natural gas, hydrogen, and any other fuel dissimilar to gasoline or diesel, the chicken-or-egg syndrome is manifested as follows: on the fuel supply side, potential fuel producers are unwilling to make large initial investments until they are assured of a large market for their product, and on the fuel demand side, automakers are reluctant to invest in the engineering and retooling of factories until they are assured that large supplies of dissimilar nonpetroleum fuel will be available to vehicle users. According to this hypothesis, fuel suppliers condition their investment decisions on the existence of a fuel market, and automakers condition their investment decisions on the supply of fuel, with the combined result of investment paralysis; no one is willing to make the initial commitment.

The chicken-or-egg conflict has been cited by government and industry as a principal deterrent to the introduction of methanol fuel.[1] This conflict—that is, who is to make the first commitment—may be characterized as arising whenever the following conditions prevail:

1. Incompatibility between the new energy material and the existing end-use technologies, thus precluding the availability of a ready market;
2. Initially, a comparable or higher price for the new fuel than the dominant fuel in the market;
3. Long lead times from when an investment decision is made until the product reaches the marketplace;
4. Strong economies of scale in energy production, energy distribution, and the manufacture of end-use technologies;
5. Homogeneity in the end-use market, which provides less opportunity for incremental market penetration.

All of these conditions prevail to some extent for alcohol fuels made from biomass, coal, and hydrocarbons, and some of the conditions hold for compressed natural gas (CNG). The first condition applies insofar as the physical and chemical characteristics of alcohol and methane (CNG) are different enough from gasoline and diesel fuel to require the modification of end-use technologies; engines and vehicles would have to be retrofitted or replaced to accommodate alcohol or CNG.

The second condition, the problem of initially higher or similar costs, would apply to all alternative fuels for very long periods of time. Methanol from remote natural gas is likely to have similar or somewhat lower costs than gasoline, but only under exceptional circumstances (and perhaps in isolated settings) would those costs ever be significantly lower. CNG, on the other hand, could have costs as low as one-half those of gasoline, and thus this condition does not always hold for CNG.

The third condition, long lead times, especially affects producers with large plants. Lead times for designing, building, and acquiring permits may be as much as ten years. The risk of carrying the costs of a large project for so long with no certainty of a market developing is huge. Long lead times also exacerbate market uncertainty, since it is difficult to predict so far in advance how market demand will develop. These lag times would be only a few years for the smaller biomass and RNG-to-methanol plants; for CNG imported as LNG lag times would be somewhat longer (in large part because of likely opposition from those concerned with safety and environmental dangers).

The fourth condition, economies of scale, is a key factor in determining the extent to which the chicken-or-egg metaphor applies. If large economies of scale are pervasive throughout fuel production, distribution, and end-use systems, then it will be more difficult to initiate change in small steps, since those small steps will have large cost penalties—and the chicken-or-egg stasis will become more dominant. As shown earlier, large economies of scale are indeed pervasive and dominant for coal and RNG-based alcohol fuel activities, not only in fuel production but also in fuel distribution and end use. Economies of scale are large for the fuel process plants, pipelines, and motor vehicle manufacturing plants. Large size is not so desirable for biomass plants, however, because of the strong diseconomies of scale in feedstock collection. Thus biomass fuel investments are less sensitive to the need for large existing or promised markets. Unlike larger multibillion-dollar coal-to-methanol plants, they can target small market opportunities without running the frightening risk of having to find a market for huge slugs of methanol when the plant comes on line. CNG imported in the form of LNG would also be based on large economies of scale in production and distribution, but the risk is less severe because it has the advantage of a huge existing nontransportation market for the gas, and so this economy-of-scale condition does not hold for CNG.

The fifth condition, homogeneity of markets, is a major factor in creating chicken-or-egg-type conflicts in the introduction of all dissimilar nonpetroleum fuels. Although some market niches exist—e.g., in the case of methanol, at petroleum refineries with high demand for octane-boosting additives, in areas with poor air quality and stringent air quality rules, and by remote users of agricultural fuel—these market segments in total constitute only a small fraction of the retail gasoline market (see chap. 13). The homogeneity in the retail gasoline market is very real. Not only is pricing highly uniform and undifferentiated by location or end-use market, but the retail fuel market is dominated by automobiles and light-duty trucks, which are designed for, and used as, general purpose vehicles. These vehicles are expected to serve in a wide range of conditions; their use is generally not specialized to a specified trip purpose, driving range, geographic area, or time slot. The homogeneity of transportation fuel markets that use spark ignition engines is highlighted by comparison with the diverse operating conditions and end-use applications of other new energy technologies (such as those for solar and nonconventional electricity-generating technologies).

Normal business practice is to target small attractive market opportunities and then expand from there. The homogeneity of fuel markets, exacerbated by slow turnover of motor vehicles, is an especially profound barrier because it deprives business of its time-tested strategy of first targeting small attractive market niches and then expanding. Homogeneity therefore acts to discourage the penetration of prospective transportation fuel markets.

Market homogeneity (and slow vehicle turnover) is a problem that faces all new fuels, but the severity of its effect and the response from different fuel producers vary.

Natural gas fuels are least affected by the homogeneity phenomenon since they have large preexisting markets in the electric utility industry and in the industrial, commercial, and residential sectors. Small biomass alcohol producers would be more greatly affected. Their response would be to content themselves with those few small market niches that are readily penetrable; they would focus their production, distribution, and marketing strategies on agricultural users and, depending upon location, vehicle fleets and octane-short petroleum refineries, as well as perhaps certain electricity-generating combustion turbines. Large-scale producers of mineral methanol would face the most severe challenge. Because other market segments are so small and dispersed, that response would likely be to aim for the much larger retail gasoline market from the beginning. This strategy is fraught with risk. The necessity of having to target the homogeneous retail fuel market exacerbates the threat of investment paralysis for methanol producers.

CONCLUSION

In conclusion, investments in large coal-to-methanol plants confront all five conditions that characterize the chicken-or-egg stasis. Indeed, the chicken-or-egg metaphor accurately represents the barriers facing coal-to-methanol production. The metaphor also applies, though to a lesser extent, to RNG-to-methanol and biomass production investments.

It is much less accurate in describing the problems facing natural gas fuels. When and where this chicken-or-egg problem exists in the context of transition paths and how it would or could be dealt with will be addressed next in chapter 18.

EIGHTEEN

Transition Strategies: Shortcuts, Detours, and Dead Ends

When should the transition to alternative fuels be initiated and what role could or should government play? In the case of those fuels similar to petroleum the important question is *when* to begin the transition; the question of *how* to do so is straightforward, since petroleum-like fuels face few compatibility barriers. The barriers and risks for dissimilar fuels are broader and therefore more difficult to deal with analytically and institutionally. This chapter presents strategies for overcoming the fuel distribution and end-use barriers impeding the transition to alcohol and gaseous fuels.

Barriers to new fuels are those that discourage industry from investing in nonpetroleum vehicles and fuels and those that discourage consumers from purchasing nonpetroleum vehicles and fuels. One may characterize the reluctance of industry to make investments in terms of the "chicken-or-egg" and "not-me-first" syndromes. The chicken-or-egg syndrome, analyzed in chapter 17, is relevant mostly when large megaprojects are the only considered options.

The not-me-first syndrome (see chap. 14) is relevant only in stagnant markets such as the transportation fuels market, not in growth markets where being first into the marketplace is important. The problem is that new fuels are replacing an existing fuel in existing applications; they do not have the opportunity to gain market share by creating new markets, as petroleum did earlier in the century (and coal before that). New fuels will succeed only by taking market shares away from an existing product. Being first at producing or marketing a new fuel, or at producing a new vehicle to operate on new fuels, does not create the potential for large new profits as it would within a growing fuel or vehicle market.

Consumers are reluctant to purchase nonpetroleum vehicles and fuels because they are resistant to making vehicle purchases under conditions of uncertainty. Consumers exhibit very conservative vehicle-buying characteristics.

TWO STRATEGIC PLANS

Two organizations have developed plans to overcome the reluctance of industry and consumers to invest in and purchase nonpetroleum fuels—both plans were for introducing methanol fuel in California.[1] These plans are reviewed as background for the development of more fundamental transition strategies.

The first plan is that proposed by the California Energy Commission (CEC), a state agency that first became a strong proponent of methanol in 1981. Given its relatively modest resources, a belief that the major barrier to the introduction of methanol fuel was market related, and the production emphasis of the well-endowed U.S. Synthetic Fuels Corporation, the CEC devoted its efforts to end-use activities. The CEC strategy, laid out in a 1982 paper,[2] was based on what will be characterized later in the chapter as a market segmentation approach. The agency financially supported the development and testing of methanol-powered heavy-duty diesel engines (especially transit buses), methanol-fired combustion turbines, and cofiring of methanol and residual oil in steam boilers. It also financed the largest demonstration of methanol vehicles in the world—over 500 automobiles. The vehicles were incorporated into the fleets of local and state governments, and a number of methanol fuel outlets (eighteen as of 1987) were established by a private oil marketing firm (subsidized with CEC funds) to serve the 500 vehicles. The methanol outlets were open to the general public.

The CEC strategy was to encourage (with subsidies) the use of methanol in fleet vehicles throughout the state and to build up over time a network of methanol outlets. They hypothesized that at some point, identified as a national demand for 10,000 vehicles, most of which would presumably be sold in California, the market would take off and that little or no additional state subsidy would be needed to sustain the transition (see fig. 35). At that point, individuals and "noncaptive" business fleets would begin buying methanol cars. At the same time, the methanol fuel market would be expanded by increased use of methanol

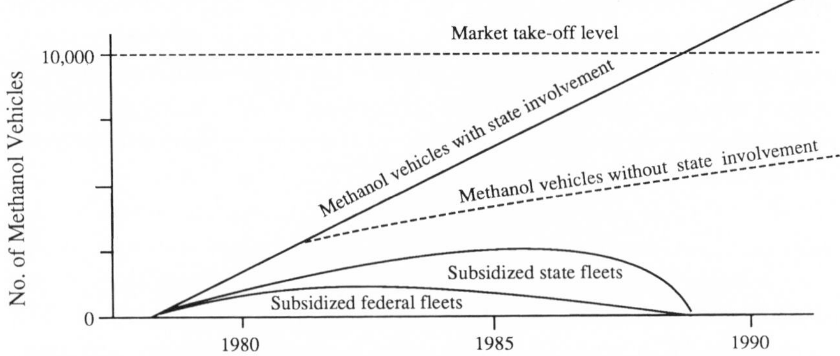

Figure 35. California Methanol Transition Strategy as Proposed in 1982

Source: K.D. Smith, A.G. Edwards, M.C. McCormack, and D.S. Kondoleon, "Alcohol Fuels for California: Establishing the Market," *Procedings of the Fifth International Symposium on Alcohol Fuel Technology*, Auckland, New Zealand, May 1982.

by electric utilities, transit buses, and other heavy-duty diesel vehicles. This increased demand would reduce distribution costs (because of economies of scale) and perhaps elicit new investments in methanol production.

While the CEC plan was in hindsight overoptimistic about methanol penetration, and naive and vague about fuel distribution costs and the willingness of fuel marketers to provide methanol outlets, it nonetheless was a coherent transition strategy.

Another plan was proposed in 1985 by the staff of the air quality management district in Los Angeles, California (though it was not approved by the district's board).[3] Following the lead of the state's Energy Commission, but motivated by air quality objectives, it proposed a schedule of activities that would create a sustainable methanol market. With a population of 12 million people and 7.5 million vehicles in the Los Angeles region, that potential market would be very large. The strategy was premised on the concept of market thresholds; in this case it presumed that when government-supported penetration of a particular market segment—e.g., transit buses—reaches 10 percent of vehicle sales, then the market will sustain itself and continue to grow. This concept implicitly assumes that until 10 percent penetration is achieved, support services (e.g., mechanics, supply of vehicle parts, and fuel outlets) will be inadequate. Another implicit assumption was that after 10 percent penetration is achieved, a significant portion of the population would perceive methanol to be an economically superior option and that sufficient support services would be in place.

The strategy was to create an initial presence for methanol by demonstrations of large fleets of transit buses and by subsidizing the use of multifuel (alcohol-gasoline) vehicles in government fleets. Methanol would also be demonstrated in selected electric utility applications. Next, lobbying would be carried out on the state level to gain tax credit incentives for purchases of methanol vehicles; on the local level mechanisms would be created that allow, for instance, petroleum companies to establish methanol outlets instead of paying fines for air pollution violations. Other local actions would include more stringent requirements on fuel quality and diesel emissions, emission "offset" credits for industrial and transportation uses of methanol, and the purchase of methanol buses and multifuel government fleet vehicles. Within about two to three years, after the transit bus market began to take off, after government fleet and electric utility demonstrations were fully operational, and after initial sales of multifuel vehicles had begun, a methanol distribution terminal would be established. In another two years or so methanol sales would be great enough to shift from costly railroad shipments from the Gulf Coast, to much less expensive bimonthly tanker shipments (of about 15 million gallons each) either from the Gulf Coast or from remote gas sources in the Pacific Basin or elsewhere.

These two strategic planning activities in California have been presented as an introduction to the subject of transition strategies.

CRITERIA FOR TRANSITION STRATEGIES

The remainder of this chapter examines three fundamental transition strategies: an incremental blending strategy, end-use market segmentation, and spatial market segmentation. Each of the three transition strategies will be evaluated in terms of its efficiency and effectiveness in accelerating the transition. The following evaluation criteria are used:

1. Stimulates significant and growing fuel production and supply;
2. Contributes to the establishment of an extensive and efficient fuel distribution infrastructure (e.g., pipelines, storage depots);
3. Contributes to the related establishment of a network of retail fuel outlets;

4. Creates demand for the production of vehicles designed for the new fuel;

5. Generates industrial experience in the handling and marketing of the fuel;

6. Utilizes fuel in high-value applications so that its economic value is fully exploited; and

7. Generates political and economic constituencies that would work to modify discriminatory laws and regulations and to create those price-determining environments that would favor the widespread introduction of one or more alternative fuels.

BLENDING—INCREMENTAL APPROACH OR DEAD END?

The most straightforward approach to introducing alcohol fuels is to circumvent system compatibility problems by blending small proportions of alcohol in gasoline (or possibly diesel fuel). (This option does not apply to gaseous fuels, which cannot be blended easily with a liquid.) By adding limited quantities of alcohol, the effect is sufficiently small so that the resulting alcohol-gasoline blend is compatible with existing infrastructure and vehicles. In this way, alcohol fuel gains a market without encountering fuel distribution and end-use barriers. This strategy is favored by fuel producers and distributors because it requires little or no change or accommodation. Blending is widely espoused as the initial stage of an incremental strategy—as a logical first step in the introduction of alcohol fuels. But a blending strategy has major disadvantages.

Blending only partly satisfies the first criterion of stimulating significant fuel production. While fuel producers argue that blending is indeed a desirable interim strategy because it creates additional demand for alcohol fuel, in fact the potential market for alcohol in blends is only about 75,000 to 150,000 b/d when alcohol is used for its octane-enhancing benefits (see chap. 13). When used as an extender, as was biomass ethanol through the mid-1980s in the U.S., market potential is somewhat greater; for instance, an increase of 75,000 b/d of alcohol production would have absorbed all the excess methanol production capacity in the U.S. in the mid-1980s as well as creating the need for an additional one-third increase in capacity—or it would call for a tripling of the size of the 1985 ethanol fuel industry. Thus a blending strategy indeed stimulates additional alcohol production, but that increase is probably not sufficient by itself to stimulate the significant expansion of an alcohol supply system, including the development of improved process technologies for converting coal and biomass into alcohol. And, more important, an alcohol blending strategy would not create enough demand by itself to elicit actual investments in those advanced production options.

The only criterion that blending fully satisfies is the fifth, in that it provides some experience to industry in the handling of alcohol fuels. Blending would lead to a dissemination of knowledge and would provide a learning experience for fuel distributors in how to handle alcohol safely, how to prevent and combat alcohol fires, how to protect against contamination, how to control volatility, and so on. This is not arcane or new knowledge, however; ethanol and methanol are long-standing industrial chemicals that have been safely handled, transported, and stored for decades. The benefits are not trivial, but neither are they of exceptional significance.

Blending fails to satisfy all the other criteria. The use of blends does not make a significant contribution toward establishing a distribution infrastructure that would facilitate the transition to the much larger retail market. Biomass alcohol would be either shipped by rail to nearby oil refineries or blended with gasoline at existing pipeline storage facilities and shipped as a blend in the pipelines; methanol from natural gas would be imported by ship directly to accessible refineries or storage terminals and blended with gasoline for overland transport. In neither case would alcohol pipelines be built. Indeed, as demonstrated earlier, the effect of blending on distribution systems is negative because of the large cost of dehydrating the system (which is not necessary for straight alcohol).

Blended alcohol would also not lead to the establishment of a network of fuel outlets, nor would it create demand for the production of alcohol vehicles. While it would create some minor constituencies among alcohol producers, octane-short petroleum refineries, and some fuel distributors, the constituencies would not be large or particularly aggressive since, with the exception of alcohol producers, they would not be uniformly committed to or particularly dependent upon alcohol.

Blending is also burdened with unique cost penalties and environmental problems (see chaps. 10 and 12). The phase separation problem requires the dehydration of alcohol and the implementation of costly water intrusion controls. The increased volatility of blends causes either increased air pollution or, alternatively, requires the removal of inexpensive volatile compounds (e.g., butanes) from basic gasoline stock so that volatility and evaporative emissions are reduced. Replacing inexpensive butanes with more expensive alcohol raises the fuel cost.

If an incremental blending strategy were implemented, it is unclear what the subsequent stages would or should be. The next stage could be a shift to one of the fuel strategies detailed later in the chapter (i.e., end-use or spatial segmentation), or it could be a gradual increase in the proportion of alcohol blended in petroleum fuels. Increasing alcohol blend proportions would require pervasive changes in fuel distribution infrastructure and motor vehicles, similar to the changes necessary for straight alcohol use—and continued attention to the still-present phase separation and volatility problems. Preliminary assessments of variable blend proportions suggest that by using larger proportions of alcohol, more economically attractive fuels might result, since lower-quality (and less-expensive) refinery blendstocks could be used for the gasoline part of the fuel.[4] Those analyses may be correct, but the cost benefits would be small. Moreover, the use of larger (and presumably variable) blend proportions would create additional problems for vehicle manufacturers and fuel distributors.

It may be that the optimal alcohol fuel—in terms of reconciling conflicting goals of cost, performance, air quality, ease of implementation, and safety—is one that contains 20 to 50 percent low-quality gasoline. Nonetheless, in terms of evaluating the attractiveness of a blending strategy, it makes little difference whether the alcohol is used straight or with large proportions of low-quality gasoline; the point is that higher blend proportions create all the implementation problems of straight (or near straight) alcohol, as well as retaining the disadvantages of low-level blends. From a strategic perspective, there is little difference between going to straight alcohol or a high-alcohol blend.

In summary, the technical arguments against blends are important but not decisive. The strategic arguments are more telling. They suggest that blending does not for the most part initiate a transition to alternative fuels, that it is an inefficient and, if supported by government, possibly a costly detour—perhaps even a dead end. From a strategic perspective, there is little justification for the argument that lower-order alcohol-gasoline blends are a necessary or important first step in the widespread introduction of alcohol fuels. If there are unfair barriers to the use of blends, then government should remove them so that if alcohol-gasoline blends are a profitable business, industry may proceed.

There is, however, not a strong imperative or compelling public interest that would justify the encouragement of blending by government. Public policy should promote strategies that are more efficient and effective in making a long-term difference.

MARKET SEGMENTATION—DEAD END?

As suggested earlier, marketing theory and theories of innovation diffusion are based on the market niche principle: first, introduce a new product or innovation in a market niche where the unique properties of that product will be highly valued (and valued highly enough to compensate for its negative attributes and its probably less reliable initial performance). As experience is gained and the product or technology improved, it can be introduced in still other niches where its properties are also highly valued. As sales volume increases, economies of scale are gained and the product price diminishes. If product improvements and manufacturing and distribution efficiencies continue, then the product eventually gains a large market share. This process may be characterized as an end-use market segmentation approach. It differs from the incremental blending strategy in that it does not reach a low threshold ceiling (i.e., 10 percent ethanol or 5 percent methanol content); rather, it results, in theory, in a continuing increase in market share. The problem, however, as pointed out earlier, is that unlike the residential, commercial, and industrial sectors, the transportation sector has few niches for new fuels and vehicles.

Let us review this problem of limited market niche opportunities. A crude approximation of the potential size of different market niches was presented in table 44 (chap. 13). The table indicates that the gasoline additive market is sizable (for alcohol fuels), but as just demonstrated, it does not contribute to the widespread introduction of alternative fuels, except in terms of creating a base level of fuel demand.

The agriculture and transit bus market niches are smaller and more dispersed. In both cases vehicles would be fueled from central depots (on the farm or at the transit agency's private depot). Neither of these by themselves—nor both together—would stimulate either significant fuel production or the establishment of an extensive fuel distribution system; nor would they create demand for alcohol or CNG cars and trucks nor stimulate the establishment of retail outlets. Also, market penetration of either of these niches would not have any significant side benefits that would assist in the penetration of other markets. The transit bus strategy has one socially redeeming virtue—the reduction of nitrogen oxide and particulate emissions—but from a strategic perspective, penetration of either or both of these market niches by a new fuel (e.g., CNG or alcohol) leads to a dead end; it does not create the conditions that would facilitate penetration of the broader and much larger household vehicle market.

The remaining market niche, light-duty vehicles in fleets, is large. Those vehicles that are centrally fueled at a private depot are of particular interest. A marketing strategy targeting vehicle fleets appears attractive on the surface because the potential market is large and fuel distribution problems minimal. In practice, though, these fleets are highly conservative. Fleet operators make their choices based on cost and reliability; individuals incorporate a larger number of variables in their decision and therefore are more likely to purchase a unique vehicle even if it is more expensive.

Moreover, the advantage of centralized refueling is less attractive when viewed in terms of overall strategy; vehicle fleet market penetration tends to meet most of the criteria of an attractive transition strategy, but it does not stimulate the establishment of a network of retail outlets. Because of the dispersed nature of the market and the tendency of most major cities to be located on navigable waterways, a vehicle fleet strategy also may not lead to the establishment of a pipeline-based distribution system. For instance, the market potential of vehicle fleets in a major metropolitan area such as Los Angeles would have an upper bound in the first five years or so of less than 10,000 b/d, insufficient to justify pipeline deliveries. When ocean-going tankers are used to deliver methanol from foreign locations or the Gulf Coast, there is little or no need for developing an overland distribution infrastructure that would also serve coal and biomass-based alcohol. Thus penetration of the vehicle fleet market satisfies the following criteria: creating substantial demand for new fuels and vehicles, generating industrial experience in handling and marketing the fuel, and generating a broad set of constituencies—but it fails to create a network of retail outlets and also possibly fails to create an efficient fuel transport and storage network. Moreover, because fleet operators are more cost sensitive and risk averse, and because CNG, alcohols, LPG, diesel, and electric vehicles are all competing for the same limited-range fleet vehicles, the fleet market will prove not to be a significant initial market opportunity for any of those fuels.

The one other market niche that is attractive from a strategic perspective (for alcohol but not CNG) is high-performance cars. This market niche may be the largest, most easily penetrable market for alcohol (see chap. 14). The true size of this market for alcohol will depend on the relative prices of alcohol and gasoline. This market is attractive because it satisfies all seven criteria, including inducing the establishment of a network of retail fuel outlets.

In summary, there are relatively few opportunities to implement a successful end-use market segmentation strategy. Vehicle fleets are the only significant opportunity for CNG, and vehicle fleets and high-performance cars are the only significant opportunities for alcohol. However, the fleet market may be smaller than perceived by alternative fuel enthusiasts, and the high performance car market may be more accessible through spatial segmentation market strategies.

SPATIAL SEGMENTATION

The most efficient and effective strategy is to target spatially defined markets. Two important initial conditions must be met for spatial segmentation strategies to succeed:

1. The targeted area must be relatively isolated so retail fuel marketing and distribution can be limited to that area, and
2. The targeted area must be a large consumer of gasoline.

The state of California is an example of an area that meets those conditions. It is bounded on all sides by either water or sparsely populated lands and only about 1 to 2 percent of the total vehicle miles traveled by California vehicles are driven outside California,[5] thus meeting the first criterion. The fact that California's more stringent vehicle emission standards have already created the precedent of distinct vehicles for that state further increases the viability of California as an isolated market. California easily meets the second condition: about 800,000 barrels per day of gasoline are consumed in the state. California is especially attractive for alcohol because it has a mild climate (thus minimizing the cold start problem), and for both CNG and alcohol because of a strong commitment to environmental quality (thus highly valuing the superior air pollutant characteristics of alcohol and natural gas).

The strategy for fuel marketers in a spatial segmentation approach would be to market the new fuel statewide (or in the Los Angeles area) as a replacement for gasoline and diesel fuel. A network of retail outlets could be rapidly established, with about 15 percent as many outlets as gasoline (see chap. 14) and a fuel distribution system based on one or more bulk storage centers. In California about 1,500 outlets would be needed, and fuel could be delivered by tanker to bulk storage centers in the Los Angeles and San Francisco areas.

The spatial segmentation strategy has the advantage of creating demand for methanol vehicles *and* stimulating the establishment of a network of retail fuel outlets. By focusing on targeted regions, it is much easier to create an efficient bulk fuel delivery system and to build a sufficiently dense network of retail fuel outlets. This facilitates penetration of the large household retail fuel market. Moreover, by targeting only a few regions, greater effort can be directed at marketing and supporting the sales of new fuels and vehicles, thus assuring quality control of fuel, greater expertise in handling fuel with less likelihood of accidents, and denser and better support services for the vehicles. Spatial segmentation also facilitates the creation of political support to make adjustments in various rules and regulations (for instance, in local air quality rules, both to accommodate fuel differences and to reward the fuel for its positive attributes) and to create the price-determining environment that would favor the new fuel.

Two factors work against a spatial segmentation strategy in the U.S. One is that no other region besides California meets the two criteria. Other candidate regions might be the Florida panhandle, perhaps other parts of the Pacific Coast or Southwest, or major metropolitan areas such as Chicago and New York, but none of these is as attractive as the California area. The second factor is political: concentrating activities and resources in one area runs counter to the need at the national level of gaining widespread geographical support for new programs. If the targeted regions were to receive significant federal support for the new fuel program, then there would be a difficult uphill task of convincing other regions that it is in their interest to support the program.

Arguments in favor of a California-oriented spatial segmentation program are that since the California market is so large, once major fuel activities develop there, the start-up cost to other regions would be much lower. Economies of scale in vehicle production would be attained, and the knowledge gained in California could be readily transferred. If this strategy were successful in California (or other targeted markets), subsequent expansion elsewhere would proceed more easily. That is, as fuel availability would expand and vehicles become more readily available at reasonable prices, confidence in the permanence of the fuel would grow, and vehicles could be comfortably used over a larger geographical area.

Although a spatial segmentation strategy has some limitations, it appears to be the most efficient and effective strategy for initiating a path toward widespread usage of alternative fuels. The key to its success is local commitment—to initiate and sustain it. If this commitment exists, then a spatial segmentation strategy is more likely to succeed than either the blending or end-use segmentation strategies.

In Brazil ethanol was introduced first as a low-level blend (0 to 20 percent), then as a straight fuel in the retail market. Special attention was given to taxicabs. What is instructive is that the Brazilian transition also was based on a spatial segmentation strategy. Over half the ethanol was produced in the state of São Paulo during the initial stages of the transition and most of that was consumed in the São Paulo metropolitan area (population 14 million in 1985). Presumably this spatial concentration of production activities created economies of scale in production and distribution, and concentration of consumption stimulated the increased network density of fuel outlets and of alcohol support services. Thus Brazil incorporated a spatial segmentation emphasis once it moved from the blend to straight alcohol stage.

New Zealand also used a spatial segmentation strategy in introducing CNG, in part because the total market was so small that any particular market niche would be inconsequential. As described in chapter 5, CNG was sold only on the North Island with marketing efforts and government incentives directed at the retail market.

INFLUENCE OF GOVERNMENT INTERVENTION ON STRATEGY IMPLEMENTATION

Government has at hand an array of policy instruments to support and accelerate the transition to alternative fuels via any and all of the three sets of strategies identified earlier in this chapter. What appears to be a dead end or detour could be invigorated with appropriate government initiatives.

The fuel blending strategy could be invigorated by a requirement that all gasoline contain, for instance, 5 percent alcohol. This action would create a much larger production base than a market-induced gasoline additive blending approach—325,000 to 650,000 b/d vs. 75,000 to

150,000 b/d—and therefore would stimulate new investments in the conversion of remote natural gas and coal into methanol, and in R&D and commercial investments in the conversion of cellulose into ethanol and methanol. The least costly options would prevail, and the cost of fuel to the individual consumer would increase, but not by much.[6]

Several government initiatives are possible for pushing the fuel blending strategy beyond a 5 percent requirement and toward higher alcohol penetration levels. One policy option is to require each fuel marketer to include a fixed percentage of alcohol in total gasoline sales, a requirement similar in concept to the CAFE (fuel efficiency) standards imposed on automakers. This policy action has the advantage of providing fuel marketers with the flexibility to determine when and where they would include the alcohol (for instance, in mild California but not wintertime Chicago); however, this advantage for fuel marketers is at the same time a problem for automakers, who must build vehicles to accept a range of blend proportions. A second policy option for expanding demand beyond the blending stage is to pursue a variation of the spatial segmentation strategy whereby direct subsidies are provided to those states that require or encourage the sale of larger alcohol blend proportions.

The role of government in invigorating the end-use market segmentation strategy would be somewhat different: it would target motor vehicle users and vehicle fleets. The transit bus market niche would be easiest to influence in the U.S., since the federal government, through the Department of Transportation, pays 80 percent of the cost of all urban transit buses. Government could direct the vehicle fleet market toward alcohols or CNG by having regulatory bodies require or encourage telephone, gas, and electric utilities to use clean-burning nonpetroleum fuels in their fleet vehicles, and by having the federal and state governments do the same with their vehicles. The Post Office and, to a lesser extent, local governments might also be responsive to alternative fuels. Various tax incentives and subsidies could be offered to business fleets (and farm vehicles) to encourage them to convert to alcohol or CNG. The next step for the government to take in an end-use market segmentation strategy is to encourage the household sector to purchase alcohol or CNG vehicles. This might be done with tax incentives for the purchase of nonpetroleum fuels and vehicles or by controlling fuel prices. The establishment of retail outlets for alcohol and CNG could also be mandated.

The principal role for government in supporting the spatial segmentation strategy is to make nonpetroleum vehicles attractive to households—by reducing uncertainty about future nonpetroleum fuel availability and prices and by making the nonpetroleum fuel option economically attractive. Government may offer various tax incentives to make the purchase and manufacture of nonpetroleum vehicles more attractive and may offer regulatory relief to vehicle manufacturers in the form of eased fuel and vehicle certification rules and requirements. Some of these incentives and dispensations would necessarily be provided at the national level; for the spatial segmentation strategy to work, however, additional support is necessary from state and local governments.

The importance of reducing consumer uncertainty cannot be overemphasized, as demonstrated by events in Brazil. As related in chapter 4, the 1981–1982 ethanol car market collapse was due in large part to uncertainty over future prices and availability of ethanol. Uncertainty was created by mixed signals from the government—by government leaders who expressed various doubts and concerns and by reductions in subsidies. It is important once a decision is made to promote a fuel, that a fairly long-term and guaranteed legislative and administrative commitment be made in support of that fuel. Without such a firm commitment, there is a high probability of disaster. Consumers are making a large financial commitment when they buy a vehicle; if the slightest doubt exists whether the fuel will continue to be readily available at competitive prices, then individuals can easily select a gasoline car for their next purchase. The result would be a collapse of the market for the new fuel.

The range of possible government initiatives for each of the three transition strategies is presented in table 61. At one extreme is a laissez-faire approach in which government sees its role as disseminating information and "leveling the playing field" by removing or modifying discriminatory rules and regulations (e.g., changing CAFE standards to miles per Btu instead of miles per gallon). The result of this laissez-faire approach would be to allow nonpetroleum fuels to languish. A more active role is to work within the market system and to rely on market signals to guide the transition, but to direct those signals so that investments and purchases would be made sooner than otherwise. Another approach for government is to avoid or circumvent the market system by directly orchestrating the transition. This last approach is more authoritarian but it assures a more rapid transition.

The Brazilian experience is an example of government working for

TABLE 61 GOVERNMENT INITIATIVES TO SUPPORT TRANSITION STRATEGIES

Transition strategy	Authoritarian (direct intervention)	Incentives	Laissez-faire
Blending	Require use of alcohol in gasoline (and diesel)	Fuel tax exemptions	Pump labeling requirements; generic volatility rules; eased fuel certification rules
End Use Market Segmentation	Require clean fuels for buses and fleet vehicles where air pollution standards are violated; guaranteed vehicle resale	Tax benefits for using clean nonpetroleum fuels; favorable treatment under CAFE standards; fuel tax exemptions	Fuel information dissemination; ease fuel and vehicle certification rules; "level the playing field"
Spatial Segmentation	Requirement by states that specified share of vehicle sales be able to operate on CNG or alcohol; concurrent requirement that retail outlets supply specified fuels; guarantee future fuel availability and competitive fuel prices	Same incentives as above but strengthened by local and state governments	Same as above

the most part within the market system, but doing so in a highly active manner. The Brazilian government had a history of active participation in the economy; its active role in promoting the transition to ethanol was not extraordinary in the Brazilian context, although it would be in a less politicized market economy such as that of the U.S. The Brazilian government worked within the market system by offering attractive loans and subsidies to ethanol producers and vehicle purchasers in order to redirect market signals. The government also placed a ceiling on ethanol prices (as a percentage of gasoline prices), which suggests that in this instance it went so far as to ignore market signals, although in reality the relationship between ethanol prices and costs was carefully monitored.

The New Zealand government also actively promoted alternative fuels (CNG) in a market context using various incentives. The government subsidized CNG conversion kits sold to consumers and provided subsidies to fuel marketers to establish refueling outlets. The government was also active in disseminating information about CNG to the public.

The diesel car and LPG experiences in the U.S. are examples of a laissez-faire approach in which no special government incentives were offered. Attracted by low prices, LPG was used almost exclusively by vehicle fleets and agricultural vehicles and equipment, but rarely by household vehicles because of the few public LPG fuel outlets. CNG was practically nonexistent. In contrast, large numbers of LPG and CNG outlets were established in Canada because of government incentives to fuel marketers and vehicle owners; vehicle manufacturers even began supplying production-line LPG vehicles. The difference between LPG and CNG use in the U.S. and Canada highlights the difference between a laissez-faire and active incentives approach. In the U.S., even with large potential fuel savings, LPG and CNG did not become a major fuel; in Canada, with incentives to establish fuel outlets and to convert vehicles to LPG and CNG, those fuels gained a far greater market share.

The U.S. diesel car experience further highlights the large disadvantage faced by dissimilar fuels and the likely results of a laissez-faire approach. Diesel fuel did not receive any subsidies, but it was a well-known fuel because of its use by heavy-duty trucks; a significant number of fuel outlets were already in place to serve those trucks. Even so, when diesel fuel prices increased to near those of gasoline, diesel car sales practically disappeared, despite the fact that diesel cars still provided substantial fuel cost savings (because of greater engine efficiency and higher energy content), with little difference in vehicle cost. Consumers needed a strong economic incentive or a conviction that diesel fuel was not a risky proposition. The growing number of diesel fuel outlets and the long history of diesel fuel allayed fears of risk and uncertainty. But as diesel fuel prices approached gasoline prices the economic incentive began to shrink. The diesel car market began to collapse in 1982 just as did the ethanol car market in Brazil (although on a smaller scale); the difference is that ethanol car sales were revived through renewed government incentives, while the U.S. government did not act, resulting in diesel car sales virtually disappearing.

The introduction of unleaded gasoline in the U.S. is an example of an authoritarian approach to introducing alternative fuels. Regulations were enacted that required automakers to produce vehicles that operated on unleaded gasoline and fuel marketers to supply unleaded fuel almost everywhere in the country.[7] Market forces were ignored in this accelerated fuel transition. A similar approach in the case of nonpetroleum fuels would be to require vehicle manufacturers to produce alcohol or CNG vehicles in the targeted regions (perhaps by the relevant state governments), and to require fuel marketers to sell alcohol or CNG at their retail outlets. Two DOE officials proposed a somewhat milder version of such an approach in a 1986 paper.[8] They suggested that the federal government require that vehicle manufacturers produce a certain percentage of vehicles (say 20 percent) that were multifuel cars that could operate on methanol and gasoline.

MULTIFUEL VEHICLES: A TEMPORARY SOLUTION?

The prospect of a multifuel vehicle somewhat eases the initial barriers to new fuels. A true multifuel vehicle that would perform equally well with all fuels and have similar efficiency and purchase price as fuel-specific vehicles would greatly facilitate a transition. There would be little need to worry about blending, end-use, and spatial segmentation strategies because vehicles would be equally attractive for all types of fuel. But such a situation will never exist.

Because of inherent differences in fuels, a multifuel vehicle could never perform as well on two different fuels as a vehicle specifically designed for one of those two fuels. In theory and practice, a multifuel vehicle will be optimized for a particular fuel (or fuel blend) and will run satisfactorily, but not optimally, on other fuels and fuel blends.

Multifuel cars designed for both gaseous and liquid fuels will be especially compromised; they will always be burdened by redundant fuel tanks and fuel lines, and suffer the space, weight, and cost penalties that that entails.

At first observation, therefore, the concept of a multifuel CNG/liquids vehicle is not very attractive. A multifuel alcohol/diesel vehicle is also unattractive because the fuels are so different.

A multifuel alcohol/gasoline vehicle is, relatively speaking, the most attractive. Such a vehicle could be optimized for gasoline but be able to operate on alcohol (essentially an "alcohol-compatible" gasoline vehicle), or it could be designed as an ethanol/methanol car that also could run, though not efficiently, on gasoline.

The first option, the alcohol-compatible multifuel vehicle, is less risky. The cost for developing such a vehicle is small since few changes need to be made—certain incompatible parts would be replaced and a more flexible fuel metering system would be installed. Moreover, these vehicles could be manufactured in large volume on a production line; they would be indistinguishable from a consumer's perspective from a conventional gasoline vehicle. The government could mandate that a certain proportion of each manufacturer's production be alcohol compatible—as proposed by the two DOE officials—or manufacturers could sell these vehicles at a slight cost premium (say $300). Perhaps a vehicle manufacturer could make an entire line of cars alcohol compatible, for instance, their high-performance sports cars. They might promote that feature by pointing out that if (high-octane) alcohol were used instead of gasoline, the car would respond with somewhat greater power and acceleration, and that if high-octane gasoline were ever in short supply, alcohol could be used as an alternative. In this case, the manufacturer must be convinced that the additional production costs are offset by the prospect of additional sales in the near future, and by the advantage of getting a technological and marketing head start on the competition.

The second option—to produce an alcohol-optimized multifuel vehicle—is more aggressive but still less risky than building an alcohol-specific vehicle. In this case, an initial network of retail fuel outlets must be in place before the alcohol-optimized multifuel vehicle is marketed. To reduce consumer uncertainty, government would also have to send fairly strong signals that alcohol would be the fuel of the future. Government or the vehicle manufacturers might also have to provide repurchase guarantees to consumers to reduce risk to an acceptable level. The decision to pursue this alcohol-optimized multifuel option hinges on the degree of support forthcoming from government in terms of financial incentives, regulatory benefits (via CAFE or emission control rules), and general policy commitment to alcohol.

The concept of a multifuel CNG/gasoline vehicle is, as indicated above, relatively unattractive. A natural gas transition path must shorten as much as possible the period during which multifuel (actually dual-fuel) vehicles are used and move quickly to dedicated CNG or LNG vehicles. Even so, multifuel vehicles will be a necessary though inefficient transition technology. These multifuel CNG/gasoline vehicles could be optimized for either CNG or gasoline, analogous to the multifuel alcohol/gasoline vehicle options described above.

The availability of multifuel vehicles greatly reduces the disadvantage of initially sparse networks of fuel outlets. From a planning and investment perspective, however, it is still unclear what the consumer response would be to the different multifuel vehicle options. For instance, how many consumers would be willing to pay a small premium to purchase an alcohol-compatible or CNG-compatible production-line vehicle? And if they did make this purchase, how often would they use the alternative fuel? The answers are unknown,[9] although, as indicated in chapter 14, there is substantial evidence that consumers are very conservative in their vehicle purchase behavior; but once they purchase an alternative fuel vehicle, they are likely to be fairly responsive to the alternative fuel—they will be, as suggested earlier, "personally committed." However, if the sale of alcohol-compatible vehicles is mandated and the consumer does not make a deliberate decision to purchase a multifuel vehicle, then they are likely to be conservative in using the alternative fuel. A rough guess is that the alcohol fuel would need to be five cents or so cheaper per gallon (on an energy basis) than gasoline for most consumers to consider refueling with alcohol, unless government and industry were to mount a convincing campaign in support of the fuel. In any case, the availability of multifuel vehicle technology makes the end-use and spatial segmentation strategies relatively more viable and mitigates somewhat the "dissimilarity" penalty of alcohols and natural gas.

MARKET CREATION AND TAKEOFF

The critical consideration in the introduction of dissimilar fuels is market creation. Once the process technologies are at or near commercial readiness, there are numerous firms (and cooperatives) that would invest in fuel production if they felt a market existed for the fuel. That was the case with ethanol fuel in the U.S. in the late 1970s and 1980s. Even in the case of more sophisticated and expensive process plants, there would be no shortage of willing investors—if profits seemed attainable. Large oil companies in particular now view themselves as energy companies. They insist they will provide any fuel that consumers want. Of course, they would prefer fuels with which they have greater expertise—for instance, large capital-intensive projects are more attractive than small biomass fuel projects, which require management of many small investments.

Market creation involves the expansion of fuel sales so as to reduce the cost of distributing and producing the fuel, and supplying end-use technologies to consume the fuel. Low-level fuel blending does not create a base for the development of alcohol fuel markets and fuel distribution systems, and is therefore not an attractive strategy (especially for methanol and natural gas fuels, since large production industries are already in place). The most efficient and effective strategies are spatial segmentation strategies that focus on one or more regions.

The basis for a spatial segmentation strategy is the concept of market takeoff thresholds, which both the California and Los Angeles planners implicitly or explicitly recognized. This concept, represented in figure 36, is in principle useful for developing transition strategies and policies, but in practice, considerably more research is needed to operationalize the concept. Suffice it to say that the point on the graph in figure 36 where the slope of the sales curve begins to increase sharply is roughly the point of market takeoff. Actually specifying that point is difficult and has not been attempted; it would result in a multidimensional mathematical surface of tradeoffs between fuel and vehicle cost, fuel availability, driving range, maintenance costs, storage space, performance, etc. As soon as a few fuel stations are available and a vehicle manufacturer begins marketing a nonpetroleum vehicle, there will be some buyers. As certain conditions shift, the number of buyers will increase, level off, or decrease, as vividly illustrated by the diesel car case. Therefore, while the introduction and dissemination of a new product may be represented conceptually by an S curve,[10] specification of market takeoff points is difficult and is different for each setting and set of conditions.

CONCLUSIONS

If one accepts the efficacy of a spatial segmentation strategy, one must come to the following conclusions. First, nationwide demonstrations of

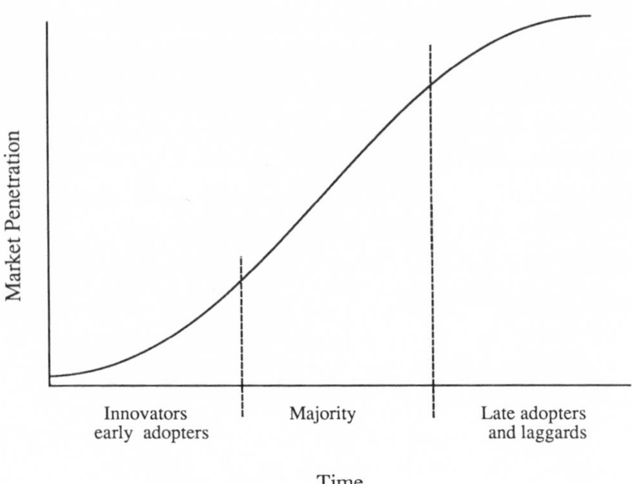

Figure 36. Pattern of Innovation Diffusion and Market Penetration

nonpetroleum vehicles, such as the 1985–1987 U.S. Department of Energy methanol demonstration fleet, are a waste of money.[11] Such dispersed demonstrations of an already proven technology made little or no contribution to the transition process. The analysis of transition strategies also makes clear the minor benefit provided by government support of ethanol or methanol additives. While there is no reason to oppose industry use of alcohol additives in gasoline as long as air quality is not threatened, there is also little reason to provide fuel excise tax rebates or any other subsidies to fuel blenders. This is especially so in the case of the U.S. ethanol fuel program, which has little long-term potential, and where subsidies benefit mostly one company.

The ultimate success of a spatial segmentation strategy is contingent in part, but not totally, upon external events. The vast array of price-determining actions that have been identified throughout the book can alter the relative attractiveness of petroleum products and alternative fuels. Brazil has altered the economics of ethanol fuel; New Zealand, Canada, and South Africa have done so for other fuels. California has contemplated doing so. Thus, even though external events such as wars and cartel actions may alter the economic attractiveness of alternative fuels, societies and governments have considerable discretion to stake out their own destiny.

The vagaries of the oil market will keep the world in an environment of uncertainty. It is the task of governments to prepare and commit themselves to transition plans. Each country and region will make different choices. Each is well advised to follow some sort of spatial segmentation approach and to focus on reducing uncertainty and risk for the consumer.

PART VIII

Making Choices

By this point, readers presumably are seeking a definitive determination of which energy options are superior, which merit public support, and what level and type of support is advisable. Those are reasonable expectations that I do not intend to evade. But the choices are not simple or straightforward, and they are sensitive to a number of uncertainties. The choice is made more difficult by the lack of consensus as to what evaluation criteria should be used, how they should be weighted, and by the imprecise measurements of costs and benefits.

The preceding chapters have specified, as accurately as possible given current knowledge and foresight, the important factors bearing upon the attractiveness and implementability of nonpetroleum transportation fuel options. In the next four chapters those various options are synthesized and reviewed.

First, in chapter 19 the question of cost is reopened. It is argued that future costs and decisions should not be based on inflation-adjusted extrapolations of current market-based prices. That would be simplistic and wrongheaded because future costs of new energy options will reflect dominant societal values, goals, and beliefs, and will be a function of the supporting technological and infrastructural environment that evolves in the intervening years. Future costs are uncertain in large part because of shifting values and beliefs. We should understand how those shifts are manifested in governmental actions and institutional and technological changes, and how those actions and changes alter the relative costs of products and technologies. With these understandings, we can identify which choices are most attractive for particular combinations of world views, values, beliefs, resource availability conditions, and societal settings.

In chapter 20 the framework of hard and soft energy paths proposed by Amory Lovins is tested to determine whether it would be suitable for evaluating transportation energy choices. Although found wanting, it does provide a model for representing and evaluating energy choices. In chapter 21 transportation energy options are sorted into five transportation energy paths. Each path represents a unique set of resource/fuel options that is plausible and that would be attractive in certain settings under certain specified conditions.

In chapter 22 emerging trends and activities in various regions and countries are examined to determine which appear to be following one or more of the five paths. In light of these developing trends and activities, the viability of the five paths is addressed.

NINETEEN

A Price-Determining View of Energy Choices

Individual initiative and private enterprise are the forces that made us an energy rich nation, which moved us from wood and whale oil a century ago to breakthroughs in electricity and petroleum and to our astounding developments in atomic energy.[1]

—*William E. Simon, 1978*
Former Director of the U.S. Federal Energy Administration and Former Secretary of the Treasury

Simon's faith in a market economy blinded him to the very major role the U.S. government has played in the development and commercialization of nuclear energy. It created and financed the Manhattan Project of World War II, legislated limitations on liability, and provided uranium enrichment and waste storage facilities, as well as huge and continuing subsidies for research and development. Regardless of its questionable veracity, Simon's statement represents a strong ideological belief in the ability of an autonomous market system to determine the appropriate choice and timing of energy investments—a belief that price signals will guide energy investments toward the most efficient allocation of resources. The objective of this chapter is to convince the reader that conventional economic analysis and passive reliance on the market system *will not and should not* by themselves determine long-term transportation energy choices.

The conventional approach to evaluating energy options is to compare cost estimates, as was done in chapter 8. Underlying such an analysis is a perception that energy choices are determined by the marketplace, and that the appropriate role of government is to assure that undesired environmental and distributional effects are softened, and to reduce institutional and other start-up obstacles that hinder investments in those options that are economically most attractive. This is a price-determined view of how energy choices are and should be made. Price determinists often speak of creating a "level playing field" so that all comers will fail or succeed purely on the basis of the marketplace. But as a former top executive of an oil company points out, "the oil sector has not generally functioned as a free transparent market in which price is determined by the interaction of many buyers and sellers."[2] As demonstrated in chapter 8, the specification of future energy production costs is subject to tremendous uncertainties and is highly vulnerable to bias. The concept of a level playing field is difficult, if not impossible, to operationalize accurately because of government's already deep and pervasive involvement and because of uncertain and erratic energy prices.

More important, a price-determined view is simplistic and often a misleading, if not incorrect, explanation of energy developments. If we are to assess future transportation energy options, we must have some understanding of why certain options prevail, and to what extent cost estimates based on price signals are or should be reliable predictors of future success.

Energy and transportation systems are what Nobel laureate Herbert Simon (unrelated to William) refers to as artificial systems.[3] They are systems that evolve over time in response to their social, physical, economic, and political setting. They evolve partly in response to available resources, but also in response to goals and values of the local society. In modern times, these local and external political preferences and systems change rapidly and unpredictably. The future will therefore be very different. There may be a drastic change. The design and structure of technologies and systems will be different; cost measurements of future energy options based on today's situation most likely will bear little resemblance to those same measurements made in the twenty-first century. Perhaps an accelerating "greenhouse effect" will cause governments to tax heavily carbon dioxide emissions; perhaps increasing cancer rates will result in a concerted campaign against toxic and suspected carcinogenic materials; perhaps a society will move toward socialistic ideals and curtail large corporate enterprises. The world has changed dramatically in the last one hundred years; the process of rapid technological innovation will undoubtedly create or instigate even more sweeping transformations in the next one hundred years. To determine in a precise manner what the relative costs of particular energy options will be in twenty, fifty, or one hundred years would be difficult and almost beside the point.

Future costs do not necessarily represent inherent values of resources and goods. In large part they represent evolving societal goals and values that are manifested in government actions, policies, and pronouncements. In this sense, the choice or pursuit of a particular energy future is determined only indirectly by market prices; those prices perceived by developers and investors and used by them as a basis for making energy choices have already been manipulated, usually in many indirect ways. In the words of Herman Daly, "the choice between . . . energy futures, is price-determining, not price-determined."[4] The transportation and energy technologies that thrive are not necessarily the most efficient, environmentally benign, or technologically superior. Rather, they are the technologies that create and sustain a supportive political environment.[5]

A price-determining view does not argue that market forces and (price-determined) cost analysis should be ignored or necessarily subverted,[6] even though the allocative mechanism of markets is flawed by many imperfections and by the failure to deal with "externalities" and distributional issues. To do so, to ignore the market system, is to place the responsibilities for resource allocation in a bureaucracy or political body. That option is not necessarily preferable because those public bodies are neither omniscient nor wholly competent.[7]

The purpose of this chapter is to sensitize the policy-making process to the many factors influencing market signals and to explain why a valid evaluation must extend far beyond an intellectually simple (albeit empirically sophisticated) examination of price-determined costs. Once the price-determining nature of energy choice is understood and accepted, one can identify and incorporate into the analysis important values and goals that could or should guide future decisions.

HOW NUCLEAR POWER BECAME COST COMPETITIVE

Examples of how government initiative guided market forces and signals to favor one or another type of energy investment are plentiful. The U.S. government's attempt to nurture the nuclear energy industry is a prime example. The U.S. government made nuclear power a reality by providing huge subsidies and by protecting owners of nuclear reactors against liability for major disasters. It was in the 1950s that the Eisenhower Administration launched the "atoms for peace" program and committed the nation to nuclear power. By 1969, nuclear energy received 93 percent of all energy R&D expenditures by the federal government.[8]

One study estimates that in fiscal year 1984 nuclear power, even

TABLE 62 FEDERAL SUBSIDIES FOR ENERGY PRODUCTION, FISCAL YEAR 1984

	Crude Oil	Natural Gas	Coal	Synthetic Fuels	Fossil Electric	Nuclear Electric	Fusion	Hydro-electric	Nonhydro Renewables	Total
Subsidies (billions of $)	8.6	4.6	3.4	0.6	5.6	15.6	0.6	2.3	1.7	43.1
Energy Output (10^{15} Btu)	21.6	17.7	19.7	0.05[a]	6.0	1.11	0.0	1.1	2.9	47.0
$ of subsidy per million Btu	0.41	0.26	0.17	12.00	0.93	14.05	----	2.09	0.59	----

SOURCE: H. R. Heede, R. E. Morgan, and Scott Ridley, *The Hidden Costs of Energy* (Washington, D.C.: Center for Renewable Resources, 1985), pp. 26-27.

[a]Estimated in original source as "negligible."

though supplying only 1.8 percent of all energy in the U.S. and already a mature technology, received $15.6 billion in subsidies, one-third of total energy subsidies in that year.[9] This subsidy about equaled the total 1984 revenues from nuclear-generated electricity;[10] in other words, about half the cost of producing electricity in nuclear power plants was covered by government subsidies. As shown in table 62, nuclear power received seven times more subsidies than any other established energy technology or energy material. Indeed, even the collapsing synthetic fuel industry performed about as well as the nuclear industry on the basis of subsidy per unit of energy output. The subsidies included in the calculations in table 62 include reduced tax expenditures (e.g., accelerated cost recovery, investment tax credits, tax-exempt pollution control and utility bonds, resource depletion allowances, and "expensing" of drilling and exploration costs and construction-period interest), federal agency program outlays, and loans and loan guarantees. The estimates do not include military expenditures to protect Middle Eastern oil interests, costs of the Strategic Petroleum Reserve, or, except where covered by federal programs, environmental and social externalities. With respect to nuclear energy, subsidy estimates also *do not* include the $8.8 billion in uranium enrichment costs still not recovered from the private sector (i.e., utilities) including the 2.5 to 3.0 billion dollars sunk in an abandoned uranium enrichment plant in Portsmouth, Ohio,[11] tax writeoffs for the many abandoned nuclear power plants, or legislated limits on the liability of owners of nuclear reactors.

Protection of the nuclear industry from liability for accidents was important in initiating and sustaining nuclear power. The Price-Anderson Act of 1957 limited liability to $60 million for any nuclear accident.[12] Any damages over $60 million and up to $560 million were covered by the federal government; no one was liable for any damages over $560 million. Liability limits were raised in 1975; according to one estimate the value of reduced liability to nuclear power resulting from the Price-Anderson Act was about 10 to 20 million dollars per reactor-year in 1984, or a total of about 0.7 to 1.4 billion dollars.[13]

The organization that conducted the preceding study is an advocate of renewable energy and an opponent of nuclear power, which suggests the possibility of bias in the analysis. However, the study is fully documented, and the point is well made that nuclear power receives exceptionally favorable treatment. The only attempt by the U.S. government to compile comprehensive estimates of energy subsidies was conducted by the Battelle Memorial Institute for the U.S. Department of

Energy in 1980.[14] That study used a much longer time period and a narrower definition of subsidies (preferring to call them incentives). When their estimates of total dollars were divided by energy output, the conclusions were the same: nuclear power had received much more government assistance than any other energy source.

It is an incontrovertible fact, therefore, that nuclear power came into being because of government actions, not because it was inherently less expensive than other options. The U.S. government decided in the 1950s to promote nuclear power for various reasons, including a need to justify military development of nuclear weaponry. It took responsibility for research and development, supplying of enriched uranium, storing radioactive wastes, protecting purchasers of the technology by limiting their liability in the case of accidents, and even helping promote sales outside the U.S. so as to reduce unit production costs (as well as to promote other strategic goals). Nuclear power became economically attractive to electric utilities because more than half the cost was being borne by the government. Active intervention by government in effect made nuclear power attractive by lowering costs; this is one example of the price-determining nature of energy choices.

Although the nuclear power case may be extreme in terms of the government resources and benefits it received, it is certainly not a unique episode.

BRAZILIAN ETHANOL FUEL

Brazil's experience with ethanol is another illustration of the price-determining phenomena. Brazil subsidized the ethanol program, protecting and nurturing it until it became more efficient and could develop a constituency that would guarantee its continuing survival. Over time Brazil adjusted its refinery mix of petroleum products, restructured its fuel distribution system, improved the efficiency of alcohol vehicles and alcohol production plants, and began exporting distillery equipment. The result was steadily decreasing ethanol costs. At the same time, expanding ethanol production increased employment and income in the agriculture, fuel-production, and equipment supplier industries, thereby strengthening ethanol's political and economic constituency.

Brazil's pursuit of ethanol at the same time world petroleum prices were tumbling was a case of bad timing, but its continued (though less aggressive) commitment to ethanol into the late 1980s suggests that major benefits are perceived to be accruing to the society, that the production efficiency has improved over time, and that these changes are great enough for Brazil to be willing to outlast the 1980s downturn in petroleum prices. Indeed, by 1986 the enormous commitments in sugar cane planting, construction of distillery facilities, redesign of motor vehicles and retooling of motor vehicle manufacturing plants, and changes in oil refining and fuel distribution were so great that turning back to petroleum would have been unthinkable.

OTHER EXAMPLES

There are many other recent examples of government actions having major effects. For example, in 1978 the U.S. Congress passed the Public Utility and Regulatory Policy Act, which required electric utilities to purchase electricity from small electricity generators (e.g., windmills and sawmills that generate electricity from unused residues) and to pay a premium price (the "avoided cost" of not having to use petroleum to generate electricity). In this way Congress effectively created many new industries and businesses for generating electricity. Similarly, a federal tax credit of 55 percent for investments in solar installations created a solar heating industry. Small oil refineries were aided by numerous programs, such as "entitlements" and special rules permitting them to continue using lead in gasoline, that together allowed them to flourish long after they otherwise would have. And, of course, federal and state gasoline tax exemptions for gasohol have by themselves created a billion-dollar-a-year ethanol fuel industry. These examples indicate the major role that government intervention and initiative may have on the apparent cost competitiveness of new energy options, and the extent to which future costs are a function of previous infrastructure and industrial investments.

ENVIRONMENTAL REGULATION

Environmental costs generally are not accounted for by the marketplace, thereby encouraging government to play a major role in protecting environmental quality. This intervention is a major source of uncertainty that could have a significant impact on which options are chosen and when. For instance, solid wastes from Lurgi gasifiers or from direct liquefaction could be declared "hazardous" under the conditions of the 1976 Resource Conservation and Recovery Act (or later legislation). If so, those process plants would be subject to stringent disposal and monitoring requirements that would limit feasible plant locations and increase costs. As suggested in chapter 16, the effect of those actions could be to raise the cost of fuel (in the case of a Lurgi coal-to-methanol plant) by as much as $20 per barrel.

Another example where government action to protect environmental quality would accelerate the introduction of new fuels is with respect to air quality. In the future, regulations might reward alcohol, CNG, and hydrogen vehicles for their less hazardous emissions. The effect would be to improve the cost-competitiveness of alcohol, CNG, and hydrogen.

AGRICULTURAL MARKET DISTORTIONS

An example of government intervention that promotes one (nonenergy) policy while unintentionally penalizing an alternative fuel is the various programs of the federal government that keep grain prices substantially higher than they would be otherwise. Thus corn feedstock costs for ethanol production are increased by government grain subsidies. If corn prices were $0.50 per bushel lower, then feedstock costs would diminish by $0.20 per gallon. The $0.50 figure is purely speculative, but it suggests the significant influence the government's agricultural support programs has on U.S. ethanol fuel prices. The U.S. General Accounting Office, a nonpartisan agency that by request of members of Congress audits government projects and programs, came to the following conclusion in evaluating government subsidies to biomass ethanol fuel:

> The alcohol fuel industry must deal with an additional element of government intervention in the marketplace which tends to increase the price of its feedstock. Through various agricultural price support programs, the government has reduced production and maintained an artificial floor under the price of corn, the ethanol industry's primary feedstock.[15]

While corn to ethanol is not an attractive fuel option even with significantly lower feedstock prices, the point is again made that current prices are often distorted or reflective of particular governmental policies and societal preferences that may be specific to a particular time and place and not necessarily relevant to estimates of future costs.

DIRECTION INTERVENTION

Government may also influence the introduction of a particular option by direct intervention. For instance, in the case of fuels that are dissimilar to petroleum, there is great uncertainty on the part of vehicle manufacturers and fuel producers over the future market of their new product (see chap. 17 for a discussion of the chicken-and-egg metaphor). In Brazil the government orchestrated the introduction of ethanol cars and ethanol fuel so that they were available at the same time in the same place. In the U.S. there are strong antitrust rules that discourage cooperation and coordination between fuel producers and vehicle manufacturers. Failure to coordinate the supply of fuel with the supply of vehicles in each region during an initial introductory phase would create additional risk for both parties. While the Justice Department and the courts have not ruled on this issue,[16] the existence of antitrust laws prohibiting collusion and price fixing renders the introduction of nonpetroleum fuels (dissimilar to petroleum) relatively less attractive. These risks could be reduced if, as with unleaded gasoline, government intervenes to orchestrate the transition.

Another way in which government actions and social preferences alter maket prices and costs is by responding to widespread public fears of a particular option, as for instance against the siting of liquefied natural gas (LNG) terminals. LNG arouses public opposition similar to nuclear power plants. During the 1970s a major LNG terminal was never built on the California coast (Point Conception) primarily because of widespread fears over the danger of explosion. As in the Point Conception LNG case, activist citizens and public interest groups may bring legal proceedings against project sponsors, and local governments may hinder such projects by refusing permits, or making permit conditions time consuming and expensive to satisfy.[17] Governments at the regional, state, and federal level, as well as the judicial system, may erect barriers and create hindrances through the use of such mechanisms as Environmental Impact Statements and coastal commission permits (in California and several other coastal states). The effect of opposition to LNG terminals is to increase the attractiveness of methanol from remote natural gas relative to the otherwise less expensive option of importing natural gas as LNG and using it as CNG.

One public policy factor working to hinder all alternative fuels is the lack of a coherent energy policy, a situation that has existed in the U.S. since before 1973. The lack of continuity and consistency from one government administration to the next—Reagan was the only U.S. president between 1973 and 1990 to be in office for more than four years—creates uncertainty over government commitment and thereby discourages risky new energy investments. If the federal government or a state such as California were to make a credible long-term commitment to methanol or CNG, then vehicle manufacturers would be much more willing to invest in the development of CNG or methanol vehicles, energy producers would be more willing to develop remote sources of natural gas, and fuel marketers and distributors would be more inclined to promote the fuel and to prepare their facilities for the new fuel. This preparation and initiation phase takes many years; ideally, some firms would like fifteen years or so advance notice so they can better plan for the future. While this longer advance planning may not be plausible, the lack of advance planning and the existence of great uncertainty have the effect of discouraging change and strengthening the status quo. Indeed, as oil prices dropped below $30 and then $20 per barrel in the 1980s, alternative energy was pushed beyond the planning horizon of almost all companies; they could no longer justify investments in R&D and demonstrations of new fuel technologies that might not be commercially viable for many years.

PAST INVESTMENTS INFLUENCE FUTURE CHOICES

While numerous government actions—or lack of actions—will strongly influence future energy and transportation choices, there are many types of direct private investments and many long-standing industry practices that will also strongly influence future energy choices. For instance, consider the material compatibility of nonpetroleum fuels with current facilities and equipment. Previous decisions regarding the design and deployment of underground fuel storage tanks, materials used in fuel lines and fuel tanks of motor vehicles, and corrosion inhibitors used in pipelines in effect favor one fuel over another. To overcome those incompatibilities, a new fuel incurs a cost penalty (initially). Similarly, the use of certain operational practices in fuel distribution systems discourages new fuels—for instance, using water in hydrostatic pipeline tests discourages investments in small fuel plants.

The enormous investments in petroleum refineries made in the past also favor certain fuels, in particular "synthetic crudes," those petroleum-like liquids produced through direct liquefaction of coal and pyrolysis of oil shale. If these petroleum-like liquids are to be used as transportation fuels, they need to be upgraded. This could be accomplished by piping them to existing oil refineries. The existence of those refineries effectively reduces the cost of producing shale oil and coal liquids, by making unnecessary the construction of special upgrading facilities. Moreover, their very existence encourages refinery owners to favor those fuels that preserve the value of their investments. In contrast, alcohol fuels and CNG would not use existing oil refineries, thus rendering the huge sunk cost in petroleum refineries superfluous and valueless.

The motor vehicle industry is going through a transformation that could also significantly influence the prospects of nonpetroleum fuels dissimilar to petroleum. Vehicle manufacturers must decide in the near future between two strategies that are to a large extent mutually exclusive: whether to transfer manufacturing activities to developing countries where labor costs are less expensive, or to invest in highly sophisticated computer-controlled "flexible manufacturing" processes.[18] If manufacturing is transferred to developing countries, then the traditional practice of routinized high volume production will be followed as a means of capturing economies of scale. There would be considerable automation, but it would be of the precomputer type in which one machine is designed and built to stamp out or assemble one particular component hundreds of thousands of times. This form of vehicle manufacturing discourages the introduction of new engines for alcohol and CNG and, for that matter, any change in which the production volume is small.

The deployment of flexible manufacturing processes in which computers control robotlike machines is much less dependent on achieving economies of scale through repetitive identical motions. Instead, the robot machines can be programmed to perform one action a fixed number of times, then perform a somewhat different action for a subsequent number of repetitions, and so on. In other words, robots could produce and/or assemble gasoline engines, then produce and/or assemble slightly different alcohol engines, then return to gasoline engines, without intervention by humans and without stoppages or machine redeployments of any type. Small batches of alcohol or CNG vehicles could be manufactured with little cost penalty and without forgoing economies of scale. If the vehicle manufacturing industry chooses to follow the developing country strategy, the barriers to alcohol and CNG vehicles are raised; if the industry favors the flexible manufacturing approach, the barriers are lowered.

The homogeneous nature of motor vehicle design and use provides a final example of how past decisions and patterns of development have created a situation that discourages the introduction of new fuels and engines. As stated earlier, vehicles are expected to be multipurpose and to perform equally well for long-distance and neighborhood trips.[19] There is not a unique road system or unique vehicles for strictly neighborhood use (e.g., specially designed low-speed roadways that are closed to all but small golfcart-like vehicles and nonmotorized traffic). Nor, for instance, are there separate roadways and unique vehicles for peak hour urban commuter trips (e.g., narrow lanes for small, limited-range personal vehicles that carry one or two persons).

If specialized vehicles were used only in local neighborhoods with traffic restricted only to those vehicles, then this would be a market niche for new fuels and engines. Likewise, the same applies for commuter vehicles on special commuter roadways or lanes. As it is, most individuals will not purchase a vehicle unless they are assured it will operate in any topography, geography, or weather that might ever be encountered. Thus alcohol and CNG cars, which tend to have a limited driving range per tankful of fuel and an initially sparse network of fuel outlets, are sharply disadvantaged by the multipurpose and homogeneous nature of the passenger vehicle market.

CONCLUSION

This chapter has provided selected examples of actions and investments from the past and governmental actions and policies in general that are inherently conservative forces that nurture some changes while sharply limiting others. Cost analyses based on market prices do not capture the effects of all these factors. For isolated and incidental decisions that is of little matter. For major changes with far-reaching and long-term implications, such as choices of new fuels, accepting current (or even projected) market prices as the basis for corporate planning and government policy making is not satisfactory. Certain assumptions and certain given conditions are implicit in any cost analysis based on market prices. Those assumptions and givens need not and should not be accepted when evaluating the attractiveness of major energy options. As demonstrated in chapter 8, production costs have a large range of uncertainty even before the many factors identified in this chapter are considered. It is intellectually (and sometimes ethically) dishonest for analysts to rely on current market-determined estimates of future costs. The evaluation of transportation energy options should be based in part on currently estimated costs but, more fundamentally, it should be based upon those basic values and goals that are most important to us. What we decide about environmental quality, distribution of wealth and power, and self-sufficiency will in turn affect the cost associated with different energy options—in the form of government regulations, purchases, and information dissemination. That is, the choices would have a price-determining nature to them. This is not a revolutionary thought. It simply explains how government and society already function. As one writer states, "if one looks at the ways our society organizes energy, one notices a number of different regimes of varying size and complexity, each one created to exploit a particular kind of energy resource, each one using appropriate instruments, techniques, and social relationships to to so."[20]

In one sense, introducing this concept of price determining adds more complexity to an already complex evaluation. But purely economic inquiry is simply not sufficient. As one scholar notes, "[an] alternative and more accurate approach would place technological change in its social and political context and explain the development and adoption of a technology as a result of social values, a specific set of economic and political payoffs, and its technical feasibility."[21] The importance of the concept of price determining is that it emphasizes the importance of thinking through the ramifications and implications of fuel options and addressing what type of future we envision and how we might direct ourselves toward that future.

TWENTY

Hard and Soft Choices?

Transportation energy options abound. For analytical and policy purposes it is useful to reduce the long list to a smaller number of categories or classifications. Also, as suggested earlier, an evaluation scheme should be based on world views and fundamental values and beliefs. Such an evaluation framework was developed by Amory Lovins in 1976.[1] Lovins argued that all energy options, including transportation energy options, could be grouped into two distinct and mutually exclusive paths of change—which he labeled "hard" and "soft." By this means of categorizing all energy options, he created a framework for proposing drastic changes in energy investment strategies and policies.

This framework is a useful but controversial way of viewing energy choices. Controversy exists at three levels. At one level one might question the structure of the framework itself; at a second level one might accept the framework but debate the attractiveness of the values that are defined as underlying the paths; and/or at a third level one might debate the "facts" used to argue the superiority of one energy path. The second and third levels—debates about values and facts—assume and accept the framework and focus on which path is preferable. This chapter addresses the first level of the debate: the accuracy and power of the framework for explaining and evaluating transportation energy options. In the next chapter, the second and third levels of the controversy—the world views, values, and "facts" (i.e., attributes of systems)—are addressed.

Various analysts have pointed out that conflict regarding this framework and energy choices in general has mostly focused on the third level of the debate—the factual accuracy of the numbers used to demonstrate the superiority of one or the other of the two energy paths.[2]

As an SRI report observes,

> It may be quite impossible, at this time, to establish which picture . . . of reality is "correct," because the available data can be fitted into more than one pattern. Each . . . has significant representation among decision makers, analysts, and the body politic. Yet each of the perceptions tends to lead to quite different actions, and when persons appear to be arguing about technical issues or choices among energy options, they may in fact be arguing from different fundamental perceptions of social reality.[3]

Energy choices, therefore, will be made not by any resolution of facts (e.g., costs), but how facts fit into world views and social preferences.

This chapter investigates the logic, robustness, and validity of the hard/soft energy framework for the study of transportation energy choices. It will be shown that the framework does not fit well with transportation energy choices. Why not? Is it because the hard/soft conceptualization is somehow not adequate for categorizing the range of choices?

I suggest that the criteria and indicators for organizing hard/soft choices are not as generalizable as has been generally assumed and that transportation energy choices may be more usefully characterized by different criteria. The purpose is not to pass judgment upon the general validity of the hard/soft energy paradigm, but rather to suggest that the transportation energy sector differs significantly from other energy sectors and is best treated within a distinct or at least modified framework.

The hard/soft energy paradigm is not unique to energy policy. It is analogous, in a way explicitly recognized by Lovins,[4] to the dichotomous Jeffersonian and Hamiltonian ideologies of governmental structure and organization. The soft and hard paths can be thought of as a revival of the conflict between the Hamiltonian concepts of centralized control and governance and the Jeffersonian concepts of greater decentralization and participation.[5] The intellectual and ideological roots of the hard and soft paths are therefore deep and strong.

Thus the controversy over energy futures brought ideological considerations into the energy debate. The debate became "preoccupied with arguments over policy in which ideologies are entangled with economic interests and technical judgments."[6] Lovins's style of presentation, in which he wove together a story that implanted vivid but not necessarily sharply defined images, encouraged this entanglement.[7] This chapter (and book) attempts to untangle the ideologies, economic interests, and technical judgments by taking one strand of a supposed soft path, biomass transportation fuels, and analyzing each of its components, individually and as linked subsystems.

There are three parts to this chapter. First, the hard/soft paradigm is described; criteria and indicators formulated by Lovins are presented to distinguish between what is defined as hard and what is defined as soft. In the second part of the chapter biomass transportation fuel systems are scrutinized to determine whether they truly embody soft path precepts as claimed by soft path advocates. The last part is an inquiry into the relevance and usefulness of a hard/soft paradigm for formulating and analyzing transportation energy policy options.

Before embarking on the substance of this chapter, some explanation of the methodological treatment of the subject matter is necessary. Soft and hard energy paths are most appropriately viewed in terms of processes of change; they were not conceived or treated as specific end states or steady state situations. The distinction is crucial. As argued earlier in chapter 2, an end-state orientation is not useful for analyzing policies related to dynamic, rapid-changing systems; it does not help define societal goals or decision criteria, does not facilitate consensus formation, and is usually inaccurate. Thus the approach taken in this chapter is to investigate energy futures in terms of processes of change and transition, *not* as some speculative snapshot of the future. Lovins in his landmark book quickly sketches a soft energy future (i.e., end states), but is more illustrative than definitive and emphasizes direction of change rather than end-state specifications.

The following analysis is conducted in a near- to medium-term time frame. The risk of this approach is that the analysis adheres to conventional wisdom—for instance, that liquid-fueled internal combustion engines and personal vehicles are likely to be dominant into the foreseeable future. It seems, however, that the greater risk is to specify a far-off future that is arbitrary, based on large amounts of speculation, and probably inaccurate.

THE SOFT/HARD PARADIGM

This paradigm categorizes energy options as either hard or soft. For analytical purposes, the paradigm may be treated as an index, with a continuum ranging from soft to hard. Choices at the hard extreme are premised on sustained expansion of energy production to meet a growing (and inefficient) use of energy. The synthetic fuels program promoted by the Carter Administration is an example of a hard path. A hard path strategy means a rapid expansion in coal utilization, accelerated search for oil and natural gas reserves, and continued growth of nuclear power.

The soft/hard paradigm asserts that the continued implementation of hard energy technologies would lead to a more centralized and concentrated economy, and a more centrally controlled society. Centralization of technological facilities would create a vulnerability to sabotage and disruptions and breakdowns of other types; it would also lead to the creation of elitist and dehumanized social control in which people would depend *not* on

> understandable neighborhood technology run by people you know who are at your own social level, but rather on an alien, remote and perhaps humiliatingly uncontrollable technology run by a faraway, bureaucratized, technical elite who have probably never heard of you.[8]

Thus it is claimed that a hard path creates an elitist technocracy, concentrated political and economic power, and technological vulnerability, resulting in various inequities and other social, political, and economic distortions.

In comparison, soft energy choices are premised on a more restrained production of energy and a more efficient use of energy sources. Soft energy choices are based on solar and other renewable energy. A soft energy path would lead to a future where small, decentralized systems form an increasingly large component of energy production and utilization. Soft energy technologies would be resilient, sustainable, and benign.

Criteria put forth to distinguish between hard and soft technologies are renewability of energy flows, diversity of energy supply, flexibility and sophistication of technologies, as well as matching of scale, spatial distribution, and energy quality between energy production and end use. Other criteria that relate to the effects of deploying the technologies include initial capital cost, ability to respond quickly to changing economic and political conditions, tractability and reversibility of environmental effects, existence of diseconomies of centralization, and level of economic participation.[9] These criteria are applied to biomass fuels later in the chapter.

Soft technologies would tend to be renewable, diverse, flexible, matched with end-use needs, have low initial (sunk) costs, be responsive to changing conditions, have tractable and reversible environmental effects, lack diseconomies of centralization, and engender high levels of economic participation.

Hard technologies would be characterized by the converse: they would be highly centralized and uniform, poorly matched with end-use needs, rely on nonrenewable resources, be characterized by inflexibility, low levels of economic participation, and high capital costs, and create intractable and nonreversible environmental effects.

Lovins creates the soft path as a means for arguing that there exist the opportunities not only to reduce national energy investments, improve system efficiencies, and reduce environmental risk but also to facilitate a more democratic society. The soft path, based on the proliferation of soft technologies, would be resilient, tolerant, and encouraging of social diversity and alternative energy strategies, and would enhance freedom of choice. It would preclude the growth of a technological and economic elite that would likely divorce social power from political accountability.

HOW GOOD IS THE FIT WITH TRANSPORTATION ENERGY?

Through careful empirical investigation, one can specify within a fairly narrow range the attributes of biomass transportation fuel options. This was done in earlier chapters. It was indicated that biomass fuel options would comprise mostly—but not exclusively—small, independent activities and investments. Process plants would be small to medium-size because of large diseconomies of scale in biomass collection. Trucks and railroads would be the principal means of delivering fuel. End-use markets would be located close to process plants and would tend, at least initially, to be in rural areas. Biomass fuels would come into being through attempts to exploit market niches; local inexpensive feedstock sources would attract initial production investments. On the end-use side, however, fewer niches exist; enthusiastic participation by the auto industry and its willingness to take risks may be critical for expansion of liquid biomass fuel activities.

Soft path advocates have presumed that biomass fuels would be produced by soft technologies and would be one component of a soft energy path. Lovins states that "many genuine soft technologies are now available and are now economic. . . exciting developments in the conversion of agricultural, forestry, and urban wastes to methanol and other liquid and gaseous fuels now offer practical, economically interesting technologies sufficient to run an efficient U.S. transport sector."[10] However, little or no systematic evidence is provided to support that hypothesis. Lovins, for instance, devoted only a few pages to transportation fuels in his 1977 and 1982 books.[11]

It is argued here that biomass transportation fuels would not fit neatly into a soft path and that a pure soft path of the type envisioned by Lovins, even if attainable, would have a very minor effect on the supply of transportation energy. Biomass transportation fuel systems would have some features of a soft path but would also differ in important ways. Only by deviating from a pure soft path could biomass make a major contribution to the supply of transportation energy.

SOFT TECHNOLOGY CRITERIA

The concept of softness is tested and operationalized by applying Lovins's definition of a soft path to the biomass transportation fuel subsystems formulated in earlier chapters. Lovins[12] defines soft energy technologies as having the following characteristics:

1. They rely on renewable energy flows . . .
2. they are diverse . . . so that energy supply is an aggregate of very many individually modest contributions . . .
3. they are flexible and relatively low technology . . .
4. they are matched in scale and geographic distribution to end use needs . . .
5. they are matched in energy quality to end use needs.

It is shown that biomass fuels fit some of these criteria but not others. The first criterion, for instance, is easily met: biomass fuels are renewable by definition.

The second criterion of diversity would also be met, by nature of the many feedstock materials that would be used, although at this time almost all liquid biomass fuel in the U.S. is from fermented corn, and in Brazil from fermented sugar cane. If energy prices rise, other feedstock materials and biomass process technologies will undoubtedly be lured into the marketplace.

The second criterion also calls for an "aggregate of very many individually modest contributions." The atomistic nature of biomass fuel process plants would satisfy this requirement, but not necessarily in the sense that Lovins suggested. Some process plants, those that are mass produced, would meet the criterion of making individually modest contributions. It is highly uncertain, though, whether these small mass-produced plants would ever make a significant aggregate contribution. If a ten-percent share of the gasoline market was considered to be significant, and assuming gasoline demand stays constant, then roughly 4,000 mass-produced thermochemical plants (each with a 300-barrel-per-day capacity) or 900,000 bioconversion plants (with capacities of one barrel per day), or some combination of the two, would be needed.

An inventory of feedstock opportunities that would indicate the realism of this possibility is lacking. My educated guess is that many opportunities to ferment small quantities of waste and surplus material exist, but that many of these opportunities would not be financially attractive—because fermentable feedstocks are highly perishable and would be available only periodically, usually during harvesting season. The potential market for mass-produced thermochemical plants is much greater. If biomass fuels are to become an important source of transportation energy, mass-produced cellulose-to-methanol (or cellulose-to-ethanol) plants are likely to be major fuel suppliers. However, these so-called "small" plants are small only in comparison to petroleum refineries and chemical process plants, not compared with typical small business investments. As stated earlier, these "small" biomass plants would cost at least $15 million each; they are certainly not suited to new start-up businesses. In summary, the response to the "individually modest" criterion is that contributions to total supply would tend to be individually modest, but where modest is defined as entailing investments generally in at least the tens of millions of dollars.

The third criterion is "flexibility and relatively low technology." Here the biomass fuel option deviates even further from the soft technology criteria. (Only production technologies are addressed here since end-use technologies would be essentially unchanged.) Bioconversion and plant oil conversion processes would satisfy this third criterion, but as suggested earlier, their production potential is limited by the high cost of feedstock crops and relative paucity of low-cost fermentable waste materials. The fermentation-distillation technologies are also flexible in the sense that they are portable; small package plants carried on truck beds have reportedly already been used commercially. However, thermochemical process technologies (to produce methanol) are not very flexible, nor can they be considered "low technology." Unless dramatic and unexpected improvements are made, these thermochemical process technologies will handle only one feedstock type, for in-

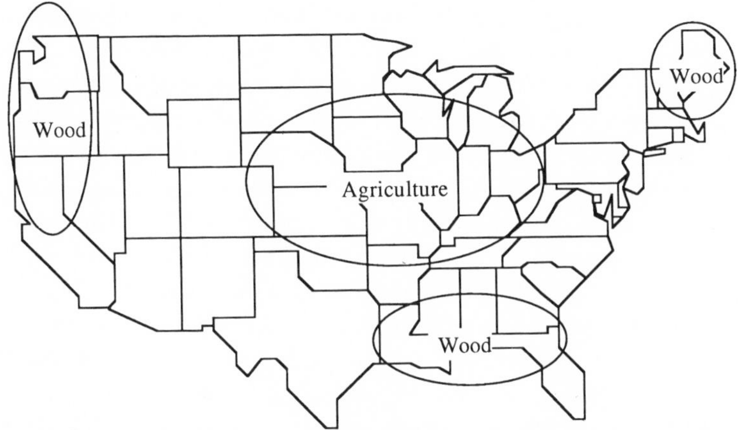

Figure 37. Attractive Areas for Large-Scale Low-Cost Biomass Fuel Production

stance cornfield residues, and would not accept even small variations within one feedstock type, such as different types of wood. Even if these more sophisticated process technologies were of the small mass-produced variety, they would not be physically mobile. They would be neither flexible nor low technology, as fermentation-distillation and plant oil pressing plants would be.

The fourth criterion, "matched in scale and geographical distribution to end-use needs," is generally satisfied by biomass fuel plants, but only in the sense and to the extent that they serve certain regions of the vehicle fuel market. Up to a certain level of production (perhaps about 30 percent of national gasoline demand), biomass fuel would satisfy fuel needs in or near the regions in which it is produced. At higher levels of production major structural changes would be required in uses of agricultural and forest land; it would be necessary to make long inter-regional fuel shipments to distribute fuel to locations where it was demanded. If one posited a future of very energy-efficient vehicles, then indeed total transportation fuel needs could be met with biomass fuel without extraordinary disruption of food and forest product markets. Figure 37 identifies areas especially suited to cultivation and harvesting of wood, crop residues, and energy crops. Excluded are those forested areas characterized by sparse and slow growth, low rainfall, or steep slopes unsuited to intensive cultivation and harvesting; excluded farm-land areas are those characterized by aridity (requiring irrigation systems that are costly and energy-intensive and also may accelerate soil erosion), poor soil quality, or precipitous slopes. Thus one observes from figure 37 that those areas with high biomass production potential tend to be located far from the populated areas of the country where the fuel would be consumed. Thus production is not well matched spatially with end-use demand under this scenario of high biomass production. Likewise, a hub-and-spoke distribution network would not fit this high-production scenario. (A factor that could mitigate these shortfalls in urban areas is fuel produced from municipal solid waste [MSW], but as noted earlier, the diverse composition of MSW materials renders them unsuitable for the feedstock-sensitive process technologies that would be used to produce liquid and gaseous transportation fuels.) Thus the fourth criterion is satisfied only so long as biomass fuels are considered an important but *not* a primary source of transportation energy.

The fifth criterion, matching of energy production technologies in energy quality to end-use needs, is the most difficult to evaluate. At issue is the appropriateness and efficiency of motor vehicle technology. A narrow view of this issue of matching energy quality to energy use would be to accept the general structure of the transportation system and land-use patterns as a given. One could then readily demonstrate the superiority of liquid fuels (including biomass fuels) and internal combustion engines in most current settings. A more profound but far more complex system-level formulation of the issue would question whether activity patterns and land uses can be restructured so as to increase the attractiveness and efficiency of transportation technologies and transportation network configurations.

In any case, the use of biomass fuels as opposed to "hard" synthetic fuels would *not* change the relationship between energy quality and end-use needs. A transition to biomass fuels would tend to bring about "soft" technologies for producing the fuels, but motor vehicles would be essentially unchanged from those that consume petroleum and coal- and shale-based fuels. Thus the introduction of biomass fuels would tend not to affect current patterns of land use, vehicle production, trip making, and vehicle usage (except to the extent that their price may be different from that of petroleum fuels and thereby may induce changes in price-elastic activities), but neither would biomass fuels be matched in energy quality to end-use needs any more or less than petroleum or transportation fuels derived from coal or oil shale.

It may be concluded, then, that biomass fuel production technologies deviate somewhat from the definition of soft technologies; they tend to be less flexible, more technologically sophisticated, and less matched in scale and geographic distribution than Lovins suggests they would be. And they are not matched in energy quality to end-use needs any more than coal- or shale-based fuels. They are, however, more diverse and atomistic and sometimes technologically simpler than petroleum refineries and "hard" synthetic fuel plants.

EFFECTS OF DEPLOYING SOFT TECHNOLOGIES

A subsequent test for determining how well a set of activities fits into a soft energy path is to compare the effects of technology deployment—in this case, biomass fuel technologies compared to the large-scale process technologies for producing liquids from coal and oil shale. The major effects claimed by Lovins of deploying soft technologies are paraphrased below:[13]

1. lower initial (sunk) cost;
2. prompt incremental construction (so that new capacity is built only when and where it is needed);
3. small environmental effects that are tractable and reversible;
4. reduction of diseconomies of large scale and centralization;
5. trend toward wider economic participation.

Again, there is not a very good fit. Biomass-based fuel production does not for the most part meet the first test of lower initial cost. Although estimates of capital cost vary considerably and are somewhat speculative, initial evidence suggests that biomass gasification/synthesis plants and large coal and oil shale synthetic fuel plants have similar capital costs per unit of energy output. Capital costs are somewhat lower for bioconversion plants (to produce ethanol), but they are also lower for large gas-based methanol plants. Nonetheless, it cannot be generalized that biomass plants have lower sunk costs. Listed below are crude mid-range estimates of capital costs for building a commercialize plant, measured in terms of cost per oil-equivalent barrel per day of capacity (see chapter 8):

Coal-to-methanol	\$102,000
Oil shale-to-gasoline	\$78,000
Wood-to-methanol	\$107,000
Corn-to-ethanol	\$62,000

Apparent savings in capital costs that would result from mass production of small process plants (e.g., bioconversion, plant oil processing, and thermochemical plants) may be somewhat illusory. Yet it is clear that there would be large cost savings for mass-produced small process plants relative to similar-size one-of-a-kind plants built on site. That is, a small mass-produced plant would have a capital cost per unit of output lower than that for a similar-size one-of-a-kind plant. However, a small mass-produced plant would have similar or greater capital costs per unit of output than large one-of-a-kind plants. Thus mass production would not necessarily result in lower capital costs.

Capital cost comparisons would also not be greatly affected by inclusion of transport costs. Atomistic biomass fuel systems, because of their minimal use of pipelines, would have lower sunk (capital) costs associated with fuel distribution than would "hard" synthetic fuel plants. However, even for remote sites in the Rocky Mountain area, pipelines would rarely account for more than about 10 percent of the total fixed investment in production and distribution facilities of synthetic fuel plants. Thus, on an overall energy output basis, biomass fuel plants would require capital investments roughly comparable to those of "hard" synthetic fuel plants.

The second criterion, prompt incremental construction, is partially satisfied for biomass fuel process plants. Small installations are easily built and can procure permits relatively easily. However, the more sophisticated thermochemical process technologies require a longer lead time for production of about three to five years unless "small" 15- to 20-million-dollar plants are mass produced. But even with these small mass-produced plants, there are likely to be only a few manufacturers (because of the complexity of the technology and the limited market), thus limiting the responsiveness of suppliers to major increases in demand. From an industry perspective, construction would be incremental, since each plant contributes a relatively modest quantity to overall supply. But it would be overstated to claim the benefit of "prompt incremental construction."

The third claimed effect is possibly the most controversial and is the effect most difficult to specify: the effect on the environment. Lovins suggests that the environmental effects of deploying soft technologies would be small, tractable, and reversible. To provide a useful frame of reference, the environmental effects of biomass fuel production can be compared with the effects of other types of fuel production. For instance, fuel plants using coal and shale feedstocks would typically be ten to one hundred or more times larger than biomass fuel plants. Thus it follows that environmental effects of mineral fuel plants would tend to be more concentrated; the effects of mining would be concentrated, and large amounts of solid waste, air pollutants, and water effluents would all be generated essentially from a single source. Biomass fuel plants would be smaller and dispersed; accordingly, the environmental effects would be more dispersed. But the effects would not be small. Although there are significant differences in pollution rates between mineral and biomass fuel production processes, as demonstrated in chapters 15 and 16, the differences are even more significant within a particular process—depending primarily on the choice of external energy source, if any, and specific feedstock characteristics. For the purposes of this discussion, the generation of air pollutants, water effluents, and solid waste may be considered roughly comparable on an energy output basis.

Small scattered facilities would seem to have an advantage in that they could rely on the assimilative capacity of the environment. For example, small plants may be able to dispose of liquid wastes on surrounding lands. On the other hand, there is a disadvantage: the costs for incorporating advanced control technologies used by larger plants into small biomass plants may be excessive. Another disadvantage, from a societal perspective, is the difficulty and cost of instituting and enforcing a regulatory program for many small scattered sources. It is much easier to promulgate and, usually, to enforce emission and effluent limits for a few very large mineral fuel plants than it is for hundreds of small biomass plants.

Possibly the most devastating environmental effect of biomass fuel production is soil erosion. Soil erosion leads to siltation and nutrient buildup in waterways, a form of (nonpoint) pollution that is presently not regulated for the most part; but the more important threat of soil erosion is the soil loss itself. Crop lands in the U.S. and elsewhere are losing fertile topsoil—some claim at an alarming rate, although this is not a consensus view (see chap. 16). In any case, more intensive and extensive use of land for fuel production would exacerbate the soil erosion problem. That same phenomenon would hold for forests. Forests generally have a thin layer of fertile topsoil. If intensive management practices were extensively applied using shorter harvest rotation periods, irrigation, and heavy machinery, then soil erosion rates would increase.

It does not necessarily follow, however, that the harvesting of plants for fuel production would inevitably accelerate soil erosion. As indicated in chapter 16, by judiciously restricting intensive land utilization to more resilient areas and by appropriate use of soil conservation and plant management practices, soil erosion on farm and forest land need not be a problem. The reality, though, is that "existing economic and regulatory incentives for biomass suppliers and users to protect the environment are weak."[14] In general terms, one cannot state unequivocally that the environmental effects of biomass fuel production are likely to be less severe than those from larger mineral fuel plants, or that they would be more tractable. And if soil erosion were to accelerate, the effect might be irreversible in any relevant time frame. Thus the environmental effects of biomass fuel production would tend *not* to be small, tractable, or irreversible.

A fourth effect claimed for soft technologies is the reduction of diseconomies of centralization and large scale. While it is true that biomass fuel systems would tend to be atomistic and dispersed, it is not clear that this atomism avoids diseconomies of centralization and large scale. Lovins cites two types of diseconomies of centralization and large scale: direct effects of loss of reliability and of higher distribution costs; and indirect effects of long lead times for deployment. Long lead times imply "increased exposure to interest and (cost) escalation during construction, mistimed demand forecasts . . . wage pressures by strongly unionized crafts well aware . . . of the high cost of delay, . . . and (vulnerability) to changes in regulatory requirements."[15]

With biomass fuels, it is also not obvious that distribution costs for more decentralized facilities would be lower; in fact, as indicated in the next section, distribution costs might be higher with decentralization because the scale economies of pipelines are not attainable. In so far as the reliability issue is concerned, I was unable after prolonged investigation to locate any evidence suggesting that small liquid fuel plants would operate more reliably than large liquid fuel plants. Also, in terms of distribution, pipelines are the most efficient and reliable transportation mode for transporting large quantities of high-viscosity liquids (see chap. 10).

It is possible, however, that biomass plants do not suffer from the "indirect" diseconomies of centralization that plague larger mineral fuel plants. In particular, biomass plants are not subject to the risk that results from long lead times. The vulnerability of large projects to changing energy and economic conditions is amply illustrated by the dismantling of new oil supertankers, underutilization of the new offshore oil port in Louisiana, and the suspension and cancellation in recent years of many mineral fuel plants and nuclear power plants.

Thus biomass fuel plants avoid some diseconomies of centralization and large scale, but the overall effect is mixed. Not enough evidence is available to state unequivocally and with confidence that biomass fuel plants have fewer diseconomies of centralization and large scale. Therefore, the cost and performance of biomass fuel plants do not necessarily benefit from decentralization and small scale.

The fifth and last effect is a fundamental element of the soft path: greater and more widespread economic participation. This effect is probably the most telling in ultimately defining the difference between hard and soft paths, and it is here that biomass fuels come closest to satisfying the test.

The atomistic scale of biomass fuel plants, induced by scale diseconomies in feedstock collection, suggests that ownership would be dispersed. Indeed, relatively fragmented ownership of agricultural and forest land and the relatively modest size of agricultural and forest product companies support a hypothesis of widespread economic participation. In a 1980 unpublished survey by the U.S. National Alcohol Fuel Commission (cited in chap. 3), it was found that of more than 300 proposed and planned ethanol fuel fermentation-distillation plants, apparently only five would have been owned by oil and gas companies, and of those companies none ranked as a major energy corporation. The prospective owners were mostly small companies; a few were farm cooperatives. The predominant business of almost all prospective owners was agriculture related.

Wood-based fuel activities may also be characterized by atomistic ownership and widespread economic participation. First, roughly 28 percent of all commercial forest land in the U.S. is owned by government;[16] thus substantial land rents would return to government and the general population. Second, the prime areas for growing biomass for fuel, the lands east of the Mississippi River, have dispersed land ownership and land management,[17] which suggests extensive participation, at least in terms of economic benefits. In the southeastern U.S. forest land is owned by relatively small landowners, though it is managed by larger corporations of the forest products industry. This fragmented ownership of land resources contrasts sharply with the control and ownership of nonrenewable hydrocarbon and mineral resources and the expected participation of very large industrial corporations in coal and shale oil fuel production.

Economic participation in biomass fuel activities would be expanded by the many new opportunities available to aspiring entrepreneurs and local businesses—retrofitting vehicles, building and operating small fermentation-distillation plants that use food wastes, and brokering the sale of feedstocks from landowners and managers to process plant operators. This latter brokering service represents an entirely new layer of business activity.

Although strong evidence indicates that ownership of biomass fuel production will tend toward atomistic patterns, there is evidence and reason to believe that large economic enterprises would also participate. Already Texaco, Chevron, and Ashland oil companies have entered joint ventures in the U.S. to produce ethanol from agricultural crops. This pattern of limited partnerships by large energy companies is likely to prevail as a means for those companies to hedge their other energy investments, and to assist smaller firms that have difficulties in raising capital and marketing fuel. However, large energy companies are capital-intensive firms that have little experience with dispersed-type land management activities (though there are exceptions), and are therefore unlikely to dominate a biomass fuel producing industry.

In the forest products and agricultural industries, large companies are likely to be major participants in biomass fuel production. According to industry sources, one food processing and grain trading firm (ADM) already accounts for over 60 percent of ethanol fuel production in the U.S. However, the largest agricultural and forest products firms are much smaller than the energy companies that would be the predominant investors in nonbiomass fuel production. In 1985 in the U.S., the largest company with any food- or wood-related activities was ranked thirty-third (by sales).[18] The highest ranked forest products firm was forty-seventh. In contrast, petroleum refining and chemical manufacturing companies accounted for seventeen of the largest twenty-seven industrial companies in the country. These facts provide strong evidence that biomass transportation fuel activities would tend to have more widespread economic participation, or at least not contribute to the concentration of economic resources to the same extent as would nonbiomass fuel activities.

HOW SOFT?

In summary, biomass transportation fuel technologies are diverse. Some are flexible and relatively simple, and to some extent they are matched in scale and geographical distribution to end-use needs. They tend to be suited to incremental and rapid construction, to have some diseconomies of centralization and large scale, and to encourage more widespread economic participation than coal and petroleum-based systems. In these ways they meet the test for soft technologies and a soft path.

Some of the process technologies, however, are not flexible or simple; large-scale production would result in mismatches in geographical distribution, the sunk cost would probably not be less than for "hard" mineral fuel alternatives (i.e., for those based on coal and shale oil), and environmental effects would threaten ecosystem sustainability and would not necessarily be small or tractable.

One must conclude that biomass transportation fuel systems would tend toward an atomistic scale and have other features that meet the criteria of softness, but that biomass fuel activities do not fit well into a soft energy path. The soft/hard energy paradigm provides a weak "model" for describing and studying biomass transportation energy activities.

ROBUSTNESS AND VALIDITY OF THE PARADIGM

What are the inadequacies of the paradigm? One gains a disquieting sense that this soft path conception of biomass transportation energy misses the mark, that it does not provide a useful framework for analyzing and planning future options, and even more important, that it obfuscates rather than clarifies important causal relationships. Is the problem one of methodology in that the criteria used to define soft technologies and the indicators used to characterize the effects of deploying soft technologies are somehow invalid and deficient—that those criteria and indicators are not generalizable and robust enough to address both portable-energy issues of the transportation sector and electrification-related issues of the residential, commercial, and industrial sectors?

Earlier in the chapter, key attributes and features of soft technology systems were defined. By thinking of these attributes and features as indicators that can be used to sort and classify energy choices along a continuum from soft to hard, the hard/soft paradigm may be characterized as an index. For the index to be valid and accurate, the individual indicators that constitute the index—i.e., criteria for defining soft technologies, and effects of deploying those technologies—must be consistent with each other and with the composite index rankings. That is, successive scores or rankings on the soft/hard paradigm index must be consistent with rankings by each indicator. As demonstrated in this chapter, the principal biomass energy choices do not satisfy many of the indicators of softness. Thus the index, and therefore the paradigm, are not valid in the case of biomass transportation fuels.

An inherent feature of transportation systems which undermines the validity of the hard/soft paradigm for transportation energy applications is the large economies of scale in pipeline costs. As pipeline size increases, capacity increases at an accelerating rate, and the cost per unit of throughput decreases sharply. In other words, larger size translates to much lower cost per unit. These economies of scale have the effect of increasing pipeline's cost advantage over other transport modes and of encouraging consolidation of shipments in larger pipelines. The scale economies are so great that not only can other modes generally not compete, but the shipment cost becomes a very small, almost negligible, part of the total energy cost. A gallon of gasoline can be transported a thousand miles by pipeline for only two or three cents, while the cost by truck would be thirty to forty cents (see chap. 10). Given this dominance by pipeline and pipeline's steep economies of scale (relative to truck and rail), the geographical matching of energy production technologies with end-use needs becomes less attractive.

The matching in scale of production and end-use technologies also becomes less important. Transportation energy end-use needs are dispersed and individually small, but the advantages of large scale in pipelines are so great that, instead of encouraging correspondingly dispersed energy production, they discourage it. Thus the overall dominance and superiority of pipelines undermine the desirability of matching the scale and geographical distribution of production and end-use technologies.

Liquid (and gaseous) fuel distribution is inherently different from electricity distribution, making it difficult for a single paradigm or framework of analysis to apply to both electricity and liquid fuel energy. Liquid fuels are easily stored; excess production capacity to meet instantaneous transportation energy peaking demands is not necessary. Also, distribution costs are a much smaller component of total fuel cost for liquid fuels than for electricity. According to industry sources, electric transmission and distribution costs are roughly the same as those for electricity generation (excluding fuel, based on California data in early 1980s), whereas for gasoline, distribution accounts for about 10 percent of the cost of retail outlets and about one-fourth the cost to individual consumers (including retail markups). Thus liquid fuel distribution costs, using a pipeline-dominated system, are relatively modest. Any direct diseconomies of scale resulting from excess capacity are much less extreme for liquid fuel systems than for electrical grid systems.

In summary, two supposedly desirable features associated with soft paths—matching of scale and end use between production and end use, and technologies that are small and decentralized—are not desirable in the case of transportation energy because of pipelines. While pipelines would be characterized as a hard technology—they have large sunk costs and are inflexible, centralized, highly automated, and vulnerable to sabotage[19]—they are not necessarily undesirable, as suggested by the hard/soft paradigm. Pipelines experience little or no downtime, require minimal repair and maintenance, have few deleterious effects on the environment, and have relatively short lead times for construction (rarely more than two years). Although pipelines are vulnerable to sabotage and unable to adapt easily to changes in market conditions, overall they represent a highly desirable option. Pipelines confound the hard/soft paradigm by contradicting the presumed effects of hard technologies and by contradicting what are considered to be desirable features of soft paths.

CONCLUSION

The study of energy systems and energy policy, partly because of the relative newness of the field, often suffers from fuzzy analysis of the tangled role of ideologies, economic interests, and technical judgments. This book attempts to unravel these tangled roles.

In this chapter it has been demonstrated that the hard/soft energy paradigm is deficient when applied to biomass fuels and, by extension, to transportation energy in general. More precisely, it was found that many of the criteria and indicators that define and categorize energy options as soft or hard were inaccurate, inconsistent, and not robust when applied to transportation energy. Some criteria, for instance, were able to stratify only a small subset of the options, such as the criterion of matching energy quality between production and end use. Others distinguished little or not at all between options, and, for instance, in regard to capital (sunk) cost. And other criteria and indicators were inaccurate—for instance, in regard to environmental effects. Thus the soft/hard index was internally and externally invalid, and a generally poor composite measurement.

These criticisms are not meant to apply generally to the hard/soft energy paradigm or even to suggest that the basic precepts of the paradigm are somehow invalid or irrelevant. These criticisms are meant to suggest that the building blocks of definitions, assumptions, criteria and indicators upon which the paradigm is built are not appropriate in dealing with transportation energy. The paradigm may be robust, accurate, and highly relevant when applied to electrification activities related to the residential, commercial, and industrial sectors. However, it is not robust, accurate, or relevant for the transportation sector. In the next chapter, a set of five transportation energy paths are defined, based on a set of goals and criteria that are more directly relevant to transportation energy choices.

TWENTY-ONE

Five Paths

In this chapter transportation energy options are organized into five pathways (see table 63). The five pathways are designed and defined in terms of a matching of scale and experience between institutions and technologies, and in terms of a set of internally consistent values, goals and beliefs (see table 64). The approach is inherently technological, not in the sense of treating technology in isolation, but by relating techno-

TABLE 63 SALIENT FEATURES OF TRANSPORTATION ENERGY PATHS

	Fuels	Feedstocks	Size and Cost of Process Plants	Marketing-Distribution Area	Environmental Effects
Path I	Alcohols	Biomass	Small to medium	Localized	Mixed
Path II	Petroleum-like liquids	Coal, oil shale, oil sands	Large	National	Negative
Path III	Methanol	Natural gas, coal	Large	National, international	Mixed
Path IV	CNG, LNG	Natural gas, coal	Medium to large	International	Mixed
Path V	Hydrogen, electricity	Solar energy, water	Small to medium	National	Very positive

TABLE 64 VALUES AND BELIEFS UNDERLYING ENERGY PATHS

Paths	Energy Supply	Economy and Society	Environment
I. Atomistic Biomass Fuels	Biomass provides a permanent solution and has potential to supply a large proportion of transportation fuel needs; reduces dependence on foreign energy	Jobs and widespread economic participation are more important than economic efficiency; local control is important; concern for world peace	Environmental quality is important; belief that soil erosion will be effectively controlled
II. Petroleum-like Mineral Fuels	Ultimately, the least expensive means of producing liquid fuels; best opportunity to reduce foreign dependence in medium term	Least disruptive; preserves existing investments	Environment is less important than concerns
III. Mineral Methanol Fuels	Provides potential for gradual transition to more plentiful (but expensive) feedstocks; provides attainable permanent solution; near-term foreign dependence is not worrisome	Low near-term fuel cost; belief that cost of modifying distribution system and motor vehicles is compensated for by higher quality, cleaner fuel	Methanol will clean up urban air; belief that pollution from methanol production can be adequately controlled

Table 64 (Continued)

IV. Methane from Mineral Feed-stocks	Opportunity eventually to develop vast supplies of coal and unconventional domestic gas; utilizes existing natural gas distribution system; near-term foreign dependence is not worrisome	Less costly than methanol production; cost savings compensate for gaseous fuel's greater inconvenience to driver, lower vehicle performance (with dual fuel vehicles), and greater cost of establishing fuel outlets	Major urban air quality benefits; acceptable disruption at production site
V. Hydrogen	Faith that vastly improved processes will be developed to produce hydrogen inexpensively; clean low cost electrical energy will be available to produce hydrogen; provides potential for permanent solution; battery vehicles might play complementary role	Environmental benefits compensate for high production costs; hydrogen can become a universal fuel with widespread benefits	Environmental quality and long term sustainability are important criteria

logical developments and industrial investments to the institutional and societal environment in which they are imbedded. These five paths are intended to be distinct (but not necessarily mutually exclusive) development models that represent the range of available choices.

ATOMISTIC BIOMASS FUEL DEVELOPMENT PATH

A distinguishing feature of biomass fuel production activities is their relatively small scale. Strong economic, geographic, and institutional forces encourage smallness in feedstock-production activities; this smallness tends to carry through into fuel distribution and, to a lesser extent, into the end-use system. Although it could veer in several different directions, the model biomass fuel path presented here is designed to embrace—and is motivated by—the values of smallness and widespread economic participation, and is based on the principle of matching technologies and institutions. This path builds upon the fundamental values and principles of Lovins's soft path, but adapts them to the specifics of biomass transportation fuels.

The principal characteristic of this path is the smallness of individual investments and path participants. The path is defined as having an atomistic scale in the sense that feedstock growers, fuel producers, distributors and marketers, and vehicle suppliers would all react independently to market conditions, making decisions in response to prices set in the marketplace (or set elsewhere by political actors, but independently of growers and producers). No single participant would unilaterally influence national (or international) prices or fuel consumption. The unfolding of this path would be based on decentralized market and political decisions. Government initiatives in this decentralized biomass fuel path would come from local and state governments, not the national government (except in geographically small countries); the local and state governments would provide incentives and remove barriers to help reduce start-up costs and to stimulate local and regional economic development. The nature of a biomass fuel path, based on small atomistic investments, encourages active local participation. No central government or major economic entity manipulates or directs. The unfolding of events is not manipulated by a major economic entity or a central government (except in small countries).

The transition to new fuels would be slow along this path. Initial fuel activities would come into being as small (public or private) economic enterprises exploited small market niches and local inexpensive feedstock sources. Path participants would patiently have to negotiate the twists and turns posed by regulatory and technical obstacles and attempt to overcome entrenched corporate investments in petroleum fuels. Participants would slowly expand in number and size as obstacles were avoided or eradicated, as market uncertainty faded, and as previous investments in petroleum refineries and other petroleum-related equipment were depreciated and not modernized or replaced. Because of its dispersed and decentralized character, there is little danger of major financial or technological disasters (although undoubtedly there would be many small failures).

Along this path entrepreneurs identify and exploit opportunities in which alternative fuels can be produced and marketed for a profit. Initially they pursue those opportunities in which barriers are least formidable, technological changes are small and unobtrusive, and large profits are most attainable. Biomass opportunities that might meet these criteria of low cost and minimal disruption include fermentation of food-processing wastes and excess and culled food crops, and thermochemical conversion of clustered residues at mill sites and elsewhere into methanol. In these instances feedstock costs are minimal or even negative and allow alcohol to be produced at exceptionally low cost. The alcohol could be added to gasoline in small quantities or used straight by local farmers and possibly some rural electric utilities. Total production potential from these sources is small, however. In the U.S. the only large opportunity during the early stages of this development path might be the diversion by the national government of surplus grain, mostly corn, to alcohol production.[1] Apart from that possibility, excess and culled crops and mill and food-processing wastes would be capable of contributing fewer than about 50,000 barrels per day (b/d) of alcohol in the U.S., not enough to meet even 1 percent of the country's gasoline demand.

Expansion of the feedstock base beyond mill and food processing residues and culled and excess crops is essential for a strong and robust biomass path to develop, but the economics for this second stage are far less attractive. Thus some government subsidies would be needed to expand the feedstock and production base beyond food processing residues and excess crops.

Based on the criteria of production costs (feedstock plus processing cost), the second group of feedstocks that would be elicited into alcohol production is forest residues, especially in the southeastern U.S., but also elsewhere. The third wave of options would include surplus wood from more intensively managed forests, crop residues, and grass and legume crops, all of which would be converted to alcohol, followed finally by silvicultural farms and new agricultural and aquacultural crops. In all cases these biomass fuel activities will be region specific, depending on local feedstock opportunities and the inclination of local and state governments to promote those fuel activities.

In the U.S. alcohol from biomass has gained a tarnished reputation as a result of ill-advised government support of corn fermentation processes. But biomass presents a truly remarkable opportunity in the United States. It is well documented that commercial forests, pasture land, and, of course, idle crop land and are vastly underutilized. Commercial forest lands, for instance, could easily provide millions of barrels of methanol daily with only minimal disturbance of conventional wood and pulp product markets.

LINKAGES BETWEEN FUEL PRODUCTION, DISTRIBUTION, AND END USE

A fundamental feature of biomass feedstocks is their dispersed availability of biomass materials, which, as indicated earlier, encourages smaller fuel production plants. An especially attractive option, where small but concentrated quantities of feedstock are available, would be to deploy mass-produced process plants of about 300 to 600 b/d for methanol (and cellulosic ethanol) plants; ethanol fermentation plants would be even smaller.

The rural location of biomass plants, their small size, and their reliance on agricultural and wood materials suggest that ownership of those process plants would be dominated by companies and cooperatives much smaller than major oil companies. As a result, investors in biomass process plants would have relatively limited access to capital, reinforcing the likelihood that biomass plants will tend to be of small to moderate size.

The small size of production plants and the probable lack of fuel marketing expertise by owners would lead to dependence on truck and rail transport for fuel distribution. The high cost of long-distance fuel shipments and the high cost of developing and maintaining business relationships outside the local area would influence small producers to focus on local markets.

Short distribution distances would allow shippers to forsake the seductive but demanding features of pipelines for the flexibility of truck and rail while still keeping distribution costs low. The key feature of fuel distribution systems in an atomistic path would be flexibility; the preferred modes would be truck and rail.

Alcohol in an atomistic path would tend to be marketed (at least initially) by independent and small fuel retailers or sold directly to small users; the major customers would be farmers, independent retail outlets (unaffiliated with vertically integrated oil companies) and perhaps some small rural electric companies and cooperatives. These small businesses would be lured into the atomistic path where and when they perceived the opportunity to further their self-interest. Smallness heightens their sensitivity to market niche opportunities and their entrepreneurial instinct, and increases the likelihood that atomistic investments and activities would be forthcoming.

The forces restraining the size of production plants and constraining distribution and marketing options have less influence when it comes to end-use activities. One source of end-use technologies for alcohol fuel, which is least disruptive to existing manufacturing activities, is retrofits of conventionally powered engines. However, this retrofitting option is unattractive except as a last resort because of the unreliable performance of retrofitted vehicles and the substantial cost to the user. A second possible source of alcohol vehicles would be vehicles manufactured in Brazil, but only if they were specially designed to meet U.S. emission standards.

The third and most promising option is production runs of "alcohol-compatible" vehicles. Automakers would need to replace incompatible materials in fuel and engine systems and modify the microprocessor-controlled fuel metering mechanisms to permit the use of alcohol. With additional engineering development, these same vehicles could be given a multifuel capability. In these cases of alcohol-compatible and multifuel vehicles, engine operation and configuration would still be based on gasoline properties. As a result, the special qualities of alcohol would not be exploited. Further into the future when alcohol market penetration became significant, automakers could develop and manufacture genuine alcohol-efficient vehicles that would be designed specifically for alcohol and would operate much more efficiently. Increased use of flexible manufacturing processes would make the production of multifuel and alcohol-specific vehicles viable sooner and in smaller quantities.

RESILIENCY

A salient feature of atomistic feedstock-production and fuel-distribution activities (and end-use activities, to the extent that retrofit and import strategies are pursued) is flexibility and responsiveness to external changes and contingencies. Responsiveness is aided by low overhead and flexibility in the deployment of capital investments. Some alcohol process plants are especially flexible; for instance, fermentation-distillation facilities, with only minor modifications, may convert almost any type of sugar or starch material into alcohol.

Flexibility and responsiveness are even more significant in fuel distribution activities. Reliance on truck and rail transport, as opposed to pipelines, enhances the ability of producers and marketers to adapt their marketing strategies as conditions change. The use of truck and rail transport allows an alcohol producer to respond to shifting markets, an important consideration especially in the early years of the alcohol fuel industry, when markets would be especially unstable and uncertain.

ROLE OF GOVERNMENT

The concept of an atomistic path implies that the national government plays a minor role because participants are small, clustered in only a few areas of the country, and without a historical connection to one another. The heterogeneous and atomistic nature of participants diminishes the ability of these participants to organize effective lobbies to influence national legislation and regulation.

Some attention from the federal government is essential, of course. As suggested earlier, the national government will be called upon to alter vehicle and fuel emission rules that hinder the introduction of non-petroleum fuels, and to preserve and extend small tax concessions that favor biomass and alternative fuels (e.g., fuel excise taxes, investment tax credits). These efforts to modify rules and extend tax concessions would require only a minimal commitment and a small share of the government's resources.

More important, regulatory accommodation and government support would be found mostly at the state level, at least initially, since the principal locus of public support and industry lobbying would be regionally based. In a particular region, alcohol would be produced using only one or two feedstock-production options. In the midwestern cornbelt area, alcohol supply would come from crops and crop residues; in forested areas of the country, it would come from wood. On the state level, corn farmers and forest product companies have more influence and are in closer communication with each other because of ongoing business relationships. The unity and political power of path participants could be brought to bear on state legislatures and regulatory bodies. They could lobby for tax concessions and other incentives such as easing the cost of vehicle conversions, eliminating or reducing sales and fuel excise taxes on locally produced alcohol, easing evaporative emission regulations, requiring the purchases of alcohol fuel by local governments and public utilities, and deploying alcohol pumps at retail fuel outlets.

For alcohol fuels to gain significant market penetration, government must at least "level the playing field" so that the new fuels are not unduly penalized by start-up costs. But in practice government must go much further: it must actively promote and facilitate the transition. These governmental initiatives will reflect the societal goals, values, and beliefs of its dominant constituencies.

A fundamental goal motivating the pursuit of an atomistic biomass fuel path is self-sufficiency. It is a goal pursued not necessarily for its own sake but to stimulate economic development and to create the opportunity locally for greater security, democracy, and equality. This goal of self-sufficiency is not an archaic notion; it is not a relic of nineteenth-century visions of isolation and insulation from an unknown and threatening world. It represents an attempt to promote greater participation in the local economy and thereby to create a more egalitarian society and a society to which more people are economically and philosophically committed. It is consistent with international trade and economic growth strategies that focus on domestic economic and social development (such as the "front-yard strategy" advocated by various economists).[2]

Expansion of local economic activity and local employment in economically depressed agricultural and forested areas through an active biomass fuels industry would at least partially compensate a state (or local government) for local subsidies that might be offered to the new fuel industry. From an economic efficiency perspective, these states might not be making the optimal choice, but that is not the point. There are other important goals, such as lower unemployment and a more democratic and healthier society. And if oil prices rise unexpectedly, then the state or region pursuing an atomistic biomass fuel path may even find itself with a more efficient energy sector than if it had continued along a petroleum path.

Another reason for pursuing biomass fuels is reduced world petroleum consumption, because that in itself will contribute to world peace. U.S. motorists, who by themselves consume about one-tenth of all petroleum produced in the world, are able and willing to pay considerably more for petroleum than are consumers in less-developed countries. In effect, they are bidding up the price of petroleum, creating difficulties for the poorer nations that rely on petroleum as an easily transported and versatile fuel that is ideal for heating, electricity, cooking, and industrial purposes. In less-developed countries, petroleum serves as a transition from firewood and other subsistence fuels to more permanent (i.e., less flexible) long-term fuels.[3] Over time petroleum can be phased out as the local economies develop and the distribution infrastructure for electricity and natural gas is improved, as new energy sources are developed for electricity generation, and as industrial, residential, and commercial equipment is adapted to other less-expensive fuels. If U.S. motorists reduce their consumption of petroleum, poorer countries will face a less-steep oil price trajectory and will have more time and flexibility to convert their economies and infrastructures to nonpetroleum energy.

Providing poorer countries with more flexibility (and more petroleum at lower prices) assists their economies and eases their frustration with the world economic order, thereby reducing international tension and contributing to world peace. This concern for world peace might be shared by advocates of other alternative fuels, but is more closely associated with the social development goals of an atomistic biomass fuel path.

On the negative side, undermining the long-term growth and societal commitment to biomass fuels is the potential problem of soil erosion. If farmers, foresters, and government regulators continue their lackadaisical attitude toward soil erosion, intensified biomass cultivation could lead to environmental disaster, thereby depriving the proponents of this path of one of their major arguments—that of long-term sustainability. Soil erosion can be readily controlled, but there must be a commitment to do so from both growers and regulators.

In conclusion, a biomass fuel path would be based on values of self-sufficiency, economic democracy, and possibly a commitment to world peace. Working against these ideals would be the threats of soil sustainability and high production costs.

PETROLEUM-LIKE MINERAL FUEL PATH

The fuel options included in this path have three important characteristics: large process plants, few opportunities to exploit small feedstock or market niches, and a strong negative impact on the environment. The fuels are virtually identical to those produced from petroleum. The similarity of these fuels to petroleum fuels is both an advantage and disadvantage. The advantage is that the fuels are compatible with existing distribution and end-use technologies. The disadvantage is that they offer no improvements in fuel quality (in fact, they are inferior because of their emissions). While average costs for biomass fuels are roughly similar to those for petroleum-like coal liquids, and perhaps slightly greater than those for oil shale production, a major difference is that the range of opportunities and costs for biomass fuels is much greater. Thus some biomass fuel will be much cheaper than any petroleum-like coal and shale liquids. Because of the large fixed investment in coal and oil shale plants, the long lead time for construction, and the lack of any market niches where these fuels have an advantage, their timely development and deployment depends on massive government intervention. Even for the most dense and concentrated oil shale deposits in the world—those found in the Colorado River Basin—the cost initially for extracting and converting them into liquid fuel would be about $60 to $70 per barrel, and that does not include the additional upgrading necessary to refine them into gasoline and diesel-like fuels. The cost for converting coal into petroleum-like liquids is likely to be even greater.

As stated earlier, in the atomistic biomass path decisions regarding production and consumption are made by individuals and small firms in response to signals from the marketplace, with government mostly playing a secondary or indirect role, generally not making a direct commitment to particular investors and generally not attempting to reduce the risk for individual investors. However, because of the high cost, large size, and therefore high risk of all oil shale and coal plants, it is unlikely that any commercial-scale plant would be built without extraordinary government support—or at least not until well beyond the time when the market would otherwise indicate that such investments are profitable. This lag time could be substantial. Added on to that lag time is the additional time taken up in developing and standardizing a reasonably efficient process technology—five to eight years for an initial plant, followed by a period to learn from that experience, and then by second- and third-generation plants. Altogether, the lag time due to risk and technological development could grow to thirty years after the time when the market first signaled that these investments are likely to be profitable.

The argument for extraordinary government support is premised on this long lag time in establishing a mineral fuels industry, and on the expectation that petroleum prices will increase in an unpredictable manner. Because the effect of supply disruptions and price increases on national economies is devastating if petroleum alternatives are not readily available, there will be pressure on government to intervene in the marketplace if and when the belief becomes widespread that oil prices will rise sharply and that disruptions are likely.

A consensus that the marketplace cannot and should not be relied upon as the elicitor of new energy investments is necessary before government would intervene to support multibillion-dollar investments. The formation of that consensus provides the legitimacy for government to claim that it is asserting the public interest in bypassing the market system and the legitimacy for assuming the high risks of doing so. If a strong consensus evolves, government action will be more vigorous in reducing start-up costs and uncertainty and in providing funding and various forms of financial support to make the new fuels competitive with hydrocarbon products. Sustained support by government for these private sector investments would yield rapid buildups of fuel production capacity.

Intrinsic to the establishment of a mineral fuel path is the deployment of large technological components supported by correspondingly large economic units. Process plants cost upward of $2 billion; a large Canadian oil sands project was to cost $20 billion. Only a handful of organizations worldwide would be capable of financing such huge projects—mostly just a few large petroleum companies and some national governments. Participation could be spread over a larger number of investors through the use of joint ventures, such as with the multibillion-dollar Great Plains coal gasification project, which started production in 1985 and was co-owned by subsidiaries of five major corporations. Generally, participation would be limited to major petroleum companies because they have the capital and the self-interest to support their existing investments in refineries and pipelines, and because they have been the principal developers of the process technologies for the production of petroleum-like liquids.

There is a major risk in investing in these projects, but the risk is limited to cost overruns, performance shortfalls, and low fuel prices. Since the fuel would be mingled with other petroleum products in the petroleum product distribution system and would be indistinguishable from gasoline and diesel fuel, there would be no compatibility problems and no risk of a market not appearing for that fuel. Thus the market risks are well known and precisely defined—though huge. Government's role is straightforward: to reduce that risk, for instance, through price and purchase guarantees. Government need not be called upon to alter regulations or be responsive in any other way. This path constitutes, from a government and policy perspective, the most straightforward approach to expanding domestic fuel production because it does not threaten or disrupt investments in fuel distribution, fuel outlets, or motor vehicles.

This path is founded on a belief that abundant resources are available but that the nation is overdependent on insecure foreign petroleum supplies. Its proponents would argue that by not having to adapt the fuel distribution and end-use systems, this path becomes superior to all others. Those who support this path would likely have a stake in the status quo and have little interest in sociopolitical or environmental issues.

MINERAL METHANOL FUELS DEVELOPMENT PATH

This path includes methanol fuel produced from coal and remote natural gas (RNG). The fuel production plants are similar to those for the production of shale oil, oil sands, and other coal liquids in that very large investments in individual process plants would be required, and only the backing of very large energy companies can be expected.

The RNG-to-methanol option of this path creates less stress than the coal-to-methanol option; indeed, it has some features that are similar to those of an atomistic biomass path: low-cost feedstock niches that could be exploited to produce methanol at relatively low costs and barge-borne RNG-to-methanol production plants that are flexible in the sense that they can be moved. RNG-to-methanol investments will precede coal-to-methanol investments.

The introduction of methanol produced from RNG and coal confronts a series of major barriers and start-up costs. While methanol is a high-quality, relatively low polluting fuel and, in the case of RNG methanol, relatively inexpensive, it requires a large number of changes in government regulations and in fuel distribution and end-use systems.

Although methanol is one of the most economically attractive early sources of alternative fuels, it faces a major market development problem. Beyond a very low threshold it runs into a major problem of no readily available market. Initially it may be delivered to oil refineries based near ports and blended into gasoline in small proportions as an octane-boosting additive, but beyond that market niche, there would be no significant market opportunities. The transit bus market would be negligible while aggressive development of the electric utility market in areas with air pollution would provide only minor opportunities. The major problems are absence of a distribution system to deliver methanol to inland markets, of retail outlets to sell methanol, and of motor vehicles to consume the methanol.

The large start-up costs in this path are also an overwhelming barrier; without government intervention, methanol prices would have to be not just comparable to gasoline but lower, in order to elicit investments. Whereas investments in petroleum-like fuels are delayed by risk and unavailable production technologies, methanol investments are delayed by risk of an unavailable market.

TRANSITION ISSUES

The transition to a methanol-fueled motor vehicle system would involve a diverse set of end uses and feedstocks. This suggests a greater resilience, but also a higher order of complexity. As already indicated, the feedstock sequence would be first natural gas and then coal. Methanol from natural gas would mostly originate in remote areas of the world and in areas with low levels of natural gas consumption. Methanol would arrive at ports; thus the early markets would be coastal cities, thereby obviating the need initially for methanol pipeline transport. Later, as inland natural gas and coal feedstocks become more important, the petroleum product distribution system would be transformed into a complex petroleum-methanol system with diverse fuel origins. It would be a difficult and possibly costly transition process, but it would not be abrupt.

The feedstock transition from natural gas to coal would likely be bootstrapped onto the earliest attractive use of alternative coal fuels—

coal gasification plants whose primary output is electricity. As argued in a 1985 book by two officials of the U.S. Environmental Protection Agency, the best use of high-sulfur coal from a political, economic, and environmental perspective is to gasify it.[4] As part of the gasification process, the sulfur is completely removed from the gas products, which are then synthesized into methanol or used to generate electricity. The Coolwater gasification plant in California demonstrated that this technology could be cost competitive with direct coal combustion processes that used add-on control technologies to remove sulfur. Methanol production could be bootstrapped onto these plants by diverting some of the gas products in the coal gasification plant to an annexed methanol facility. This coproduction option is attractive to electricity producers because it allows them to run the gasification units at full capacity and to divert gas products to methanol whenever electricity demand is slack. This option is also attractive from a methanol production perspective because it makes the timely development of a transportation market less critical and because economies of scale benefits are gained without the risk of having to market the entire plant output as methanol. By creating a double market for the coal gases, the overall risk to coal gasification investors is substantially lessened.

Care must be taken in the transition process to avoid a problem of imbalance in fuel supply. In Brazil ethanol's replacement of gasoline but not diesel fuel resulted in a continuing high demand for petroleum imports in order to maintain a supply of diesel fuel. Gasoline continued to be produced, even though it was not needed. This proved to be a major problem initially since petroleum imports did not diminish and the low-quality excess gasoline was difficult to sell in other countries.

Methanol production need not create an imbalance problem. In the U.S. there is more flexibility than elsewhere because gasoline accounts for a disproportionate share of petroleum demand—about 50 percent of the petroleum products in the U.S. as compared with about 20 percent or so in most other countries. Reducing gasoline demand would reduce the need for investments in the expensive new refinery equipment that is needed to produce high gasoline yields. These investments in more sophisticated and expensive refinery equipment (e.g., hydrocracking) increased in the 1980s as lead additives were phased out and high-quality crude oil became less available. Gradual introduction of methanol to replace up to about half the gasoline in the U.S. would not disrupt the petroleum industry; indeed, refining costs might be significantly reduced.

When methanol replaces more than half the gasoline, two strategies could be followed. One strategy is to produce shale oil liquids that can be used to replace diesel fuel (indeed, the U.S. Defense Department contracted with Union Oil Co. in the 1980s to use the shale oil from the Parachute Creek Plant as a middle-distillate diesel fuel). A second option is to phase out diesel engines and replace them with a high compression spark ignition engine (essentially a hybrid of spark and compression engine technologies) that would be suited to methanol.

CRITICAL TECHNOLOGIES

Methanol production facilities are the key component in this pathway. They have huge economies of scale: the processing costs for a 50,000 b/d coal-methanol plant are $0.15 to $0.20 per gallon lower than for a similar 10,000 b/d plant, which on a gasoline-equivalent basis translates to $0.30 to $0.40 lower per gallon (see fig. 16). To attain full economies of scale an RNG-to-methanol plant must have a design capacity of at least 10,000 to 15,000 b/d and a coal-to-methanol plant a capacity of at least 40,000 b/d. Plants of this size cost from $500 million to $5 billion or more each.

The second key technology is pipelines. When point-to-point shipments reach 5,000 b/d or more, which would be the case with these coal and RNG plants, pipelines become feasible. Pipeline transport costs are generally much lower than those for truck and rail transport and thus provide huge economies of scale; as pipeline capacity increases, unit costs drop several fold. The initial (sunk) cost of pipelines is very high, however; only about 20 to 25 percent of the lifecycle cost of pipelines is for operating costs.

The large size of coal-to-methanol and natural gas-to-methanol production plants is well matched with the similarly large capacity of pipelines.

The third key technologic component, the vehicle production plant, is another determinant of the pathway's basic structure. Large outputs of methanol fuel need large markets. For instance, one 50,000 b/d plant would supply fuel for almost 1 million vehicles. Motor vehicles are the only end-use technology that forms a market large enough to serve a coal-based or RNG-based methanol industry. Because of high costs and uncertain vehicle performance, retrofitting conventional gasoline-powered vehicles on a large scale would be unsatisfactory. The motor vehicle market is most attractive if vehicles are designed and built specifically for methanol.

The cost of research and development and retooling to produce efficient alcohol vehicles would be very high on a per-vehicle basis if only a few vehicles were sold. The sense of the auto industry is that at least 100,000 vehicles or more of a single engine model must be produced to reduce production costs to levels equivalent to those for conventional gasoline-powered vehicles (see chap. 12).

Economies of scale are significant, but few firms would have the resources and market power to produce and market large volumes of alcohol vehicles. Economies of scale in vehicle production favor large production runs of only a few engine models rather than small production runs of many different models (although the introduction of "flexible" manufacturing processes may reduce economies of scale). As with pipelines and methanol production plants, cost efficiencies gained through economies of scale push vehicle manufacturing plants and vehicle production runs toward larger sizes and larger outputs.

The critical feature of this path, therefore, is the large size of investment in individual facilities, a feature that derives from two phenomena: the ability to capture economies of scale in production and the tendency for interfacing systems to operate efficiently when they are matched in size and/or operating behavior. Firms and technologies seek out similar-size units to interact with. This path proceeds most smoothly and efficiently when technologies and firms in fuel distribution and marketing and motor vehicle manufacturing are all large.

Large size means concentration of capital in relatively few projects—billions of dollars in process plants and hundreds of millions of dollars in pipelines and vehicle manufacturing plants. These large investments encourage a matching in scale and financial resources between participating firms. Large size of firms implies that firms operate in national and international arenas. The national (and international) orientation of participating firms, the imbalances between production and demand in major producing regions, and the relatively modest cost of using high-volume large-diameter pipelines encourage nation-wide marketing and distribution of alcohol. Distribution and marketing of mineral fuels would tend to expand beyond the boundaries of the regions where the feedstock is located almost from the very beginning. Methanol would be marketed in areas far from coal-producing regions and, of course, far from remote sources of natural gas.

UNCERTAINTY

Conceivably the large investments implied by this path could come about through the normal workings of the marketplace, but if so the timing would be delayed considerably due to the risks of uncertainty. Uncertainty exists because of the unpredictability of world oil prices and interest rates but also because of consumer behavior. It is unknown when oil prices will rise and by how much, whether oil supplies will be abruptly or even gradually curtailed, and what the future cost of capital will be. The major cause of this uncertainty may be characterized as market uncertainty; it derives in part from unpredictable petroleum price and availability conditions, but above all is rooted in erratic responses by consumers. As learned in the 1981–1982 collapse of the ethanol car market in Brazil, consumer purchasing behavior can be highly erratic. Consumers can readily switch from one vehicle type to another; an ethanol or methanol or even a CNG vehicle is not much different from a gasoline vehicle. If consumers perceive that a particular fuel is likely to be in short supply or is likely to become relatively more expensive than other fuels, then they can readily turn toward or away from an alternative fuel, as the case may be. In Brazil ethanol car sales dropped from 73 percent to 9 percent in only seven months. In this environment of uncertainty, investors are reluctant to commit large sums of money into single projects.

The problem of uncertainty is found whenever there are multiple systems and participants depending upon each other. In this path that uncertainty is even more profound because of the large size and restricted number of individual technology units. That uncertainty creates the "chicken-and-egg syndrome" described earlier: investments in large plants that produce methanol will not proceed without assurance of inexpensive transport services, and large alcohol pipelines will not be built until a stable supply and stable markets are assured, and automakers will not produce alcohol vehicles until they are certain of the availability of widespread fuel supplies. New investments in each of these three major activities await the commitment by others to invest in the other two complementary activities.

The challenge is to bring into being alcohol systems that are not only well matched but well coordinated. Firms and individuals acting independently to supply and use alcohol and alcohol technologies will not be drawn into an equilibrium between production and demand. That is because of the long lead time required to bring a large methanol plant on line; an investor must begin construction of a process plant long before gaining a sense of market potential, and without knowledge of what the price of energy will be when the product is finally available. If investors sense that a market may not develop fast enough to match the productive capacity of their facility as it comes on line, they are unlikely to proceed with construction.

Process plants, pipelines, and vehicle manufacturing plants must all come into being together, because the increment of fuel output represented by even one coal-to-methanol plant would be so large as to overwhelm local markets and existing distribution systems. Timely unfolding of this aggressive path relies on coordinated deployment of all three key technologies. Left to itself, the market system will hinder, not assist, this unfolding. The public sector would be called upon to counter the stasis of inaction.

ROLE OF GOVERNMENT

The public sector would have several important roles to play in this path. It cannot be overemphasized, however, that government will act forcefully only when it perceives a unity of purpose in the country. This unity of purpose is especially critical in the United States and other countries with decentralized political systems and a proliferation of governmental units on the local, state, regional, and national levels. Governmental initiatives are most successsful when consensus and unity of purpose exist, leading to and allowing collective action at all levels and branches of government.

In this path the first and most important function of government is to reduce start-up costs by reducing risk and uncertainty. In part, risk results from the unproven nature of some of the vehicle and fuel production technologies, but these are not the major sources of risk. The major risk in this pathway, as just suggested, would be the failure of alcohol fuel markets to develop in a timely fashion—that is, at a rate that matches expansion in fuel production capacity. Resolution by government of these chicken-or-egg problems of market uncertainty and of the matching of fuel production, distribution, and end use is hindered in the U.S. by two factors: the highly decentralized nature of the political system and restrictions on cooperation between firms. Antitrust rules specifically restrict joint decision making; in this case it would restrict cooperation between alcohol producers and vehicle manufacturers. Alternative competitive paradigms do exist in the U.S., however, having been adopted for similar situations in which laws and rules adopted to promote perfect competition and perfect markets proved unsatisfactory. The public utility concept is one example of an alternative model; in this case the firm is granted a monopoly in return for allowing its prices to be determined by a governmental body. Other examples include the cases of nuclear energy and unleaded gasoline, in which government deliberately and forcefully intervened in the marketplace in order to promote these two fuels.

As indicated by the 1981–1982 alcohol market disaster in Brazil, the most difficult task of government in reducing risk is to reduce consumer uncertainty. This requires giving strong, consistent signals regarding the government's commitment to a particular fuel. It involves assuring consumers that fuel will continue to be available at an attractive price and that the vehicle will retain some resale value. Fuel and vehicle suppliers can help in reducing consumer uncertainty, but the role of government is of utmost significance. Government is not always successful in orchestrating major system-level changes. Yet without a strong government commitment, transitions to dissimilar fuels will be delayed many years.

A second function of government would be to review systematically all existing rules and regulations affecting alcohol fuel and to modify or rescind them when they are unintentionally discriminatory—and to promulgate others where regulatory gaps exist. For example, alcohol is a higher-quality fuel and has generally lower air pollutant emissions than gasoline. Yet a slate of rules and regulations at both the state and federal levels that were initially enacted to promote cleaner air and higher fuel efficiency do not reward alcohol for being superior and may even perversely penalize it. Existing air quality and safety rules have to be modified, to deal with the safety and pollutant characteristics of alcohol fuel, and to establish pollution standards for the processing plants.

A third function would be to establish and enforce a set of uniform standards and specifications. Brazil was deficient in this respect; it did not tightly regulate mechanics who performed vehicle conversions or enforce fuel quality. As a result, some alcohol was sold with excessive water and contaminants that damaged engines, and many vehicle conversions were done poorly. Early and firm establishment of specifications and standards for fuel quality and pollution control is critical because it allows automakers and fuel producers to standardize their technologies. Standardization reduces costs, especially for automakers who otherwise would have to build a proliferation of engine types to run on different fuels.

In closing, I would like to expand upon one more important feature of this path: the intentional nature of investments. Large investments in this path are intentional in the sense that they would be deployed as part of an overall coordinated strategy to introduce methanol fuels on a large scale. An intentional system is one that results from collective decision making and that strives to meet some agreed-upon objectives. An intentional system does not operate according to the rules of free and perfectly competitive markets. It is protected and supported by the public sector because of an overriding public interest. (Examples of intentional systems are the Soviet gas pipeline to Western Europe, nuclear powerplants, and ethanol distilleries in Brazil.) It is clear this path must comprise intentional systems; otherwise there is no hope of this path being realized until and unless petroleum prices increase *above* the estimated production cost of methanol fuels, at which time a decade or more would pass before substantial methanol production would be on line.

Intentional systems are deployed at least in partial disregard to market signals. The "invisible hand" is not present to steer away from mistakes; its absence increases the potential of white elephant projects. It is a risk inherent in the type of systems that would be deployed in this (and the previous petroleum-like mineral fuel) path. Intentional systems are not likely to disappear, however, even if the decision-making process is faulty. Firms and governments would be prepared to absorb the risk of mistakes because of their large sunk costs. The federal government could assure that no one unit of government or single corporate enterprise exposes itself to excessive risk. Although the prospect of large mistakes casts a shadow over these investments and, as it should, has a sobering influence on any proposals involving governmental liability (such as loan guarantees), it would not be a unique occurrence in U.S. history. Risk and responsibility for financial disasters could be dispersed (as has been the case in the past with other enterprises) through tax relief, purchase guarantees, and treatment of enterprises as public utilities.

As with the petroleum-like mineral fuel path, the decision to branch onto a methanol mineral fuel path would be founded in part on a belief that the nation is overdependent on insecure petroleum supplies. The rationale for shifting from one foreign energy source (petroleum) to another (remote natural gas) is that the number of suppliers would be larger and more dispersed, and that dependency is a long-term and not a short-term problem.

The methanol path is based on a stronger environmental ethic than is the petroleum-like mineral fuels path and on the belief that the added cost of adapting fuel distribution and end-use systems to methanol would be offset by the production cost savings of methanol over shale and directly liquefied coal fuels. It is also based on a belief that a gradual transition from petroleum to methanol produced from natural gas, coal, and perhaps later biomass provides a permanent and relatively palatable long-term solution for transportation energy.

METHANE FUELS FROM MINERAL FEEDSTOCKS

The initial feedstocks in this path are domestic and remote foreign natural gas, to be superseded later by unconventional domestic natural gas and coal gases (and perhaps biomass gases). This path is similar to a methanol path in that it relies on natural gas and coal and requires substantial changes in motor vehicles. It is different in that some gas would be imported as LNG and that the existing natural gas pipeline system could be utilized.

The advantages of using natural gas and coal feedstocks as methane rather than methanol are lower production costs, milder environmental impacts, and access to the large existing natural gas pipeline distribution system. The principal disadvantage is the relative inferiority of the dual-fuel vehicles that would be used initially. The problem is that separate fuel systems must be developed: one for gasoline, the other for CNG. Alcohol multifuel vehicles can mix the gasoline and alcohol; gases and liquids cannot mix. The result is a highly compromised vehicle, in terms of cost, performance, interior space, and emissions. As suggested earlier, the demanding customers of affluent countries would be especially resistant to these transition vehicles, despite their cost attractions.

Another obstacle to this methane path would be widespread opposition to LNG port terminals. Indeed, efforts to build an LNG terminal in California in the 1960s and 1970s were rejected after prolonged opposition by local governments and environmental groups. Most communities in the U.S. are likely to be opposed to the construction of nearby LNG port terminal because of safety fears.

Because of the high cost of building LNG terminals (and ships), their vulnerability to shifting demand and prices, and the need to develop broad local support for the LNG facilities, LNG port terminals have an "intentional" nature to them. Nonetheless, LNG imports may prove to be the least expensive means of expanding natural gas supplies in the

U.S. Rejection of LNG imports may limit the market penetration of methane transportation fuel.

A third factor inhibiting the introduction of methane fuels in the U.S. is the lack of investors and promoters. The automobile companies have a decided lack of interest in methane because dual-fuel vehicles would be unattractive to consumers; all the major auto companies have indicated either in public pronouncements or by their R&D activity that they prefer methanol to methane. The obvious sponsors of methane fuels—natural gas utilities and pipeline companies—have shown a remarkable lack of interest in using natural gas for transportation. For instance, in 1987 only about 11 percent of all gas utility vehicles in the U.S. had been converted to natural gas, which represents virtually no increase over the previous five years.[5] Another indication of lack of interest is that of about 400 business and government people who attended a 1986 international conference in Vancouver, Canada, on "Gaseous Fuels for Transportation," only a handful were from U.S. natural gas companies. The lack of interest is explained by a combination of the following factors: relative insularity of natural gas companies resulting from decades of government regulation, complete ignorance of the transportation energy market, diminishing regulation which is creating greater competition for industrial gas users and diverting utilities attention to short-term concerns of profitability and market retention, absence of large low-cost domestic gas supplies, and absence of government support for natural gas as a transportation fuel.

In contrast, Canadian gas producers and utilities have been much more aggressive in pursuing the transportation energy market. The Canadian utilities benefit from greater regulation, which protects them from unsuccessful experiments, from access to large domestic gas supplies and, most important, active government support for natural gas use as a transportation fuel.

In conclusion, a gaseous fuel path is superior to a methanol path from an economic and, to a lesser extent, an environmental perspective. However, the transition process is more difficult. As a result, the end-state attractions of a methane path may never be realized in the U.S. because of the barriers and the lack of proponents and self-interested participants. The difficulty of getting from here to there may mean the U.S. may never get there. The getting there part of the process may prove to be much easier and the attractions much greater in those less developed countries with natural gas reserves.

HYDROGEN FUEL PATH

This path would not become viable until well into the twenty-first century. This motivates the question: should we delay making commitments to other fuels until more is known about hydrogen or should we proceed tentatively with one of the first four fuel paths with the understanding that we may want to allow that transitional path to atrophy sometime in the next century? If one feels confident that serious petroleum shortages will not occur until the mid–twenty-first century and that OPEC will not be able to raise petroleum prices above $45 or so per barrel (1986 $) until that time, then it may be worth the risk to stay with petroleum until 2010 or so and then to reassess the situation.

If hydrogen were to be the ultimate fuel, then the choice of an interim path should be based on compatibility with hydrogen. In fact, no fuel or path is vastly superior to others in terms of reducing the subsequent barriers to the start-up of a hydrogen path. The methane path is a somewhat better predecessor for hydrogen than other fuels because hydrogen could probably use the natural gas pipeline network (see chap. 11) and because methane would accustom vehicle users to gaseous fuels. But these benefits are not great enough to be a principal factor in choosing among the first four paths.

The transition process to hydrogen would be difficult for both technological and institutional reasons. Technologically, the difficulties are the development of an efficient electrolysis process and photovoltaic cell and more efficient and effective means of storing hydrogen on board vehicles. Institutionally, mechanisms must be created to reward the environmental benefits of hydrogen—its clean production process and very low emissions from combustion, including the absence of carbon dioxide. Also, there would have to be a willingness to devote large areas of land to photovoltaic cells to generate the energy for hydrogen production.

The hydrogen path would be motivated primarily by concern for urban air pollution and the greenhouse effect and possibly by a desire to decentralize energy systems and increase economic participation in energy activities. This latter factor is similar to the motivations of economic democracy and decentralization that underlie the biomass path.

Significant support for the introduction of hydrogen fuel would depend on definitive evidence that the greenhouse effect is significant. Hydrogen is the only transportation fuel other than solar-generated electricity and biomass fuels that does not emit carbon dioxide and other "greenhouse" gases into the atmosphere. If the greenhouse effect begins creating disastrous climatic shifts and hydrogen fuel is not acceptable for some reason (such as costs), then dramatic wrenching changes will be needed in transportation systems. There might have to be a shift toward mass transit and away from the mobility of personal vehicles.

An accompaniment to hydrogen fuel would be battery vehicles. While battery vehicles are not likely ever to provide the same combination of driving range, power, and interior space as internal combustion vehicles (including hydrogen), with continued improvements in battery technology, they could prove to be viable where long-distance travel is not necessary. For instance, they could be used as dedicated commute vehicles (for work trips), secondary household vehicles, and in many fleets. Battery vehicles are more likely to gain acceptance as part of a hydrogen path than any other path because they, like hydrogen vehicles, could reduce environmental pollution sharply, especially if the electricity were generated from solar energy or some other clean renewable source. Thus battery vehicles would be highly valued in this environmentally motivated hydrogen path and their shortcomings more readily accepted.

TWENTY-TWO

The 1980s: Where Are We? Where Are We Going?

Nonpetroleum transportation fuels are not fuels of the future. They are fuels of the present. They are already here. As indicated in earlier chapters, ethanol is the dominant transportation fuel in Brazil, coal liquids are the major source of liquid fuel in South Africa, CNG has made significant market inroads in New Zealand, and ethanol and methanol have been important gasoline additives in the U.S. What do those experiences imply about the future of alternative fuels? Is each a special case? How do they relate to the concept of energy pathways?

CASE NO. 1: IS BRAZIL FOLLOWING AN ATOMISTIC BIOMASS FUEL PATH?

The initial decision in 1975 to replace petroleum imports with ethanol was not a commitment to branch away from petroleum onto an alternative path. More accurately, it was an effort to utilize existing sugar cane mills and distilleries more effectively, stabilize the domestic sugar cane market, and reduce petroleum imports. If petroleum prices had not spiraled up in 1979–1980, then the ethanol effort probably would have remained a marginal endeavor with half-hearted support. As it happened, the price escalation and supply disruptions resulting from the Iran-Iraq war invigorated support for alternative fuels and created the political environment that allowed the government to take the crucial next steps—financial support of specialized "autonomous" ethanol plants, distribution of straight alcohol, and commitments to automakers to support and encourage alcohol vehicles. Only then, in 1979 and 1980, did Brazil truly commit itself to nonpetroleum transportation fuels.

Until 1979 Brazil was following what resembles the initial stages of the atomistic biomass development path described in the previous chapter—it was an incremental process, and government played a modest role in reducing the risk and uncertainty of new investments. Around 1979 that governmental role changed; the central government became far more aggressive about financing new fuel production and subsidizing fuel prices and alcohol car purchases. At that point Brazil was shifting onto a more directed path resembling the third development path, the one based on widespread consensus with projects having an "intentional nature."

That consensus was precarious, however. When problems with fuel contamination and converted vehicles were publicized in 1981, consumer demand collapsed and threatened the fuel program with disaster. It was through the persevering power of the central government and residual support for the goal of energy independence that the country reaffirmed its commitment to ethanol fuels.

Brazil's commitment to ethanol fuel was based on only a subset of the values and beliefs associated with the atomistic biomass fuel path. The commitment was not based on environmental concerns, and (despite early proclamations) only weakly on social goals. Little attention was given to the goal of economic participation and the food versus fuel conflict of feeding the cars of the middle and upper classes rather than the stomachs of the more numerous lower class. The Brazilian ethanol program was, however, strongly rooted in a belief that self-reliance and self-sufficiency were important and that a high priority should be placed on developing national resources and stimulating national economic activity. In a sense they were pursuing the "front yard strategy" suggested in the discussion of the biomass fuel path, but without the attendant pursuit of social goals. One wonders how long this more narrow commitment will retain national support in the face of low petroleum prices.

CASE NO. 2: ETHANOL FUEL IN THE U.S.

The U.S. experience with ethanol fuel is difficult to characterize. Biomass ethanol activity in the U.S. closely parallels the early ethanol activity in Brazil—flexible production capacity tied to existing industrial operations and the use of small blend proportions so that the fuel is compatible with existing distribution and motor vehicle technologies. In both cases the government was subsidizing farmers, fuel producers, and marketers. However, while U.S. subsidies of biomass ethanol reflect in principle a philosophical commitment to small businesses and renewable fuels, the reality is very different. Most of the biomass fuel is produced in a few fairly large plants, much of it by a single large corporation. While some of the fuel is produced from waste materials and takes advantage of local feedstock opportunities, most is produced from corn in large integrated industrial facilities. Moreover, the fuel is distributed over long distances, not just locally—some of it as much as 2,000 miles away (e.g., Illinois to California).

This discrepancy between the concept of an atomistic biomass development path and the reality of highly centralized industrial activities raises the question whether the concept of an atomistic-style path is viable in the U.S. This question has never been diligently addressed, and there is little published data to test the proposition. The absence of small fuel producers is apparently due to their lack of experience in the fuel industry, their inexperience in marketing and utilizing coproducts (e.g., protein-rich materials, carbon dioxide gas), and their lack of market power in selling the fuel. Ethanol producers are operating in an energy industry where even the smallest independent gasoline marketers are fairly large—handling much more fuel than would be produced even by a group of small biomass fuel producers.

Most of the biomass ethanol fuel being produced in the U.S., even that produced by large agricultural corporations, is being sold through independent marketers such as 7-eleven. One wonders how many small ethanol fuel producers failed and how many prospective producers withdrew because of their inability to find marketers and retail outlets to sell their fuel. The complexity of fuel quality regulations (e.g., vapor pressure) and tax refund procedures, as well as ignorance of how to do business in this new industry, surely provided additional barriers. My suspicion is that analysts and researchers underestimate the difficulties faced by a small new business with a unique product trying to get started in such a well-established industry as fuel marketing and distribution. It is unclear how large and significant those difficulties are, but it is probably safe to say that the initial institutional barriers to a true atomistic biomass path are considerable.

In conclusion, biomass energy production and fuel distribution in the U.S. have grown in a direction that is just the opposite of what an atomistic biomass fuel path represents. The U.S., halfheartedly, and some of the farm states, more enthusiastically,[1] have promoted biomass ethanol, but so far with disappointing results. The case of the U.S. economy—and of energy production industries in particular—is so strong that a shift to small atomistic activities could only be achieved with revolutionary changes. One scholar of the politics of technological change observes that

> for more than a century utopian and anarchist critiques of industrial society have featured political and technical decentralization. While it has wonderful appeal . . . [h]ow can it have any importance in a society thoroughly enmeshed in centralized patterns.[2]

Earlier it was argued that long-term energy choices are best viewed as part of a price-determining process. In the case of biomass fuels it is clear that a major shift in societal values and a strong sustained commitment to the promises and attractions of a biomass fuel path are necessary if such a plan is to become a reality. Without a shift in values and without public commitment to atomistic decentralized activities, whatever biomass fuel activities come into being are likely to resemble centralized electric and hydrocarbon activities.

CASE NO. 3: SOUTH AFRICA'S PETROLEUM-LIKE MINERAL FUELS PATH

The Republic of South Africa has branched onto the second development path identified in the previous chapter: the production and use of petroleum-like coal liquids. South Africa's experience fits that conceptualized path almost perfectly. In South Africa's case international trade sanctions provided the motivation to build a domestic fuels industry. The central government instigated the establishment of an energy company, Sasol Ltd. (30 percent owned by the government), to produce liquid fuels from domestic coal reserves. As detailed in chapter 5, by 1985 about one-third of liquid fuel demand was met by synthetic liquids, a total production of over 70,000 barrels per day.

South Africa had sought a reliable source of energy that was compatible with conventional motor vehicles. By producing compatible fuels, the government avoided the responsibility of having to maintain an equilibrium between production and end-use markets; it was able to substitute coal fuels for imported petroleum fuels without having to modify distribution and end-use systems.

The cost of reducing risk and of preserving investments in existing infrastructure and technology was large; it involved more expensive and less efficient process plants, and lower-quality and more-polluting fuels. Nonetheless, the country is apparently willing to accept those costs in order to achieve its primary objective of fuel independence.

The decisive commitment to a coal-based fuel path in South Africa was possible only because of the apparently strong consensus that existed in the country that energy security was an extremely important goal and that the most attractive means for attaining that goal was via a coal liquids program.

CASE NO. 4: PETROLEUM-LIKE MINERAL FUELS IN U.S.—A MISSING INGREDIENT?

In 1979 the Carter Administration had proposed the creation of a huge fund for alternative fuels, a public corporation to disburse those funds, and a high-level commission to expedite the approval and construction of fuel production projects (see chap. 3). Those proposals, if fully implemented, would have been excellent examples of a petroleum-like mineral fuels development path such as described in the previous chapter. The intent of the proposals was to encourage investment in alternative fuels by reducing risk. Biomass and heavy oil were eligible, but the intended emphasis was on the production of liquids and gases from coal and oil shale. It was expected that a number of very large process plants would be built. The entire process would have had an "intentional" nature to it; the plan was to offer large price, loan, and purchase guarantees in order to transfer risk from the private investors to the federal government. In 1979 it seemed that the urgency to reduce petroleum imports and to reduce dependency on unstable and erratic suppliers would translate into widespread support for Carter's aggressive proposals. Clearly, the Congress would have had to perceive a strong consensus before it would have been willing to make an $88 billion commitment and to establish an organization that could override environmental laws and various project approval prerogatives of state governments.

The consensus proved to be fragile and transient. The apparently strong support for energy independence and domestic fuels was not felt very deeply. As the sense of crisis resulting from supply disruptions and price increases in 1979 dissipated, so did the urgency for developing domestic energy sources. The 1980 legislation resulting from President

Carter's proposals established the government-financed Synthetic Fuel Corporation with authority to commit only $17.5 billion by 1985 (with "promises" to consider appropriating an additional $68 billion for a second stage), and included no provision for the Energy Mobilization Board. The final legislation paled in comparison to the original proposals. Enthusiastic initial support in Congress had dissipated; the necessary consensus had fragmented.

Given the decentralized political system of the U.S., it is unlikely that a strong consensus to promote alternative fuels will form until another disruptive event occurs and, judging from recent history, perhaps not even then. Thus, for the U.S., it seems that the more extreme elements of mineral fuel paths—those premised on an assertive and aggressive role by the federal government—are unlikely. Direct intervention to assist very large individual investments does not seem a likely possibility. The U.S. government (and electorate) tends to prefer an indirect role; the example of the nuclear energy industry, whereby the government provided numerous indirect subsidies (e.g., restricted liability, waste storage facilities and uranium enrichment and reprocessing facilities, and massive research programs), illustrates this preference. In the case of nuclear energy the government could reassure itself that it was only facilitating market forces without directly subsidizing particular investments and investors, even though in hindsight this indirect approach is seen as having resulted in several financial disasters for the government—e.g., multibillion-dollar losses each on the Clinch River breeder reactor and uranium enrichment and reprocessing facilities.[3]

CASE NO. 5: A METHANOL FUEL PATH IN THE U.S.?

Beginning in 1982 in the U.S. methanol began to be used in increasing quantities as a gasoline additive. In 1985 about 250 million gallons (16,300 b/d) of methanol, representing about 0.25 percent of gasoline consumption, were used for transportation energy. Only about 75 million gallons of that methanol were used directly as a gasoline additive; the other 175 million gallons were used to produce MTBE, a chemical which is not soluble in water and is therefore more compatible with the petroleum product distribution system than methanol (but also more expensive). The trend is to use methanol for MTBE and less as a gasoline additive itself. In 1987 total methanol fuel use increased to about 375 million gallons, although 350 million of those gallons were for

TABLE 65 U.S. METHANOL MANUFACTURING CAPACITY, MILLION OF GALLONS PER YEAR, 1985

Company	Plant Location	Capacity	Feedstock	Remarks
Air Products	Pensacola, FL	60	Natural Gas	
Allemania*	Plaquemine, LA	130	Natural Gas	Shut down temporarily (7/84)
ARCO	Houston, TX	200	Natural Gas	
Borden	Geismar, LA	210	Natural Gas and "Off Gas"	110 million gallons/year portion of capacity shut down temporarily
Celanese	Bishop, TX	145	Natural Gas	
	Clear Lake, TX	230	Natural Gas	Closed
DuPont	Deer Park, TX	200	Residual Fuel Oil	Shut down temporarily (9/84)
	Beaumont, TX	250	Natural Gas	
Georgia-Pacific	Plaquemine, LA	130	Natural Gas	
Getty	Delaware City, DE	100	Raffinate	
Monsanto	Texas City, TX	100	Natural Gas	
Tenneco	Houston, TX	150	Natural Gas and "Off Gas"	Closed
Tennessee-Eastman	Kingsport, TN	60	Coal	
Total Available Capacity		1965		

SOURCE: U.S. Dept. of Commerce, A Competitive Assessment of the U.S. Methanol Industry (Washington, D.C.: GPO, 1985), p. 2.

*A joint venture of Ashland Chemical and International Minerals and Chemicals.

MTBE and only about 25 million were for direct methanol use. Methanol also gained wide usage as a gasoline additive in Western Europe, especially West Germany, where, as of 1985, almost 70 percent of all gasoline contained 3 percent methanol.[4] This use of methanol as an additive is likely to grow in the U.S., Europe, and elsewhere as tetraethyl lead is phased out.

Unlike ethanol fuel, methanol penetrated the U.S. transportation fuel market with no special government subsidies or unusual government intervention. In one sense it is an anomaly of an oversupply of methanol in chemical markets; in another sense it represents the most likely genesis of widespread production and use of alternative fuels.

TABLE 66 WORLD METHANOL CAPACITY

	Billions of Gallons Per Year
U.S.	
Existing, 1983	1.53
Projected, 1986	1.90
Shut down (as of 1984)	0.36
World (including U.S.)	
Existing, 1983	5.41
Projected, 1986	8.36

SOURCE: Chevron, Inc., The Outlook for Use of Methanol as a Transportation Fuel (Richmond, Calif.: 1985).

Note: Plants in the following countries and states were completed between 1983 and 1986: Bahrain, Saudi Arabia, New Zealand, Indonesia, Malaysia, Burma, U.S.S.R., East Germany, Trinidad, Libya, India, Texas (ARCO), Delaware (Getty), and Tennessee (coal-to-methanol).

The increasing use of methanol and the pressure for expanding its use is directly related to dramatic changes in the world methanol industry. Until 1982 the U.S. was a net methanol exporter. In 1985 U.S. methanol production capacity was about 2 billion gallons per year (130,000 b/d), almost all based on natural gas and natural gas liquids (see table 65). In the 1980s, however, a number of countries with large amounts of unused or flared natural gas decided to gain economic value (and foreign currency) by converting the gas into methanol for the export market (see table 66). These new plants were built on the expectation of strong chemical demand for methanol, which did not materialize, and the perception that their low feedstock costs would allow them to produce methanol far less expensively than plants in the U.S., which used high-priced domestic natural gas, or plants in Europe and Japan, which used mostly naphtha, a light petroleum product. In some cases major U.S. and international chemical companies hold an equity stake or are managerial participants in the projects, while in other cases the national oil company is the sole owner and operator.

Methanol supply and demand became highly imbalanced in the 1980s. In 1984 world production capacity was about 5.4 billion gallons, but within two years methanol production capacity had increased to over 8.3 billion. Since worldwide demand (for all uses) was only about 4.6 billion gallons in 1986, many plants were forced to close or operate at less than capacity, especially those without access to low-cost remote natural gas. By early 1985 four U.S. methanol plants were shut down. Despite this huge oversupply, continued investment in new methanol plants is likely, especially if there is even a glimmer of hope that methanol will become a transportation energy reality. The driving force is the desire of gas-rich countries to exploit their natural gas reserves to improve their trade balance and acquire foreign currency to finance their debt payments. The trade press in 1985 and 1986 carried notices of numerous plans for new methanol plants in countries with no previous history of methanol production. The ease of purchasing state-of-the-art methanol plants from international construction companies makes these plants easily realizable, as indicated by the methanol plant construction craze of the early and mid-1980s.

Natural gas production costs will remain low in those countries where gas is a by-product of petroleum production. Table 67 lists countries with large gas reserves but low domestic gas production. Most of the gas in those countries was found associated with oil, since there has been little or no incentive to explore for gas in those countries. Future world supply of methanol will increasingly shift to these low-cost countries, not unlike the shift to Middle East oil that took place in the 1950s, 1960s, and 1970s. New methanol plants are not likely to be built in the U.S. for at least several decades unless the U.S. government adopts a policy to promote domestic production.

In any case, methanol prices are likely to remain low worldwide as long as large amounts of easily accessible natural gas are available. These untapped resources are huge. If, for instance, only 10 percent of the reserves of the countries listed in table 67 were converted to methanol, then 35 billion barrels would be produced, enough to last nineteen years if consumed at the rate of 5 million barrels per day. Clearly, there is a huge amount of natural gas readily (and cheaply) available that at present has almost no market or value.

The developing use of methanol fuel in the U.S. and Western Europe is following a very different path from that of ethanol in Brazil and the U.S., and petroleum-like fuels in South Africa and New Zealand. One difference is that by increasing the usage of methanol, the U.S. and Europe will become more dependent on foreign supplies (though many of the foreign methanol plants are partially or totally owned or operated by U.S. corporations).

A second unique aspect of the methanol fuel experience in the U.S.

TABLE 67 COUNTRIES WITH HIGH POTENTIAL FOR METHANOL PRODUCTION BASED ON THE AVAILABILITY OF NATURAL GAS

Country	1983 Est. Gas Reserves (10^{15}Btu)	Gas Consumption (10^{15}Btu/yr)	Reserve/ Production Ratio	% of Reserves Necessary for One Large Methanol Plant*
Australia-New Zealand	22.4	0.51	44	2.2
Brunei-Malaysia	55.3	0.33	167	0.9
Burma	5.1	?	High	9.8
India	14.8	0.073	203	3.3
Indonesia	30.5	0.694	44	1.6
Pakistan	16.2	0.328	49	3.1
Bahrain	7.8	0.182	43	6.4
Iran	369.5	0.256	1443	0.1
Iraq	25.1	0.036	697	1.9
Kuwait	32.9	0.219	150	1.5
Oman	5.9	?	High	8.5
Qatar	59.8	0.146	410	0.8
Saudi Arabia	114.0	0.511	223	0.4
United Arab Emirates	99.8	0.474	210	0.5
Algeria	130.0	0.766	170	0.4
Egypt	9.7	0.073	133	5.1
Libyia	19.8	0.146	14	2.5
Nigeria	48.9	?	High	1.0
Tunisia	6.4	?	High	7.8

SOURCE: U.S. Dept. of Commerce, A Competitive Assessment of the U.S. Methanol Industry (Washington, D.C., GPO, 1985), p. 11.

*Assumes 20-year life, 25 billion cubic feet (25×10^{12} Btu) of gas per year, and production of 20,000 barrels of methanol per day.

has been the absence of assistance from the government. In fact, the only significant action by the government has been to hinder and constrain methanol's use as a fuel through various rules and regulations. It represents the only significant market situation, possibly worldwide, where an alternative fuel has been economically attractive without government intervention, even at very low gasoline prices. In that sense, methanol has neither an "intentional" nor a market-ignoring nature.

The continued and growing use of methanol fuel produced from natural gas has some commonalities with the atomistic biomass path. Although individual plant investments are large—around a half billion dollars each—they are significantly smaller than full-size coal, oil shale, and oil sands plants. They do not require (or at least do not receive) government support and are not coordinated or orchestrated in any way. They exploit local feedstock opportunities and also are somewhat resilient in that there is a fair amount of opportunity to expand methanol's chemical market and to expand local market opportunities in the countries where the plants are located. And the development of methanol fuels provides opportunities for wide and diverse economic participation.

Thus the early stages of a methanol fuel path reflect many of the principles underlying the atomistic biomass path. The key difference is that these principles apply on an international, not a domestic, level. It is a crucial difference because the key values and beliefs underlying the atomistic biomass path—those articulated by proponents of a "soft" path and "front yard" strategy—mean controlling one's own destiny. Relying on methanol from foreign suppliers, many with erratic and unstable governments, is not a way to control one's destiny.

PART IX

Conclusions

It is inevitable that petroleum transportation fuels will be phased out eventually; indeed, that process has already begun, albeit in fits and starts. The relevant questions, therefore, as first suggested in Part II, are: Which fuels are the most attractive replacements for petroleum? What role should government play in the transition to those fuels? And what would be the most effective and efficient manner for initiating that transition? Although there are no definitive answers to these questions, some generalizations can be made.

WHICH FUELS?

First, which fuels are most attractive? The answer, as I have indicated, is different for each region and society. In most locations, including North America, Japan and Europe, the use of starch and sugar crops to produce ethanol is a clearly inferior option. Cellulosic biomass feedstocks have much more potential, but they would require major societal changes, including mechanisms for controlling soil erosion, in order to become an attractive and accepted option. Such revolutionary restructuring and reorganization of economies and societies is not unreasonable and may be highly desirable; it would represent a shift toward decentralization and dispersion of economic power coupled with greater regulatory intervention to assure sustainable ecological systems.

Conventional wisdom favors mineral fuels, however: first, because coal, oil shale, oil sands, heavy oil, and possibly natural gas are so abundant that each could keep the world supplied with energy for a century or more, and second, because extraction, production, and distribution patterns for those fuels would fit well with the technical, financial, and institutional attributes of current energy industries. Current energy companies would easily and relatively painlessly be able to make the transition from petroleum to these other feedstocks, especially when these mineral feedstocks are converted into fuels that are similar or identical to gasoline and diesel fuel. In this latter case of petroleum-like mineral fuels, the impact would be small not only on the energy sector but also on the transportation sector. Unfortunately, these petroleum-like mineral fuel options have strongly negative impacts on the environment. They also reinforce the current concentration of resources and power in a small number of corporations. Some amount of development of these mineral resources to make petroleum-like fuels is undoubtedly desirable, but other options are more attractive environmentally and in some cases in terms of market costs.

One somewhat more palatable use of coal from an environmental perspective would be to convert it into methanol and gaseous products. An even more attractive option in the near term from both an economic and an environmental perspective is the use of natural gas for the production of methanol and gaseous fuels. The extraction and production of these natural gas–based fuels is much less environmentally intrusive, and the fuels themselves are environmentally far superior to petroleum and petroleum-like fuels. The disadvantage for the U.S. is that natural gas is not as abundant domestically as the other mineral feedstocks.

An important debate is which of the two fuels, methanol or natural gas (as CNG or LNG), is preferable for the transportation sector. Natural gas, especially in compressed form as CNG, is generally less expensive than methanol, but it is also less compatible with existing pertroleum fuel distribtuion systems and relatively less attractive during the transition period (because dual-fuel CNG vehicles are more inferior to dedicated CNG vehicles than are dual-fuel methanol vehicles relative to dedicated methanol vehicles). Natural gas is therefore more attractive where consumers place greater weight on price than performance and where countries have large domestic supplies of natural gas but not massive sunk investments in petroleum distribution systems. Indeed, the potential benefits of converting entire energy systems, not just the transportation sector, from petroleum to natural gas are huge for those countries. All developing countries with large supplies of natural gas, which may include as many as fifty countries, should be seriously pursuing this idea of a natural gas economy.

In summary, the preferred choices in the near- and medium-term future are fairly obvious. Assuming that more natural gas will be found, especially in Asia, Africa, and South America (which are less explored than North America and Europe), the following choices seem most attractive. First, as suggested above, countries with large gas reserves should switch most or all motor vehicles to CNG. This list includes many countries in Asia and several in Africa and Latin America, as well as Canada, the Middle East and the Soviet Union.

Second, many of those countries without gas reserves of their own would be well served by also diverting part of their motor vehicle fleet to CNG. Japan is already a major importer of LNG and could continue along that path by phasing natural gas fuel into the transportation sector. Europe has access to nearby gas in the Soviet Union, the Middle East, and North Africa and could easily use that gas as CNG motor fuel.

The one risk these nongas countries would want to avoid is undue reliance on natural gas imports because it is costly and time consuming to secure new gas supplies and to build new LNG facilities, both in the originating and the receiving countries. If, for instance, Iran were to expand and then for some reason curtail natural gas exports—whether because of sabotage, natural disaster, or political upheaval—then the receiving countries, for instance in Europe, would have considerable difficulty replacing those gas imports. These nongas countries would have to build new pipelines to other adjacent supplies or build an LNG system for imports from more remote locations. Natural gas supply and transport systems are much less flexible and much more costly than liquid fuel systems. It would be irresponsible to become heavily dependent on natural gas imports.

Japan and Europe might therefore safely divert part but not all of their transport sector to CNG. For Europe this presents somewhat of a quandary because natural gas imports could come by pipeline and therefore be considerably less expensive than methanol (or imported LNG). The U.S. and Japan, on the other hand, in order to increase significantly their natural gas consumption, could not rely on pipeline imports. They would have to rely on LNG imports, which would not be significantly less expensive than imported methanol. Thus, if Japan and the U.S. had the desire to divert a large share of their transport sector to a new fuel in the near- to medium-term future, one would expect them to prefer methanol imports to natural gas imports. That decision would lead to somewhat higher energy prices and result in slightly greater pollution, but it would reduce vulnerability.

Farther into the future is hydrogen. While hydrogen appears to have somewhat larger market costs than other options, it may prove to have the lowest total social cost. Into the foreseeable future, the cost of producing hydrogen from water with photoelectricity will be considerably greater than the cost of manufacturing other fuels; hydrogen will also continue to be more expensive to transport and more difficult and expensive to store on board vehicles than other fuels. However, when produced with solar electricity from water, hydrogen represents a clean, sustainable energy source. Hydrogen vehicles would not pollute urban areas, and hydrogen production would not generate carbon dioxide and other gases that cause greenhouse warming of the planet. If the greenhouse effect proves to be as serious as some scientists claim, then it may be advisable to shift to hydrogen as quickly as possible, notwithstanding its higher private cost. As with biomass, photoelectric hydrogen production is also amenable to small production modules, thereby creating the conditions for a more decentralized economy and society.

Whether hydrogen's promise will ever be realized is uncertain. It depends to what extent production efficiencies can be improved and how highly societies value its environmental and social benefits. There may be a convergence of values, beliefs, and economic and political forces that make hydrogen inevitable, but that time is not likely to be in the near or even medium term future.

The preferred choices in the longer-term future, beyond the year 2020 or so, therefore are less obvious. As illustrated in chapter 21 ("Five Paths"), depending on one's vision of the future and how one weighs different values, one can justify a preference for any of the five paths. Even if one preselects a particular long-term option—for instance, solar hydrogen—that does not strongly influence near-term choices, mostly because there is not a strong technological or institutional linkage between different sets of fuel options. Methane, for instance, even though it is a gaseous fuel, does not significantly ease or facilitate the transition to hydrogen.

An idealist would hope that in the future greater emphasis would be placed on sustainable clean energy sources and a more democratic society, and that fuel choice decisions would be made accordingly. For that idealized future to be realized, however, major shifts in societal preferences are required. That future may be attainable. But how does a nation shift onto that or any other path?

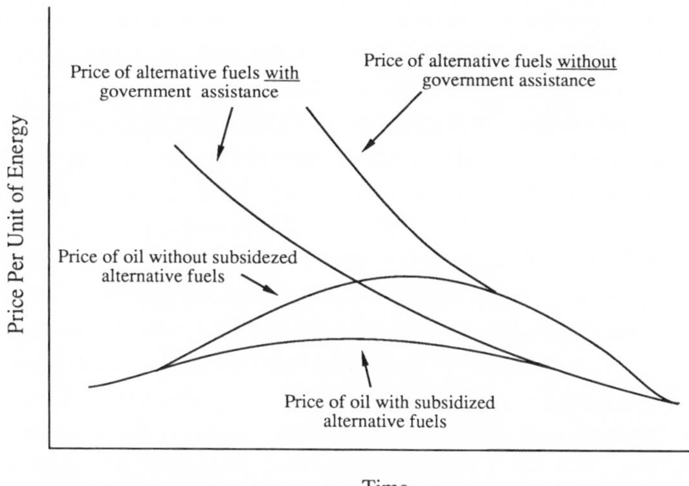

Figure 38. Conceptual Representation of Relationship Between Fuel Prices of Petroleum and Alternative Fuels

ROLE OF GOVERNMENT

Government plays an absolutely critical role. The petroleum market has demonstrated for many years its vulnerability to political manipulation and its inability to operate in an orderly and efficient manner. The maldistribution of petroleum resources and their concentration in a few sparsely populated countries suggest that there is no reason to believe that the market will operate any more effectively in the future.

If governments of the U.S. and other oil-importing countries remain aloof, they relinquish control and power to a few underpopulated Middle Eastern nations; they remain vulnerable to price and supply disruptions and allow petroleum prices to follow a steeper trajectory than is otherwise possible. If petroleum importing countries actively pursue alternative fuels, then they reduce the possibility and severity of disruptions and they suppress the possibility of oil price escalations. Government participation could assure the timely introduction of substitute fuels and assure that the price of transportation fuels never exceeds about $45 per oil-equivalent barrel (1987 $), even for environmentally attractive options. This benefit of flattened fuel prices is illustrated in figure 38.

There are other justifications for government intervention. First, as suggested above, government intervention helps assure that the socially preferable alternatives are the ones that are chosen. Government assures that environmental, national security, and sociopolitical considerations are incorporated into the decision process.

Another motivation that is especially relevant to the U.S., a motivation based on moral principles and a desire for world peace, is to ease the cost of energy for less affluent countries in order to facilitate their development.[1] High oil prices have a severe impact on the less developed countries. Motorists in affluent countries are willing to pay $2 or more per gallon, and the nontransportation sectors of those countries have the resources and expertise to switch to other fuels as oil prices rise, but poorer countries do not have the same resources and flexibility. Petroleum is a highly attractive fuel to less developed countries because it is easy and relatively inexpensive to handle and transport and because it can be readily used in a wide variety of applications. Donald Hodel, U.S. Secretary of the Interior (and former Secretary of Energy), noted in late 1986 that "what may be only a ripple on the energy shores of the United States, . . . may be a tidal wave on the shores of the developing nations."[2] Those countries, such as Brazil, that have developed the institutions, human resources, and whatever else it takes to get on an upward growth trajectory can sometimes overcome the hindrance of high oil prices. Many countries cannot and have not. High energy prices deprive these countries of the flexibility they so desperately need in developing their economies and providing a better life for the population. Reducing petroleum demand in the U.S. would flatten the oil price trajectory for these poorer countries (and for the U.S. as well) and would provide them with more time to make the shift to alternative energy and to devote more resources and initiative to the pressing problems of agriculture, health, education, and industrial development.

Government involvement could take many forms, from playing an information dissemination role to full financial and operational involvement in fuels projects. The spectrum of governmental roles is indicated in table 68. In all cases the objective of governmental action is to reduce start-up barriers.

At one extreme is the clearly insufficient laissez-faire approach of only responding to the market's information dissemination deficiency.

A more active but still minimal role is to provide research support for promising new technologies that are far from commercialization and that would otherwise be ignored by private corporations. For large countries with an active private sector this means supporting research in such areas as hydrogen production and storage, cellulosic biomass

TABLE 68 SPECTRUM OF PUBLIC POLICY APPROACHES AND OPTIONS

Government Ownership and Management	Government Mandates	Regulation of Market Power	Government Incentives	Reduce Uncertainty	Government R and D	No Intervention (laissez-faire)
	Mandates to produce and/or market particular products such as methanol-compatible cars or CNG at fuel stations	Encourage cooperation between firms and industries; create energy utilities with regulated entry, exit, and prices	Tax credits, grants, loan guarantees, import tariffs and other subsidies and actions meant to alter the relative price relationships of fuels	Reduce uncertainty over future energy prices and fuel availability for consumers and industry	Support of long-term and risky technologies and those not likely to be developed in a timely manner by the private sector	Let decisions be made by market without government intervention; rely on exhortation and information dissemination
Cost to government: Can be very large	Cost to government: None	Cost to government: Negligible	Cost to government: Can be very large	Cost to government: Can be very large	Cost to government: Significant	Cost to government: Negligible

conversion, fuel cells, coal conversion, and so on. In smaller countries where the private sector has fewer resources, such as in New Zealand, government R&D might also include more applied research such as developing engine conversion kits and fuel metering devices. In all countries government R&D involves being a nurturer of knowledge, not letting knowledge disappear when interest wanes. It would have been a devastating loss, for instance, if the fully operational Great Plains coal gasification plant had been abandoned in 1986. Instead, knowledge is being retained in that cadre of engineers that was retained to continue operating and improving the plant. Having those types of people and their knowledge readily available may dramatically shorten the time needed to move on to the next stage of development—whenever that may be.

Venturing one step further into market activities, government could reduce uncertainty for both industry and consumers. Uncertainty about the future amplifies all other start-up barriers. One technique for reducing uncertainty to industry is to offer loan guarantees and price guarantees for specified fuel products, such as was done by the U.S. Synthetic Fuel Corporation. Price guarantees are generally preferable because they force a producer to stay in business instead of defaulting on loans. A technique for reducing uncertainty for consumers is to guarantee a price relationship, as in Brazil and New Zealand. The Brazilian government guaranteed that the price of ethanol would not exceed 65 percent that of gasoline, and the New Zealand government established a 50 percent limit for CNG. Other techniques include guaranteeing the resale value of nonpetroleum vehicles and a certain level of fuel availability (e.g., a certain number of fuel outlets). The credibility of these approaches is undermined by past experience in which new administrations do not honor the promises of their predecessors. This problem of reneging can be mitigated by purposely moderating the programs so that they are not too prominent or too expensive—the Synthetic Fuel Corporation is an example of how *not* to do it.

From a political perspective, a more interventionist role is to alter price signals with various incentives so as to make alternative fuels more attractive relative to petroleum fuels. These incentives could include reduced fuel and vehicle taxes, low-interest loans (such as loan guarantees to fuel producers and subsidized car loans to consumers), free fuel vouchers, grants to new fuel stations, special treatment within regulatory programs (such as fuel economy standards), and oil import tariffs. These incentives are not necessarily more expensive to government than actions to reduce uncertainty, but they are farther to the left politically because they are deliberate attempts to "tilt the playing field."

A still more interventionist role—but one that tends to be far less costly to government—is the public utility model. A government agency would regulate fuel prices and the entry (and exit) of companies into the business of producing, distributing, and marketing fuels. In this way the government is able to influence which fuels will be chosen and when and where they will be introduced. A mild form of this type of intervention is easement of antitrust restrictions to allow coordination between private firms in different industries and even in the same industry with the intention of mitigating some of the "chicken-or-egg" barriers.

The most active role would involve direct intervention in the market with the purpose of making or forcing investments that otherwise would not be made. This might involve direct investment in fuel production plants (for instance, through a public corporation), such as with the synthetic gasoline plant in New Zealand, or it could involve direct mandates without taking on any financial liability, as when the U.S. government ordered oil marketers to provide unleaded gasoline at retail fuel stations.

In summary, government has a vast range of options for promoting alternative fuels. The effectiveness and acceptability of different approaches depend on the sense of urgency, economic circumstances, and social and environmental goals and priorities. At a minimum, the timely introduction of new fuels will need some combination of initiatives that increase R&D, reduce uncertainty, and provide incentives. In most cases some of the more extreme forms of market intervention will also be needed. Keep in mind that the types of actions suggested here are not exceptional, even for the U.S., which has a strong ideological commitment to the sanctity of the free market.

BACKING INTO THE FUTURE?

When and how should the transition be initiated? The salient question for oil-importing countries such as the U.S. is how long they are willing to accept the risk of dependence on a finite resource that is subject to erratic supply and price manipulations. So far the U.S. has been bumbling along with no coherent strategy. It has played mostly an information dissemination role, with minimal investments in R&D. Sometimes the laissez-faire approach works; indeed, the U.S. is much better off in the short run having invested relatively little in new fuels during the 1970s and 1980s. In contrast, Brazil and New Zealand acted aggressively in a fairly directed manner; they have been penalized by financial losses, at least in the short term. Whether in the long term the Brazilian ethanol and New Zealand synthetic gasoline investments pass into history as economic disasters or become examples of brilliant foresight will depend upon what happens with future oil prices.

Thus the question of when and how the transition should be initiated is the most difficult to answer of all those posed at the beginning of the chapter. The answer depends only in part on knowledge and the specification of societal priorities; mostly it depends on divining a very uncertain future—for instance, forecasting future oil prices, predicting major political changes, and specifying future greenhouse effects. And thus no human, no matter how knowledgeable, can know what the optimal choice and timing should be.

Several general principles and guidelines can and should be observed, however. First, fuel production, distribution, and end-use activities should be made more flexible. For instance, materials in motor vehicles, storage tanks, and fuel pumps that are not compatible with methanol should be replaced. Flexibility could also be enhanced by making government rules, regulations, and testing procedures generic rather than specific to gasoline and/or diesel fuel.

Second, R&D investment in advanced energy conversion processes should be expanded; some of that effort should go into coal and biomass conversion, but the emphasis should be on hydrogen production and storage technologies. Given the possibility that petroleum may be available worldwide at modest prices well into the twenty-first century and that the greenhouse effect may become severe, it may prove attractive to skip from petroleum directly to hydrogen. More knowledge is needed before such a decision can be made, and much more work is needed for hydrogen to become an attractive fuel.

Third, cost-sharing arrangements between government and industry for demonstrating new production technologies should be expanded. Over 200 billion dollars are spent on petroleum fuels each year in the U.S. It would seem that devoting at least 1 percent of that amount (about $2 billion) to demonstrating the technology for producing alternative fuels (and for controlling their negative environmental impacts) is reasonable, especially considering that this cost would be roughly the same as that of only a $1 per barrel increase in imported petroleum. It might even be that a sustained investment of this scale might be sufficient to restrain petroleum prices enough to more than cover on a national basis the full cost of such activities (although it would be difficult if not impossible to know whether this is true). In any case, the risks of investing in demonstration and commercial-scale production plants are so great that government must be willing to play a major financial role initially.

Fourth, greater attention should be given to environmental considerations. The human race is poisoning water supplies with chemical wastes, producing carbon dioxide and other gases that appear to be causing global climatic changes, discharging carcinogenic materials into air and water, acidifying water and soil, and in general causing extensive damage to the planet. The production and combustion of energy is arguably the greatest threat to the sustainability of global ecological systems. Future energy supplies—especially coal, oil shale, and oil sands—will create exceptional stresses having far-reaching and possibly devastating effects on the health and welfare of the planet's inhabitants. In the future a saner strategy would be to err on the side of placing too much emphasis on ecological sustainability instead of, as in the past, too little.

Similarly, too little attention has been given to the complex relationship between the deployment of technology and the structure of society. Modern society has undergone a radical transformation in moving from an agrarian to an industrialized economy. Another radical transformation could occur. Because energy plays a large role in regional, national, and international economies, energy choices can have major fundamental impacts on the distribution of political and economic power and even on the physical distribution of people. It is incumbent upon policy makers to evaluate energy choices not only with economic and environmental criteria but also with a broad range of social criteria.

LEADING TECHNOLOGY INTO THE FUTURE

The process of technological change—in this case the many changes involved in introducing new fuels—does not, as indicated earlier, obey invariant laws. There is not one evolutionary path that is *the* optimal path. So it is with transportation energy choices. The transportation and energy systems we create are artificial systems whose designs respond to their social, political, economic, and physical settings. The systems we now have are artifacts of a previous time. They served us well. But they will not serve us well in the future. The time has come to recognize this and to begin nudging the existing systems toward a more appropriate configuration for the twenty-first century. Those changes might seem revolutionary now, but the same would have been said one hundred years ago regarding a petroleum-highway system. Let us not constrain our imagination. Let us not let current cost analyses dictate future choices; the future design of our transportation and energy systems is and should be price determining. Let us accept that premise and stretch our imagination. Perhaps the most conservative choices—petroleum-like fuels that preserve the status quo—will prove to be the preferred choices. I doubt that and hope it will not be true. New transportation fuels should not be treated as second-best alternatives to petroleum, but as opportunities to make a better world.

Notes

PART I: LEARNING FROM THE PAST

1. Vice President George Bush argued in favor of both methanol and ethanol in an editorial published in Des Moines, Iowa, newspapers on March 5, 1987. Ford Motor Company CEO Donald Petersen has consistently supported methanol since the early 1980s (see, for instance, Petersen, *Proceedings of the International Symposium on Alcohol Fuel Technology*, vol. 3, Ottawa, Canada, May 1984, pp. 41–44). California Energy Commission Chairman Charles Imbrecht and his predecessor in that position, former astronaut Rusty Schweikert, have promoted methanol in California since about 1980. For statements by Toyota and the U.S. Environmental Protection Agency, see U.S. Congress, House of Representatives, Energy and Commerce Committee, Subcommittee on Energy Conservation and Power, *Methanol as Transportation Fuel* (Hearings, 98th Congress, 2d session, 4 and 25 April 1984, GPO, Washington, D.C., 1984, Serial No. 98–145). For General Motors, see Richard L. Klimisch, Testimony before U.S. Congress, House of Representatives, Committee on Energy and Commerce, Subcommittee on Energy and Power, 17 June 1987.

1: IS THERE A TRANSPORTATION ENERGY PROBLEM?

1. Charles Masters, "World Petroleum Resources—A Perspective," U.S. Geological Service, Report 85–248, Reston, Virginia, 1985.

2. Production cost data are not divulged, but various estimates suggest that excluding sunk costs, most of the Persian Gulf oil can be produced for less than $2 per barrel. Because many of the development and infrastructure costs were incurred long ago, and because oil fields are large and easily accessible, this $2

per barrel figure is plausible. In contrast, the production costs for new oil fields are about $7 to $20 per barrel. For instance, the cost just to transport Alaskan North Slope oil to continental U.S. is about $6 per barrel. See Donald Woutat, "Big Oil Still in the Black," *Los Angeles Times*, 5 February 1986; and Douglas Martin, "Has Canada lost its oil gamble?" *New York Times*, 9 February 1986, p. 4.

3. The exception may be Iran. The condition of oil wells reportedly has deteriorated to the extent that without considerable new investment oil production could return to only one-half the 1978–1979 production levels.

4. Forecasts of energy prices and energy consumption have been notoriously inaccurate since the early 1970s. For an overview of energy forecasting inaccuracies see William Ascher, *Forecasting: An Appraisal for Policy-Makers and Planners* (Baltimore, Md.: Johns Hopkins University Press, 1987), pp. 93–141.

5. Estimate for 1900 computed from U.S. Dept. of Commerce, Bureau of the Census, *Historical Statistics of the United States, Colonial Times to 1970, Part I* (Washington, D.C.: 1975), p. 588. Estimates for 1978 are from U.S. Energy Information Administration, *Annual Energy Review, 1984* (Washington, D.C.: 1985).

6. Computed from U.S. Energy Information Administration, *Monthly Energy Review* (Springfield, Va.: NTIS, 1986, DOE/EIA-0035).

7. For an analysis of fuel switching worldwide in the residential sector, see Lee Schipper and Andrea Ketloff, "The International Decline in Household Oil Use," *Science* 230 (December 1983): 1118–1125. For an analysis of fuel switching in Europe, see Mans Lonroth, "The European Transition from Oil," *Ann. Rev. Energy* 8 (1983): 1–25.

8. A. Altshuler, M. Anderson, D. Jones, D. Roos, J. Womack, *The Future of the Automobile* (Cambridge, Mass.: MIT Press, 1984), p. 50.

9. American Petroleum Institute, *Basic Petroleum Data Book*, vol. 6, no. 1 (Washington, D.C.: 1986).

10. Research has shown that changes in urbanization, population density, and gasoline prices have little or no effect on auto ownership rates. See J. C. Tanner, *International Comparisons of Cars and Car Usage* (Crowthorne, U.K.: Transport and Road Research Laboratory, 1983, Report 1070), pp. 8–9; and A. Altshuler et al., *The Future of the Automobile* (Cambridge, Mass.: MIT Press, 1984), pp. 107–111.

11. U.S. Energy Information Administration, *1981 Annual Report to Congress*, vol. 3 (Washington, D.C.: 1982).

12. Energy Information Administration, *Annual Energy Review 1984* (Washington, D.C.: 1985), p. 79. See also EIA, *Annual Energy Outlook 1984* (Washington, D.C.: 1985), p. 104.

13. "Fill the Oil Reserve, Academy Report Says," *Science* 232 (April 25, 1986): 441–442.

14. Marianne Millar, Margaret Singh, Anant Vyas, and Larry Johnson, *Transportation Energy Outlook Under Conditions of Persistently Low Petroleum Prices* (Argonne, Ill.: Argonne National Laboratory, 1987). For forecasts by DRI, Chase Econometrics, National Petroleum Council, and others, see U.S. Energy Information Administration, *Annual Energy Outlook 1986* (Washington, D.C.: 1987).

15. If estimated "undiscovered recoverable" reserves were included, the percentage would drop to 56 percent. Production figures are from U.S. Dept. of Commerce and United Nations data reports; reserve estimates are from Energy Information Administration.

16. Cost data are closely guarded by oil producers; my estimates are based on newspaper interviews with energy industry experts and newspaper accounts of private industry studies, including a study carried out by Texas Eastern Corp. in 1985. A 1985 Argonne Lab report estimated that the cost of finding and extracting new petroleum in the U.S. was about $14 to $18 per barrel. See M. K. Singh, M. Millar, and L. R. Johnson, *Dominant Trends in Worldwide Petroleum Supply and Demand* (Springfield, Va.: NTIS, 1985, ANL/CNSU-TM-175). Note that oil companies started closing oil wells when oil prices dropped below $20 in 1986. See D. Martin, "Has Canada Lost Its Oil Gamble?" *New York Times*, 9 February 1986, p. 4; "Texaco Closes 1,500 Oil Wells," *San Francisco Chronicle*, 6 March 1986, p. 29. Recent evidence indicates that with modest and gradual increases in petroleum prices into the $30 to $40 per barrel range the U.S. could sustain its current level of oil production. See William F. Fisher, "Can the U.S. Oil and Gas Resource Base Support Sustained Production?" *Science* 236 (26 June 1987): 1631–1636. Of course, even if that is true, oil imports would still be substantial.

17. U.S. Congress, "Energy Policy and Conservation Act," Public Law 94–163, December 22, 1975, and amended.

18. Committee on Strategic Petroleum Reserve, Commission on Engineering and Technical Systems, National Research Council, *The Logistics of the U.S. Strategic Petroleum Reserve in the World Petroleum Market: 1990–2000* (Washington D.C.: National Academy Press, 1986).

19. Ibid., pp. 1–30. Some of the improvement was due to larger planes, increased load factors, and other operational factors, but most was due to engine improvement (and to a lesser extent, weight reductions).

20. Ibid., pp. 1–8.

21. Charles L. Gray and Frank von Hippel, "The Fuel Economy of Light Vehicles," *Scientific American* 244, no. 5 (1981): 48–59.

22. Estimates range from about 1.4 to 1.6 passenger miles per vehicle mile. See Congressional Budget Office, *Urban Transportation and Energy: The Potential Savings of Different Modes* (Washington, D.C.: 1977), p. 19.

23. U.S. Dept. of Transportation, *National Transportation Statistics* (Washington, D.C.: 1984).

24. Eric Hirst, Robert Marlay, David Greene, and Richard Barnes, "Recent Changes in U.S. Energy Consumption: What Happened and Why," *Annual Review of Energy* 8 (1983): 193–245.

25. U.S. Congress, "Energy Tax Act of 1978," Public Law 95–618, November 9, 1978.

26. Richard Shackson and H. James Leach, *Maintaining Automotive Mobility: Using Fuel Economy and Synthetic Fuel to Compete with OPEC Oil* (Arlington, Va.: Mellon Institute, Applied Energy Services, 1980); and U.S. Congress, Office of Technology Assessment, *Increased Automobile Efficiency and Synthetic Fuels: Alternatives for Reducing Oil Imports* (Washington, D.C.: GPO, 1982).

27. For summary of report see James Brinkley, Wallace Tyner, and Marie Mathews, "Evaluating Alternative Energy Policies: An Example Comparing Transportation Energy Investments," *The Energy Journal* 4, no. 2: 91–104.

2: A CONCEPTUAL AND ANALYTICAL FRAMEWORK

1. See R. E. Gomory, "Technology Development," *Science* 220 (1983): 576–580.

2. See William Ascher, *Forecasting: An Appraisal for Policy-Makers and Planners* (Baltimore, Md.: Johns Hopkins University Press, 1978), p. 166.

3. M. J. Cetron and T. I. Monahan, "An Evaluation and Appraisal of Various Approaches to Technological Forecasting," in J. R. Bright, ed., *Technological Forecasting for Industry and Government* (Englewood Cliffs, N.J.: Prentice-Hall, 1968), p. 151.

4. See W. L. Garrison, "Thinking About Public Facility Systems," *The National Research Council in 1978* (Washington D.C.: National Academy of Science, 1978), pp. 244–265; and W. L. Garrison, ed., "Technology Development" (special issue), *Transportation Research* 18A, no. 4 (1984).

5. For an elaboration see D. Sperling, "Assessment of Technological Choices Using a Pathway Methodology," *Transportation Research*, 18A, no. 4 (1984): 343–353.

6. Peter Hall, *Great Planning Disasters* (Berkeley, Los Angeles, London: University of California Press, 1980).

7. See J. D. Lewis, "Technology, Enterprise, and American Economic Growth," *Science* 215 (1982): 1204–1210; and Charies E. Lindblom, "The Science of Muddling Through," *Public Administration Rev.* 19 (1959): 79–89.

8. See Chalmers Johnson, *MITI and the Japanese Miracle: The Growth of Industrial Policy, 1925–75* (Stanford, Calif.: Stanford University Press, 1982).

9. W. Ascher, *Forecasting*, p. 165.

10. Lawrence H. Tribe, "Technology Assessment and the Fourth Discontinuity: The Limits of Instrumental Rationality," *Southern California Law Review* 46, no. 3 (1973): 659.

11. Edward Merrow, K. Phillips, and C. Myers, *Understanding Cost Growth and Performance Shortfalls in Pioneer Process Plants* (Santa Monica, Calif.: RAND Corp., 1981, R-2569-DOE), pp. 47–48.

12. Ibid., p. 2.

13. Ibid., p. 31.

14. Tribe, "Technology Assessment and the Fourth Discontinuity," p. 659.

15. Amory Lovins, *Soft Energy Paths: Toward a Durable Peace* (New York: Harper & Row, 1979). Originally published by Ballinger, 1977.

16. See for instance John B. Robinson, "Energy Backcasting: A Proposed Method of Policy Analysis," *Energy Policy* (December 1982): 337–344.

17. Richard R. Nelson and Sidney G. Winter, *An Evolutionary Theory of Economic Change* (Cambridge, Mass.: Harvard University Press, 1982).

18. Ibid., p. 19.

19. Robert H. Hayes, "Why Strategic Planning Goes Awry," *New York Times*, 20 April 1986, p. F-2.

20. Herbert Simon, *Sciences of the Artificial* (Cambridge, Mass.: MIT Press, 1981).

21. Tribe, "Technology Assessment and the Fourth Discontinuity," p. 660.

PART II: HISTORICAL REVIEW OF ALTERNATIVE TRANSPORTATION FUEL EXPERIENCES

3: U.S. FLIRTATIONS WITH BIOMASS AND MINERAL FUELS

1. Bureau of the Census, Department of Commerce and Labor, *Census of Manufactures: 1905, Automobiles and Bicycles and Tricycles. Bulletin 66* (Washington, D.C.: GPO, 1907), p. 19.

2. H. Bernton, W. Kovarik, and S. Sklar, *The Forbidden Fuel: Power Alcohol in the Twentieth Century* (New York: Boyd Griffin, 1982), p. 11.

3. Arnold Krammer, "An Attempt at Transition: The Bureau of Mines Synthetic Fuel Project at Louisiana, Missouri," in L. J. Perelman, A. W. Giebelhaus, and M. D. Yokell, eds., *Energy Transitions: Long-Term Perspectives* (Boulder, Colorado: Westview Press, 1981), p. 65.

4. For a historical description of energy production and use in the U.S., see Sam H. Schurr and Bruce C. Netschert, *Energy in the American Economy, 1850–1975: An Economic Study of Its History and Prospects* (Baltimore, Md.: John Hopkins University Press, 1960).

5. For a brief review of early coal gas history, see Ronald F. Probstein and R. Edwin Hicks, *Synthetic Fuels* (New York: McGraw-Hill, 1982), pp. 6–7; and Jonathan Lash and Laura King, eds., *The Synfuels Manual* (New York: Natural Resources Defense Council, 1983).

6. For a detailed history of German fuel activities between World War I and World War II, see Kenneth S. Mernitz, *Progress at a Price: Research and Development of Liquid Motor Fuels in the U.S. and Germany, 1913–1933* (Ann Arbor, Mich.: University Microfilms Int., 1984); also see Arnold Krammer, "An Attempt at Transition: The Bureau of Mines Synthetic Fuel Project at Louisiana, Missouri," pp. 65–108.

7. Mernitz, *Progress at a Price*, p. 204.

8. German synfuel production data are somewhat uncertain.

9. These estimates were based on experimental work carried out by the U.S. Bureau of Mines after the war.

10. Krammer, "An Attempt at Transition."

11. This and the following information and data on research activities of the U.S. Bureau of Mines come from ibid.

12. See Borking, 1978, cited in Bernton et al., *The Forbidden Fuel*, p. 30.

13. A hostile oil industry trade organization, the National Petroleum Council, prepared an apparently biased report that had estimated the costs at 41.4 cents per gallon, but it inflated the costs by including all funds spent on the research project, even housing costs for all scientists and staff; the Bureau of Mines had estimated the costs for this hypothetical plant to be 10.2 cents per gallon.

14. Cited in Bernton et al., *The Forbidden Fuel*, p. 99.

15. Richard H. Vietor, *Energy Policy in America Since 1945: A Study of Business-Government Relations* (New York: Cambridge University Press, 1984), p. 163.

16. Ibid., p. 164.

17. Ibid., p. 171.

18. For reviews of oil shale developments outside of the U.S., see U.S. Congress, Office of Technology Assessment, *An Assessment of Oil Shale Technologies*, Vol. I (Washington, D.C.: GPO, 1980), pp. 108–111; Paul Russell, *History of Western Oil Shale* (E. Brunswick, N.J.: Center for Professional Advancement, 1980), pp. 2–4; and Jonathan Lash and Laura King, eds., *The Synfuels Manual* (New York: Natural Resources Defense Council, 1983), pp. 12–14.

19. Lash and King, *The Synfuels Manual*, p. 13.

20. Ibid.

21. U.S. Congress, OTA, *An Assessment of Oil Shale Technologies*, Vol. I., pp. 108–110.

22. Paul Russell, *History of Western Oil Shale*, p. 5.

23. Ibid.; The remainder of this paragraph comes from pp. 6, 75, and 135 of Russell, 1980, and Lash and King, p. 14.

24. R. Vietor, *Energy Policy in America Since 1945*, p. 178.

25. P. Russell, *History of Western Oil Shale*, pp. 107–114.

26. R. Vietor, *Energy Policy in America Since 1945*, p. 182.

27. Ibid.; and Edward W. Merrow, *Constraints to the Commercialization of Oil Shale* (Prepared for U.S. DOE) (Santa Monica, Calif.: Rand Corp., 1978, R-2293-DOE), p. 126.

28. R. Vietor, *Energy Policy in America Since 1945*, p. 182.

29. Ibid., p. 183.

30. See ibid. for thorough discussion of land-leasing activities of the U.S. Dept. of Interior.

31. Ibid., p. 187.

32. Address to Congress, 4 June 1971, cited in ibid., p. 324.

33. Description of oil shale activities from 1973 to 1980 in the next few paragraphs is based on ibid., pp. 327–329, 334–335.

34. The following history of legislative actions and administration proposals is based on Ann Pelham, *Energy Policy*, 2d ed. (Washington, D.C.: Congressional Quarterly, Inc., 1981); and R. Vietor, *Energy Policy in America Since 1945*, pp. 333–334.

35. R. Vietor, *Energy Policy in America Since 1945*, p. 333.

36. Ibid., pp. 333–334.

37. Ibid.

38. A. Pelham, *Energy Policy*, p. 210.

39. Ibid., p. 210.

40. Ibid., p. 212.

41. R. Vietor, *Energy Policy in America Since 1945*, p. 338.

42. The following history of activities by the U.S. Synthetic Fuels Corp. is based on annual reports of that organization from 1981–1984, reviews of those activities in the *New York Times* and *Wall Street Journal*, and Ernest J. Yanarella and William C. Green, eds., *The Unfulfilled Promise of Synthetic Fuels: Technological Failure, Policy Immobilism, or Commercial Illusion* (Westport, Conn.: Greenwood Press, 1987).

43. N. A. Holt, T. P. O'Shea, and J. McDaniel, "The Coolwater Project —Preliminary Operating Results," Paper presented at American Institute of Chemical Engineers Spring National Meeting, Houston, Texas, March 1985.

44. Personal communication with Bill Waycaster (Dow Chemical), Houston, Texas, August 1985.

45. Statement of Robert Long to U.S. Congress, House of Rep., Subcomm. on Fossil and Synthetic Fuels and Subcomm. on Energy Cons. and Power, *Methanol as a Transportation Fuel*, 98th Cong., 2d sess., April 4 and 25, 1984, pp. 321–330. Serial No. 98–145.

46. Andy Pasztor, "Fading Dream," *New York Times*, 9 August 1984, p. 1.

47. U.S. Synthetic Fuel Corp., *Annual Report 1984* (Washington, D.C.: 1985), p. 1.

48. Information provided by G. Hedstrom, U.S. DOE, February 1987. See also U.S. General Accounting Office, *Great Plains: Status of the Great Plains Coal Gasification Project* (Gaithersburg, Md.: GAO, 1985, GAO/RCED-86-49FS).

49. The following historical review is based on four sources: Kenneth S. Mernitz, *Progress at a Price: Research and Development of Liquid Motor Fuels in the U.S. and Germany, 1913–1933*; Bernton et al., *The Forbidden Fuel*; Krammer, "An Attempt at Transition"; A. W. Giebelhaus, "Resistance to Long-Term Energy Transition: The Case of Power Alcohol in the 1930s," in L. J. Perelman, A. W. Giebelhaus, and M. D. Yokell, eds., *Energy Transitions: Long-Term Perspective*, pp. 35–64.

50. B. H. Markham (Director, API Committee), "Memorandum re: Alcohol-Gasoline Blends," April 15, 1933, cited in Giebelhaus, "Resistance to Long-Term Energy Transition," pp. 40–41.

51. Bernton et al., *The Forbidden Fuel*, p. 32.

52. See Herman F. Wilkie and Paul J. Kolachov, *Food for Thought* (Indianapolis, Ind.: Indiana Farm Bureau, circa 1942), cited in John W. Lincoln, *Driving Without Gas* (Charlotte, Vt: Garden Way Publishing, 1980).

53. Giebelhaus, "Resistance to Long-Term Energy Transition."

54. Gasoline prices at the retail outlet were about $0.30 per gallon in 1920 but dropped to about $0.12 in the mid-1930s.

55. See Bernton et al., *The Forbidden Fuel*, p. 55. The small "stills" were encouraged by the U.S. Department of Energy—one of the worst ideas ever promoted by government because the stills only produce low-quality (e.g., hydrated) ethanol that is unsuited for blending with gasoline.

56. U.S. Department of Energy, *The Report of the Alcohol Fuels Policy*

Review (Springfield, Va.: NTIS, 1979, DOE/PE-0012), p. 8.

57. Bernton et al., *The Forbidden Fuel*, p. 105.

58. Information on DOE loan guarantees was gathered from various *Wall Street Journal* articles (esp. 17 August 1981, p. 6); Burl Hagewood, Renewable Fuels Assoc., Washington, D.C., August 1985; and U.S. Department of Agriculture, *Fuel Ethanol and Agriculture: An Economic Assessment* (Washington, D.C.: GPO, 1986).

59. Information on USDA loan guarantees was provided by Marlyn Aycock, FMHA, U.S. Dept. of Agriculture, August 1985; and ibid.

60. See U.S. General Accounting Office, *Potential of Ethanol as a Motor Vehicle Fuel* (Gaithersburg, Md: GAO, 1980).

61. Unpublished survey conducted by U.S. National Alcohol Fuel Commission. Partial results published in U.S. NAFC, *Fuel Alcohol, Final Report* (Washington, D.C.: GPO, 1981).

62. "The 500 Largest U.S. Industrial Corporations, Ranked by Sales," *Fortune*, 29 April 1985, pp. 266–285.

63. U.S. Department of Transportation, Federal Highway Administration, *Monthly Gasoline Sales Reported by States* (Washington, D.C.: January 1986).

64. USDA, *Fuel Ethanol and Agriculture.*

65. John E. Murtaugh, "Fuel Ethanol Production—the U.S. Experience," *Process Biochemistry* (April 1986): 61–65.

66. For elaboration of this point, see David Prindle, "Shale Oil and the Politics of Ambiguity and Complexity," *Ann. Rev. Energy* 9 (1984): 351–373.

4: ETHANOL IN BRAZIL

1. In the 1980s domestic oil and natural gas production started to boom as a result of greatly expanded exploration efforts in the late 1970s and early 1980s. Production grew from about 170,000 barrels b/d in 1979 to 600,000 b/d in 1985. But this was not foreseen in 1975.

2. Fred Moavenzadeh and David Geltner, *Transportation, Energy and Economic Development: A Dilemma in the Developing World* (New York: Elsevier, 1984). For U.S. data see Oak Ridge National Laboratory, *Transportation Energy Data Book*, 7th ed. (Springfield, Va.: NTIS, 1984, ORNL-5765), pp. 1–20.

3. Ministerio das Minas e Energia, *Balanco Energetico Nacional* (Brasilia: 1977), p. 119.

4. Ministerio das Minas e Energia, *Balanco Energetico Brasileiro* (Brasilia: 1983).

5. Ministerio das Minas e Energia, Conselho Nacional do Petroleo, *Anuario Estastico* (Brasilia: 1983).

6. This background discussion on the foreign debt of Brazil and the cost of oil imports is borrowed from Moavenzadeh and Geltner, *Transportation, Energy and Economic Development*, pp. 474–475.

7. Harry Rothman, Rod Greenshields, and Francisco Rosillo Calles, *Energy*

from Alcohol: The Brazilian Experience (Lexington, Ky.: University Press of Kentucky, 1983).

8. Ibid., p. 14.

9. B. Nunberg, cited on pp. 134–136 in Michael Barzelay, *The Politicized Market Economy: Alcohol in Brazil's Energy Strategy* (Berkeley, Los Angeles, London: University of California Press, 1986).

10. For a discussion of the initiation of the alcohol program, see Allen L. Hammond, "Alcohol: A Brazilian Answer to the Energy Crisis," *Science* (11 February 1977): 564. See also Barzelay, *The Politicized Market Economy: Alcohol in Brazil.*

11. William S. Saint, "Farming for Energy: Social Options Under Brazil's National Alcohol Programme," *World Development* 10, no. 3 (1982): 223–238.

12. See, for instance, Barzelay, *The Politicized Market Economy.* The following discussion of institutional conflict is based on Barzelay's review.

13. See W. S. Saint, "Farming for Energy."

14. See Andre Ghirardi, *Alcohol Fuels from Biomass in Brazil: A Comparative Assessment of Methanol and Ethanol* (Ph.D. dissertation) (Ann Arbor, Mich.: University Microfilms, 1983). See also Barzelay, *The Politicized Market Economy Alcohol in Brazil.*

15. *Veja* (Brazilian news weekly), 1979, cited in Saint, "Farming for Energy," p. 227.

16. Barzelay, *The Politicized Market Economy.*

17. Hal Bernton, William Kovarik, and Scott Sklar, *The Forbidden Fuel: Power Alcohol in the Twentieth Century* (New York: Boyd Griffin, 1982); also see Barzelay.

18. Barzelay, *The Politicized Market Economy Alcohol in Brazil.*

19. Antonio G. Novaes, "Market Response to Alcohol Fueled Cars in Brazil," Paper presented at U.S.-Brazilian Workshop on Transportation Energy, Polytechnic Institute of New York, January 12–14, 1983, p. 21.

20. *Veja*, 13 June 1979. See Bernton et al., *The Forbidden Fuel.*

21. Novaes, "Market Response to Alcohol Fueled Cars in Brazil."

22. Detailed data and analysis on the economic and social costs and benefits of the Proalcool program were never made available (and possibly never generated) when the crucial decisions were made in 1979. Nevertheless, recent studies indicate that ethanol production costs in the early 1980s were about $0.80 to $0.85 per gallon (more for anhydrous alcohol, and less for annex distilleries and distilleries in the São Paulo area), and were therefore somewhat higher on an energy basis than gasoline prices. Of course, such simple analyses ignore many important direct and indirect effects. For cost analyses of Brazilian ethanol production see Barzelay, *The Politicized Market Economy Alcohol in Brazil*; Adam Kahane, "Economic Aspects of the Brazilian Alcohol Program" (Master's thesis, Energy and Resources Group, University of California, Berkeley, 1985); Howard Geller, "Ethanol Fuel from Sugar Cane in Brazil," *Annual Review of Energy* 10 (1985): 135–164; C. Pamplona, "Proalcool: Technical-Economic and Social Impact of the Program in Brazil," Sugar and Alcohol Institute, Brasilia, 1984, cited in Geller, 1985; and Sergio Trindade,

"Brazil and Alcohol Fuels: A Multi-Sponsored Program," Rio de Janeiro, 1984 (available from S. Trindade, 121 Brite Ave., Carlsdale, N.Y. 10483).

23. *Veja*, 26 August 1981, cited in Barzelay, pp. 232–233, *The Political Market Economy*.

24. Novaes, "Market Response to Alcohol Fueled Cars in Brazil"; and Barzelay, *The Political Market Economy*.

25. Novaes, ibid.

26. Kahane, "Economic Aspects of the Brazilian Alcohol Program."

27. Andre Ghirardi, "Trends of Energy Use in Brazil: Is Self Sufficiency in Sight?" *Journal of Energy and Development* 10, no. 2 (1985): 173–191.

28. Alan Riding, "Oil Price Fall Perils Brazil Alcohol Fuel," *New York Times*, 29 July 1985, p. 21.

29. Barzelay, *The Political Market Economy*.

30. Ibid.

31. Ibid., p. 239.

32. Ibid.

33. A private multiclient study supervised by Sergio Trindade (see note 22) concludes that the effect on the balance of payments, taking into account the increased imports of chemicals, fertilizers, and fuel necessary for expanded sugar cane production, the loss of exports of molasses and other agricultural products, replacement of petroleum and petrochemical imports by ethanol, and export of chemical and fuel ethanol, was a net improvement of $834 million in 1980 and $2.022 billion in 1984. Other studies, however, conclude that alcohol production resulted in a net total loss of foreign exchange during that same period. See Kahane, "Economic Aspects of Brazilian Alcohol Program"; and S. Islam and W. Ramsey, "Fuel Alcohol: Some Economic Complexities in Brazil and the U.S.," Disc. Paper D-736G, Resources for the Future, Washington, D.C., 1982.

34. Alan Riding, *New York Times*, 29 July 1985, p. 21.

35. Ibid.

36. Kahane, "Economic Aspects of Brazilian Alcohol Program."

37. Ibid.

38. Bernton et al., *The Forbidden Fuel*.

39. Charles Weiss, "Fuel Ethanol in Brazil: Technology and Economics," World Bank paper presented to First Conference on Macroengineering, Crystal City, Virginia, 14 March 1986.

40. See Saint, "Farming for Energy"; Armand Pereira, "Employment Implications of Ethanol Production in Brazil," *International Labor Review* 122, no. 1 (1983): 111–127 (cited in Kahane, 1985); Howard S. Geller, "Ethanol from Sugar Cane in Brazil," *Annual Review of Energy* 10 (1985): 135–164; Pamplona, 1984, in Kahane, "Economic Aspects"; and Kahane, "Economic Aspects."

41. Pereira, 1983, cited in Kahane, "Economic Aspects," p. 52.

42. See Saint, "Farming for Energy"; Ghirardi, "Alcohol Fuel for Biomass."

43. Bernton et al., *The Forbidden Fuel*, p. 159; and Saint, ibid., pp. 226–227.

44. Charles Weiss, "Fuel Ethanol in Brazil."

5: NEW FUELS IN NEW ZEALAND, SOUTH AFRICA, AND CANADA

1. John Boshier, "Energy Issues and Policies in New Zealand," *Annual Review of Energy* 9 (1984): 51–79.

2. New Zealand Ministry of Energy, *Energy Plan* (Wellington, N.Z.: New Zealand Government Printer, 1982).

3. Boshier, "Energy Issues and Policies," p. 62.

4. Ibid.

5. Ibid. See also International Energy Agency, *Synthetic Fuels* (Paris: IEA, 1985).

6. The synthetic gasoline disaster was not necessarily New Zealand's most costly risk in attempting to adapt quickly to high oil prices and insecure oil supplies. Because the synthetic gasoline and compressed natural gas initiatives both were intended to replace gasoline but not diesel fuel, which accounted for about half of the transportation fuel demand, the government felt obligated to alter its refinery operations to produce a larger proportion of diesel fuel. Accordingly, it built a $1.65 billion dollar hydrocracker addition to its refinery. Unlike the synthetic gasoline plant, the hydrocracker project was apparently poorly managed, taking much longer and costing much more than had been anticipated. Given the drop in oil prices, New Zealand may have been better off not building the hydrocracker, importing less crude oil, and only importing as much refined diesel oil as needed. Some New Zealand energy experts have suggested that the operating losses are so great that, if not for the embarrassment of the government, the hydrocracker might have been mothballed in 1986.

7. Liquefied petroleum gas (LPG) is used in limited quantities on the South Island. It has limited potential and from a technical perspective has the same type of barriers to overcome as CNG. LPG is addressed briefly in chapter 12.

8. Information on government incentives comes from a packet of press releases provided by the New Zealand Ministry of Energy.

9. Government pamphlets for 1985 and 1986; G. Harris, P. Phillips, L. Richards, L. Arnoux, *CNG Market Development Study* (Auckland, N.Z.: Energy Research and Development Committee, University of Auckland, 1984), p. 86.

10. Reprint of energy topics from South African Department of Foreign Affairs and Information, *Official Yearbook of the Republic of South Africa—1983* (Johannesburg: Chris van Rensbourg Publications, n.d.).

11. S. Mulholland, "South Africa's Synthetic Fuel Program, Born of Necessity, Grows to Meet Demand," *Wall Street Journal*, 2 March 1982, p. 28.

12. Cited in Douglas Martin, "Developing Canada Oil Sands," *New York Times*, 27 August 1985, p. 31.

13. International Energy Agency, *Synthetic Fuels* (Paris: IEA, 1985), pp. 69–70.

14. Canadian Gas Association, Natural Gas for Vehicles Business Plan, May 1985.

15. B.C. Hydo, "Natural Gas to Go Marketing" (Burnaby, B.C.), 3 January 1987 (newsletter).

16. Energy, Mines and Resources Canada, Natural Gas for Vehicles Industry Survey, 1985.

17. Robert McRory, *Oil Sands and Heavy Oils of Alberta* (Edmonton, Alberta: Energy and Natural Resources, 1982).

18. For descriptions of the two projects, see James J. Heron and Elma K. Spady, "Oil Sands: The Canadian Experience," *Annual Review of Energy* 8 (1983): 137–163.

19. Douglas Martin, "Developing Canada Oil Sands," *New York Times*, 27 August 1985, p. 31.

PART III FEEDSTOCKS AND FUEL PRODUCTION

1. See American Society of Agricultural Engineers, *Vegetable Oil Fuels*, Proceedings of International Conference on Plant and Vegetable Oils and Fuels (St. Joseph, Mich.: ASAE, 1982).

2. Anaerobic digestion to produce methane is discussed thoroughly in U.S. Congress, Office of Technology Assessment, *Energy from Biological Processes*, Vol. II (Washington, D.C.: GPO, 1980).

6: BIOMASS AND HYDROGEN FUELS

1. See for instance John M. Nystrom, C. G. Greenwald, F. G. Harrison, and E. O. Gibson, "Making Ethanol from Cellulose," *CEP*, May 1984, pp. 68–74.

2. See Charles E. Hewett, C. J. High, N. Marshall, and R. Wildermuth, "Wood Energy in the United States," *Ann. Rev. Energy* 6 (1981): 139–170. For a description of hydrolysis processes, see Aerospace Corp., "Alcohol Fuels from Biomass: Production Technology Overview," in U.S. National Alcohol Fuel Commission, *Fuel Alcohols, Appendix* (Washington, D.C.: GPO, 1981), pp. 1–252; papers by E. E. Clausen and J. L. Goddy, and K. H. Vause in D. L. Wise, ed., *Liquid Fuel Systems* (Boca Raton, Fla.: CRC Press, 1983); and various papers in Proceedings of Symposium on Alcohol Fuel Technology (1980, 1982, 1984, 1986).

3. SRI International, "Availability and Cost of Agricultural and Municipal Residues for Use as Alcohol Fuel Feedstocks," in U.S. Dept. of Energy, *The Report of the Alcohol Fuels Policy Review: Raw Material Availability Reports* (Springfield, Va.: NTIS, 1979, DOE/ET-0114/1).

4. U.S. NAFC, *Fuel Alcohol, Final Report* (Washington, D.C.: GPO, 1981).

5. See for instance ibid., p. 71; see also U.S. Congress, Office of Technology Assessment, *Energy from Biological Process* (Washington, D.C.: GPO, 1980); Donald Hertzmark, *A Preliminary Report on the Agricultural Sector Impacts of Obtaining Ethanol from Grain* (Springfield, Va.: Solar Energy Research Institute, NTIS, 1979, SERI/RR-51-292).

6. U.S. NAFC, *Fuel Alcohol, Final Report*, p. 73.

7. O. M. Bevilacqua, M. J. Bernard, D. Sperling, et al., *An Environmental*

Assessment of the Use of Alcohol Fuels in Highway Vehicles (Springfield, Va.: NTIS, 1980, ANL/CNSV-14), p. 264. See also Fred H. Sanderson, "Benefits and Costs of the U.S. Gasohol Program," *Resources* (Resources for the Future) 67 (July 1981): 2–13; and U.S. Department of Agriculture, *Fuel Ethanol and Agriculture: An Economic Assessment* (Washington, D.C.: GPO, 1986). Production costs vary considerably from one plant to another.

8. U.S. General Accounting Office, *Importance and Impact of Federal Alcohol Fuel Tax Incentives* (Gaithersburg, Md.: GAO, 1984, GAO/RCED-84-1), p. 36.

9. Actual wholesale prices of ethanol fuel (before distribution, i.e., F.O.B.) varied between $1.40 and $1.80 for the years 1979 to 1985 in the United States. *Alcohol Week* (Inside Washington, Washington, D.C.) publishes ethanol fuel prices weekly. For a discussion of ethanol pricing see ibid., p. 130, and U.S. Dept. of Agriculture, *Fuel Ethanol and Agriculture.*

10. "Illegal Oranges Sold in Oakland," *San Francisco Chronicle*, 14 November 1981, p. 3.

11. Midwest Research Institute, "Availability and Cost of Grains," in U.S. Department of Energy, *The Report of the Alcohol Fuels Policy Review: Raw Material Availability Reports.*

12. U.S. Congress, Office of Technology Assessment, *Energy from Biological Processes*, pp. 67–69.

13. Ibid.

14. Abdullah et al., "The Potential of Producing Energy from Agricultural" (Lafayette, Ind.: Purdue University, 1979), Contractor's report prepared for U.S. Congress, Office of Technology Assessment. Cited in ibid.

15. See G. E. Ho, "Crop Residues—How Much Can Be Safely Harvested?" *Biomass* 7 (1985): 47–57, and David Pimentel et al., "Biomass Energy from Crop and Forest Residues," *Science* 212 (1981): 1110–1115.

16. U.S. Congress, OTA, *Energy from Biological Processes*, p. 68.

17. Energy Research Advisory Board, U.S. Department of Energy, "Biomass Energy, Draft Report," October 1981, p. 24.

18. Abdullah et al., "The Potential of Producing Energy from Agriculture." Costs and production opportunities are reviewed in Energy Research Advisory Board, "Biomass Energy, Draft Report," pp. 24–28.

19. Ibid., p. 184.

20. For analyses of MSW see "Methanol from Solid Waste," in John Paul, ed., *Methanol Technology and Application in Motor Fuels* (Park Ridge, N.J.: Noyes Data Corp., 1978), pp. 168–187; James Abert and Harvey Alter, "A Survery of United States and European Practices for Recovering Energy from Municipal Waste," in K. V. Sarkanen and D. A. Tillman, eds., *Progress in Biomass Conversion*, Vol. I (New York: Academic Press, 1979), pp. 145–214.

21. Institute of Gas Technology, *IGT World Reserves Study* (Chicago: IGT, 1986), pp. 39–46.

22. Ibid.

23. Ibid.

24. Robert Rea, "Methanol Fuel for East Coast Automotive Markets" (Presented at "Methanol as an Automotive Fuel" workshop sponsored by Mellon

Institute, Peat Methanol Assoc., Santa Fe, New Mexico, September 1981).

25. Mitre Corp, *Biomass-based Alcohol Fuels: The Near Term Potential for Use with Gasoline*, prepared by W. Park et al. (McLean, Va: 1978, MTR-7866), p. 3.

26. Energy Research Advisory Board, "Biomass Energy, Draft Report."

27. See, for instance, S. Y. Shen, A. D. Vyas, and P. E. Jones, "Economic Analysis of Short and Ultra-Short Rotation Forestry," *Resources and Conservation* 10 (1984): 225–270; and Mitre Corporation, *Silvicultural Biomass Farms*, 6 vols. (Prepared for U.S. ERDA) (Springfield, Va.: National Technical Information Service, 1977, MTR-7347.)

28. See ibid. (Vol. II of Mitre, 1977); and Intergroup Consulting Economists Ltd., *Liquid Fuels from Renewable Resources: Feasibility Study*, 6 vols. (Prepared for Government of Canada, Ottawa, Ontario: Fisheries and Environment Service and Environmental Management, 1978).

29. Mitre Corp., "Potential Availability of Wood as a Feedstock for Methanol Production," in U.S. DOE, *The Report of the Alcohol Fuels Policy Review: Raw Material Availability Reports*, p. 73.

30. For detailed analysis of biomass feedstock supply curves, see Daniel Sperling, "An Analytical Framework for Siting and Sizing Biomass Fuel Plants," *Energy* 9, nos. 11–12 (1984): 1033–1040.

31. T. Reed (SERI), Testimony before U.S. Congress, House of Representatives, Committee on Interstate and Foreign Commerce, Subcommittee on Energy and Power, *Alternative Fuels and Compatible Engine* (Hearings, 96th Congress, 2d session, 18 December 1980, GPO, Washington, D.C., 1981, Serial No. 96–237).

32. Ibid.

33. For research results and descriptions of specific biomass methanol technologies, see ibid., p. 197; U.S. Congress, OTA, *Energy from Biological Processes*, pp. 127–129, 134–142; Aerospace Corp., "Alcohol Fuels from Biomass," T. Reed, ed., *Biomass Gasification: Principles and Technology* (Park Ridge, N.J.: Noyes Data Corp., 1981).

34. A conversion efficiency of 100 gallons per dry ton was used to calculate plant output. This figure may prove conservative as technologies improve. Conversion factors in the literature range from 43 to 175 gallons per DT. The 100-gallon-per-DT factor is based on the assumption that power to operate the plant is cogenerated on site by the methanol or raw feedstock. If electricity or imported fuel is purchased, then the equivalent yield would be about 120 gallons per DT. Plant capacities are expressed in terms of barrels per calendar day as opposed to conventional engineering practice of expressing capacity in terms of a 330-day operating schedule. Thus, to translate daily plant capacities as expressed here into annual capacity, they should be multiplied by 365 days, not 330 days.

35. See Stephen Gage (International Harvester), "Packaged Plants," in Thomas Reed and Michael Graboski, eds., *Proceedings, Biomass-to-Methanol Specialists' Workshop* (Springfield, Va.: NTIS, 1982, SERI/CP-234-1590, CONF-820324), pp. 33–40.

36. The patents were inherited by a spinoff company, Syngas Systems, that

has sought and received research grants from the U.S. Dept. of Energy (SERI). In late 1985 an earlier one-ton-per-day gasifier (with output equivalent to 3.5 b/d of methanol) was scaled up to a twenty-four-ton-per-day gasifier. The scaled-up gasifier uses oxygen instead of air and is claimed to be the first gasifier designed especially for methanol production. See SERI, *Oxygen Gasifier Report, Phase II* (Boulder, Colo.: 1987).

37. G. Schutz, "Thermochemical Processes for Hydrogen Production," in *Hydrogen: Energy Vector of the Future* (London: Graham and Trotman, 1983), pp. 67–88.

38. Ibid.

39. Heiko Barnert, "Utilization of Nuclear Process Heat for the Production of Hydrogen," in ibid., pp. 206–217.

40. W. Donitz, "Economics and Potential Application of Electrolytic Hydrogen in the Next Decades," *International Journal of Hydrogen Energy* 9, no. 10 (1984): 817–821; Hartmut Wendt, "The Electrolytic Production of Hydrogen," in *Hydrogen: Energy Vector of the Future*, pp. 55–66; and Wendt, "Seven Years of Research and Development of Advanced Water Electrolysers: Technical Advances and Economic Implications," in G. Imarisio and A. S. Strub, eds., *Hydrogen as an Energy Carrier* (Boston: D. Reidel Publishing Company, 1984), pp. 91–101.

41. J. Gretz, "Conversion of Solar Energy into Hydrogen and Other Fuels," in *Hydrogen: Energy Vector of the Future*, pp. 89–105.

42. C. Carpetis, "An Assessment of Electrolytic Hydrogen Production by Means of Photovoltaic Energy Conversion," *International Journal of Hydrogen Energy* 9, no. 12 (1984): 969–991; O. J. Murphy and J. O. M. Bockris, "Photovoltaic Electrolysis: Hydrogen and Electricity from Water and Light," *International Journal of Hydrogen Energy* 9, no. 7 (1984): 557–561.

7: MINERAL-BASED FUELS

1. The definition of minerals used here is consistent with popular usage but is broader than the scientific definition. The legislation that established the U.S. Synthetic Fuels Corporation defined synthetic fuel as "any solid, liquid, or gas, or combination thereof, which can be used as a substitute for petroleum or natural gas or any derivatives thereof . . . and which is produced by chemical or physical transformation (other than washing, coking, or desulfurizing) of domestic sources of" coal, shale, oil sands (including heavy oil), and water (to produce hydrogen). See Energy Security Act, Section 112 of Title I.

2. World Bank, *Emerging Energy and Chemical Applications of Methanol: Opportunities for Developing Countries* (Washington, D.C.: 1982).

3. U.S. Energy Information Administration (DOE). *International Energy Annual,* 1984, 1985 (Washington, D.C.: GPO, 1985, 1986), p. 64.

4. Charles J. Mankin, "Unconventional Sources of Natural Gas," *Ann. Rev. Energy* 8 (1983). For a discussion of uncertainty in oil and gas estimates see J. J. Shanz, "Oil and Gas Resources—Welcome to Uncertainty," *Resources for the Future* 58 (1978): 1–16 (special issue); and Institute of Gas Technology, *IGT*

World Reserves Study (Chicago, Ill.: IGT, 1986), pp. 47–128.

5. For analysis of these unconventional sources see C. J. Mankin, "Unconventional Sources of Natural Gas," pp. 27–43. Especially good assessments are in IGT, *IGT World Reserves Study*, pp. 227–247, and U.S. Congress, Office of Technology Assessment, *U.S. Natural Gas Availability: Gas Supply Through the Year 2000* (Washington, D.C.: GPO, 1985).

6. Thomas Gold has been the principal proponent of this theory. See, for instance, T. Gold, "The Origin of Natural Gas and Petroleum, and the Prognosis for Future Supplies," *Ann. Rev. Energy*, 10 (1985): 53–77.

7. P. Abelson, "World Supplies of Natural Gas," *Science* 228 (June 1985): 4705.

8. B. Mossavar-Rahmani and S. Mossavar-Rahmani, *The OPEC Natural Gas Dilemma* (Boulder, Colo., and London: Westview Press, 1986).

9. Ronald Probstein and R. Edwin Hicks, *Synthetic Fuels* (New York: McGraw-Hill, 1982), p. 322.

10. U.S. EIA (DOE), *Annual Energy Outlook, 1984* (Washington, D.C.: GPO, 1985), p. 168.

11. Two top officials of the U.S. Environmental Protection Agency, Charles L. Gray and Jeffrey Alson, make this argument in their book, *Moving America to Methanol* (Ann Arbor, Mich.: University of Michigan Press, 1985).

12. See Everett A. Sondreal and George A. Wiltsee, "Low-Rank Coal: Its Present and Future Role in the United States," *Ann. Rev. Energy* 9 (1984): 473–499.

13. Harry Perry and Hans Landsberg, "Factors in the Development of a Major U.S. Synthetic Fuels Industry," *Ann. Rev. Energy* 6 (1981): 248.

14. Ibid.

15. Probstein and Hicks, *Synthetic Fuel*, pp. 294–301.

16. For elaboration of coal gasification theory see D. Hebden and H. J. F. Stroud, "Coal Gasification Processes," chap. 24 of M. A. Elliott, ed., *Chemistry of Coal Utilization* (2d supplementary volume) (New York: Wiley-Interscience, 1981).

17. Synthetic Fuel Associates, *Coal Gasification Systems: A Guide to Status, Applications, and Economics* (Palo Alto, Calif.: Electric Power Research Institute, 1983, EPRI AP-3109), p. 201.

18. Ibid., pp. 2–3.

19. See Jonathan Lash and Laura King, eds., *The Synfuels Manual* (New York: Natural Resources Defense Council, 1983), pp. 31–32.

20. In moving-bed gasifiers (sometimes also called fixed-bed) such as Lurgi gasifiers, the feedstock is placed in lump form at the bottom of the reactor vessel. Oxygen or steam is blown upward through the coal (or biomass) and gases exit through the top or side. These gasifiers produce large amounts of methane that may be sold directly or reformed at additional cost into synthesis gas. They are relatively simple to operate but do not handle caking coals well. A second technology is fluidized-bed gasifiers in which finely sized particles are suspended by a fairly high-velocity upward flow of gases. These tend to produce a lower-Btu gas and to have a lower thermal efficiency than fixed-bed reactors.

A new pressurized Winkler design, however, is reported to be particularly well suited to lignite and low-rank coals and to have high efficiencies with those feedstocks. Relatively small quantities of methane are formed. In entrained-bed gasifiers fine particles are blown into the gas stream. The chemical reactions take place while the feedstock is suspended in the gas stream. The particles are continuously filtered from the gas and recycled. This technology is used in Texaco and Koppers-Totzek gasifiers. They can handle a wide range of coal types but are more complex. For greater detail see Synthetic Fuel Associates, *Coal Gasification Systems*, pp. 2–7 to 2–14; and U.S. Synthetic Fuel Corporation, *Comprehensive Strategy Report, Appendices* (Washington, D.C.: 1985), chap. J.

21. World Bank, *Emerging Energy and Chemical Applications of Methanol*, p. 36.

22. For a description and historical review see S. S. Marsden, "Methanol as a Viable Energy Source in Today's World," *Ann. Rev. Energy* 8 (1983): 333–354.

23. Ibid., p. 338.

24. Jet Propulsion Laboratory, *California Methanol Assessment*, Vol. II (Pasadena, Calif.: 1983), chap. 4.

25. R. Probstein and E. Hicks, *Synthetic Fuels.*

26. Harry Perry and Hans Landsberg, "Factors in the Development of a Major U.S. Synthetic Fuels Industry," p. 245.

27. Cited in Bernton et al., *The Forbidden Fuel: Power Alcohol in the Twentieth Century* (New York: Boyd Griffin, 1982), p. 199.

28. See Probstein and Hicks, *Synthetic Fuels*, p. 449.

29. Ibid., p. 315. Because of their high aromatic content, coal liquids are inherently unsuited to the production of diesel-like middle-distillate fuels.

30. ICF Inc., "Methanol from Coal," in National Alcohol Fuel Commission, *Appendix* (Washington, D.C.: GPO, 1981), p. 303.

31. The costs would be high because of the severe refining necessary and the incompatibility problems that will arise if existing petroleum refineries are used. See ibid., pp. 248–354.

32. Ibid., p. 351.

8: PRODUCTION COSTS

1. The Great Plains plant was reportedly producing high-Btu gas (SNG) at a cost of well under $10 per million Btu in 1986, and the Coolwater plant was producing medium-Btu synthesis gas for slightly less. SNG should cost just a little less than methanol; the cost of producing synthesis gas is by far the largest cost component in producing methanol from coal (or biomass).

2. E. Merrow, K. Phillips, and C. Myers, *Understanding Cost Growth and Performance Shortfalls in Pioneer Process Plants* (Prepared for U.S. Department of Energy, Santa Monica, Calif.: Rand Corp., 1981, R-2569-DOE), pp. v–vi.

3. More recent experience is mixed. The actual cost of the Coolwater and

Great Plains SNG plants did not increase above the "definitive" estimates. The Union Oil shale plant went far over budget as it encountered various technical problems.

4. U.S. Synthetic Fuel Corporation, *Comprehensive Strategy Report, Appendices* (Washington, D.C.: 1985).

5. R. W. Hess, *Review of Cost Improvement Literature with Emphasis on Synthetic Fuel Facilities and the Petroleum and Chemical Process Industries* (Santa Monica, Calif.: Rand Corp., 1985, N-2273-SFC). Also see R. W. Hess, *Potential Production Cost Benefit of Constructing and Operating First-of-a-Kind Synthetic Fuel Plants* (Santa Monica, Calif.: Rand Corp., 1985, N-2274-SFC).

6. Charles Weiss (World Bank), "Fuel Ethanol in Brazil," Presented at First Conference on Macroengineering, Crystal City, Virginia, 14 March 1986.

7. Hess, *Review of Cost Improvement Literature*, p. 72.

8. Texas Eastern Synfuels, Inc., *New Mexico Synfuels Project: Coal-Liquid Fuels and High-Btu Coal Gas* (Springfield, Va.: NTIS, 1982, DE84 001654), 5 vols.

9. The costs in the New Mexico Synfuel study were expressed in 1982 dollars. Escalation of construction costs to 1984 dollars was based on appropriate construction cost indices published in *Engineering News Record* (209, no. 13; 211, no. 13; and 213, no. 12).

10. In simple terms $PV_x = \sum_{n=1}^{t} \frac{C_x^n}{(1+r)^n}$

where PV_x = present value of the cost stream x
C_x^n = future value of cost x in year n
r = discount rate
n = 1, corresponds to 1985.

11. E. Merrow et al., *Understanding Cost Growth*.

12. See cost indices in *Economic Report of the President* (Washington, D.C.: GPO, 1980), appendix B.

13. Based on early 1980s coal price forecasts for the New Mexico region. See California Energy Commission, *Preliminary Price Forecast for California Utilities* (Sacramento, Calif.: CEC, 1984).

14. The initial (somewhat undersize) coal gasification plants in the U.S. required about two to three years for actual construction, plus about two years for acquiring permits and completing detailed engineering plans and feasibility studies, plus additional earlier time for preliminary design work. The Union Oil shale plant had an extended construction period because of serious technical flaws that apparently were not detected until the plant began an initial aborted start-up.

15. E. Merrow et al., *Understanding Cost Growth*.

16. See for instance E. F. Ehrbar, "How to Bring Interest Rates Down," *Fortune*, 15 June 1982, pp. 66–82 (cover story).

17. Annualized Accelerated Depreciation Allowance = [(accelerated tax depreciation) × (investment tax basis/investment book basis) − (straight-line depreciation)] × [tax rate/(1−tax rate)]

where:
Accelerated Tax Depreciation = [2(capital recovery factor)] × [n− present worth factor/n(n+1) (weighted cost of capital)]
Investment Tax Credit = [capital recovery factor × investment tax credit rate/ ((1−r) (1−tax rate))] × [investment tax basis/investment book basis]

18. It may now be clear that this revenue-based method is appropriate when revenues of these "other" products can be estimated with confidence.

9: FURTHER ASSESSMENT OF FEEDSTOCK-PRODUCTION OPTIONS

1. See for instance R. S. Chambers, R. A. Herendeen, J. J. Joyce, and P. S. Penner, "Gasohol: Does It Or Doesn't It Produce Positive Net Energy?" *Science* 206 (16 November 1979): 789–795.

2. See U.S. Congress, Office of Technology Assessment, *Energy from Biological Processes*, Vol. II (Washington, D.C.: GPO, 1980), pp. 210–233; TRW, "Energy Balances in the Production and End-Use of Alcohols Derived from Biomass," in U.S. National Alcohol Fuel Commission, *Fuel Alcohol, Appendix* (Washington, D.C.: GPO, 1981), pp. 623–759; H. Perez-Blanco and B. Hannon, "Net Energy Analysis of Methanol and Ethanol Production," *Energy* 7, no. 3 (1982): 267–280.

3. U.S. Department of Agriculture, *Agricultural Statistics* (Washington, D.C.: GPO, 1982).

4. U.S. Congress, Office of Technology Assessment, *Energy from Biological Processes*, Vol. II (Washington, D.C.: GPO, 1980), pp. 21–22.

5. U.S. National Alcohol Fuel Commission (Survey conducted in 1979), Partial results published in NAFC, *Fuel Alcohol, Final Report* (Washington, D.C.: GPO, 1981).

6. Fortune, "The 500 Largest U.S. Industrial Corporations, Ranked by Sales," *Fortune*, 29 April 1985, pp. 226–285.

7. Ibid.

8. Stuart Dramond, "Synthetic Fuel Plant Scuttled," *New York Times*, 2 August 1985.

9. Greg Morris, "Biomass as an Energy Resource: An Economic and Environmental Investigation" (Ph.D. dissertation, Energy and Resources Group, University of California, Berkeley.) University Microfilms, Ann Arbor, Mich., 1982), chap. 4.

10. In some developing countries, including China, low-Btu gasifiers are used widely. They are small, simple pieces of equipment. The gas is usually used to run on-site stationary internal combustion engines (for pumps, food processing equipment, etc.).

PART IV: FUEL DISTRIBUTION

1. Richard Ottinger, before U.S. Congress, House of Representatives, Committee on Interstate and Foreign Commerce, Subcommittee on Energy and

Power, *Alternative Fuels and Compatible Engine Designs*, Hearings, 96th Cong., 2d sess., December 18, 1980 (Washington, D.C.: GPO, 1981, 96–237), p. 244.

10: CURRENT FUEL DISTRIBUTION SYSTEMS

1. The following historical description of petroleum transport is taken from George S. Wolbert, *U.S. Oil Pipelines* (Washington, D.C.: American Petroleum Institute, 1979).

2. Coastal transport is hindered in the U.S. by lack of deep ports and high labor costs. It is deprived of potential scale economies in the U.S. by shallow ports that preclude the use of larger tanker ships, while labor (and other operating) costs are increased by legislation (Jones Act) that requires all ships originating and terminating in the U.S. to use only U.S. crews (which are paid significantly more than foreign crews).

3. Joint venture systems are not separate legal entities. All owners set their own tariffs for their share of the flow independent of other owners. For an excellent description and analysis of alternative ownership forms of pipelines, see John Piercey, "The Pipeline Segment of the Domestic Petroleum Industry" (Ph.D. dissertation, University of Oklahoma, Norman, Oklahoma, 1978).

4. See U.S. DOT, *The Transport of Methanol by Pipeline*, prepared by P. Zebe and W. Gazda (Washington, D.C.: Transportation Systems Center, 1985); and U.S. Environmental Protection Agency, *Distribution of Methanol as a Transportation Fuel*, prepared by R. D. Atkinson (Springfield, Va.: NTIS, 1982, PB83-116822).

5. U.S. DOT, *The Transport of Methanol by Pipeline*, pp. 3–12.

6. Ibid., pp. 3–10.

7. U.S. General Accounting Office, *Petroleum Pipeline Rates and Competition—Issues Long Neglected by Federal Regulators and in Need of Attention*, Report to the Congress by Comptroller General (Washington, D.C.: GAO, 1979), p. 22.

8. See U.S. DOT, *The Transport of Methanol by Pipeline*, pp. 5–10; and U.S. GAO, *Petroleum Pipeline Rates*.

9. See Locklin, D. P., *The Economics of Transportation*, 7th ed. (Homewood, Ill.: R. D. Irwin, 1972), p. 614.

10. See Reduced Pipeline Rates Case 243 ICC 115 (1940), cited in U.S. DOT, *The Transport of Methanol by Pipeline*, pp. 5–10.

11. U.S. GAO, *Petroleum Pipeline Rates and Competition*, p. 23.

12. Ibid.

13. U.S. DOT, *The Transport of Methanol by Pipeline*, pp. 3–1.

14. Ibid., pp. 3–3. Also see ARCO Chemical Company, "Oxygenated Fuels Technical Bulletin," No. 8309, September 1983.

15. William H. Kite (ESSO), "Supplying Hydrated Ethanol as a Gasoline Substitute," Presented at Fifth Miami International Conference on Alternative Energy Sources, Miami Beach, Florida, 13–15 December 1982.

16. G. S. Wolbert, *U.S. Oil Pipelines*, p. 78.

17. Proration is the industry term used to describe the situation when demand for pipeline capacity exceeds the available capacity and all shippers must scale back or prorate their shipments. Proration when necessary is required by regulation. In 1977 a joint study team from the U.S. Departments of Energy and Transportation developed a national pipeline network model to predict future flows. The model predicted a 5 percent flow reduction in petroleum product trunk lines by 1990. Specific areas projected to experience significant flow reductions are some northeastern pipelines and links extending north from Texas toward Montana. The joint study is documented in U.S. Departments of Transportation and Energy, *National Energy Transportation Study*, Preliminary Report to the President (Washington, D.C.: 1980.)

18. See chapter on Alaskan methanol concept in J. K. Paul, ed., *Methanol Technology and Application in Motor Fuels* (Park Ridge, N.J.: Noyes Data Corp., 1978), pp. 188–240.

19. See figure 28 in text; and U.S. DOE, *United States Petroleum Pipelines: An Empirical Analysis of Pipeline Sizing*, prepared by Leonard Coburn (Springfield, Va.: NTIS, 1980, DOE/PE-0024), pp. II-24.

20. See National Petroleum Council, *Petroleum Storage and Transportation Capacities*, Vol. V (Washington, D.C.: 1979); and U.S. Congress, Congressional Research Service, *National Energy Transportation*, 95th Cong., 1st sess. (Prepared for Senate Committees on Energy and Natural Resources and Commerce, Science, and Transportation, Washington, D.C.: GPO, 1977), pp. 222–232.

21. U.S. Congress, *National Energy Transportation*, Vol. I, pp. 250–256.

22. See Lane W. Harold, "Distribution Options in U.S. National Alcohol Fuel Commission," *Fuel Alcohol, Appendix* (Washington, D.C.: GPO, 1981), p. 791.

23. U.S. Congress, *National Energy Transportation*, Vol. I, p. 299.

24. See P. Gibson, "Look Who's Booming," *Forbes*, 5 March 1979, pp. 52–57.

25. National Petroleum Council, *Petroleum Storage and Transportation Capacities*, Vol. IV, pp. G–11.

26. Unless otherwise noted, this and all subsequent data on gasoline marketing stations are cited from or based on U.S. Department of Energy, *State of Competition in Gasoline Marketing*, by J. Delaney and R. Fenili (Springfield, Va.: NTIS, 1981, DOE/PE-0026). While these data are rapidly becoming outdated because of major changes in the industry in the 1980s, those changes do not affect the findings of this chapter or book. Service-station data are not precise because there is no complete data base. Primary outlets are those where gasoline dispensing accounts for 50 percent or more of revenues.

27. See John Piercey, "The Pipeline Segment," p. 202.

28. American Gas Association, *Gas Facts* (Arlington, Va.: annual).

29. John Fowler, *Energy and the Environment* (New York: McGraw-Hill, 1984), p. 213.

30. B. and S. Mossavar-Rahmani, *The OPEC Natural Gas Dilemma* (Boulder, Colo., and London: Westview Press, 1986), p. 83.

31. Fereidun Fesharaki, "Natural Gas Supply and Demand Planning in the

Asia-Pacific Region," *Annual Review of Energy* 10 (1985): 463–493.

32. Ibid., p. 467.

33. U.S. Department of Commerce, pp. 32–33.

11: FUEL DISTRIBUTION COSTS AND PROBLEMS FOR NEW FUELS

1. American Petroleum Institute, *Alcohols: A Technical Assessment of Their Application as Fuels* (Washington, D.C.: 1976, 4261).

2. Union Oil Co., *Methanol Fuel Modification for Highway Vehicle Use* (Prepared for U.S. DOE) (Springfield, Va.: NTIS, 1978, HCP/W3683-18), pp. 41–42 and II-1 to II-8.

3. P. G. Edgington, "Distribution of Gasoline Containing Oxygenates," *VII International Symposium on Alcohol Fuels* (Paris: Editions Technip, 1986), pp. 497–501.

4. Union Oil Co., *Methanol Fuel Modifications*, pp. II-2.

5. See O. M. Bevilaqua, M. J. Bernard III, D. Sperling, et al., *An Environmental Assessment of the Use of Alcohol Fuels in Highway Vehicles* (Springfield, Va.: NTIS, 1980, ANL/CNSV-14), pp. 228–232; and U.S. DOT, prepared by P. Zebe and W. Gazda, *The Transport of Methanol by Pipeline* (Washington, D.C.: 1985), pp. 3–8.

6. Lane W. Harold, "Distribution Options," in U.S. National Alcohol Fuel Commission, *Fuel Alcohol, Appendix* (Washington, D.C.: GPO, 1981), p. 775.

7. U.S. DOT, *The Transport of Methanol by Pipeline*, pp. 3–10.

8. Kuhn (ARCO), testimony before Congress, November 1984, p. 68.

9. "EPA Spurs Attack on Illegal Fuels," *Automotive News*, 19 March 1984, p. 40.

10. F. Kant, A. Cohn, A. Cunningham, M. Farmer, and W. Herbst, *Feasibility Study of Alternative Fuels for Automotive Transport*, Exxon Research and Engineering (Prepared for U.S. Environmental Protection Agency) (Springfield, Va.: NTIS, 1974, PB-235580), 3 vols.

11. U.S. DOT, *The Transport of Methanol by Pipeline.*

12. Ibid., pp. 5–18.

13. John G. Holmes, "Distribution Systems and Costs for a Regional Fuel Methanol Market," *SAE* 861573 (1986).

14. See, for instance, Derek P. Gregory, *A Hydrogen Energy System* (Arlington, Va.: American Gas Association, 1973); Walter J. Jasionowski, Jon B. Pangborn, and Dale G. Johnson, "Distribution of Gaseous Hydrogen—Technology Evolution," *Hydrogen for Energy Distribution, Symposium Papers* (Chicago, Ill.: Institute of Gas Technology, 1978).

15. J. H. Holbrook, H. J. Cialone, and P. M. Scott, *Hydrogen Degradation of Pipeline Steels, Summary Report* (Upton, N.Y.: Brookhaven National Laboratory, 1984).

16. The cost for a single tank car with a 30,000 gallon capacity is about $50,000, and for a tank car in a tank-train configuration about $80,000, according to J. Dudlak, Director of Marketing for GATX in Chicago, 1987.

17. U.S. DOT, *The Transport of Methanol by Pipeline*, pp. 5–23.

18. Ibid., pp. 5–18.

19. Daniel Sperling, "Distribution Dilemmas," *Proceedings of the Fifth International Alcohol Fuel Technology Symposium*, Vol. III, Auckland, New Zealand, May 1982 (P.O. Box 5098, Wellington, N.Z.: National Organizing Committee, 1982), pp. 380–389.

20. Holmes, "Distribution Systems and Costs for a Regional Fuel Methanol Market."

21. Ibid.; and Charles Gray and Jeffrey Alson, *Moving America to Methanol* (Ann Arbor, Mich.: University of Michigan Press, 1985).

22. Mark DeLuchi, Daniel Sperling, and Robert Johnston, *Comparative Analysis of Future Transportation Fuels* (Berkeley, Calif.: Institute of Transportation Studies, 1987).

23. Anker Gram, President of Cryogas (Vancouver, British Columbia), 11 March 1987.

24. Don Knowles, *Alternative Automotive Fuels* (Reston, Va. Reston Publishing Company, 1984); The Aerospace Corporation, *Assessment of Methane-Related Fuels for Automotive Fleet Vehicles, Executive Summary* (Washington, D.C.: Office of Vehicle and Engine R&D, DOE, 1982).

25. Daniel Sperling and Jennifer Dill, "Unleaded Gasoline in the U.S.: A Successful Model of System Innovation," *Transportation Research Record* (forthcoming).

12: MOTOR VEHICLE TECHNOLOGIES

1. See L. J. Nuttall and J. F. McElroy, "Technical and Economic Feasibility of a Solid Polymer Electrolyte Fuel Cell Powerplant for Automotive Applications," *SAE* 830348 (1983); J. R. Huff, N. Vanderborgh, J. F. Roach, and H. S. Murray, "Fuel Cell Propulsion Systems for Highway Vehicles," *Proceedings of the Fourteenth Energy Technology Conference* (Rockville, Md.: Government Institutes, Inc., 1987).

2. Steven Shladover, "The Roadway Powered Electric Transit Vehicle—Progress and Prospects," *Transportation Research Record* (forthcoming). Electrified roadways may come about in conjunction with automated vehicle controls as a means of deploying the more reliable electric engine and thereby reducing the danger of relying on automated controls. See, for instance, J.G. Bender et al. (GM Transportation Systems Center), *Systems Studies of Automated Highway Systems* (Springfield, Va.: NTIS, 1982, FHWA/RD-82/003).

3. W. J. Walsh and J. B. Rajan, "Advanced Batteries for Electric Vehicles," *Transportation Research Record* (forthcoming). Paper was presented in January 1985 at the annual meeting of the Transportation Research Board.

4. See Ministry of Energy, Mines and Resources, *Propane—The Modern Automotive Fuel* (Ottawa, Ontario: n.d., Annex H, TE-84-12/E); and Kenneth Darrow, "Economic Assessment of Compressed Natural Gas Vehicles for Fleet Applications," Gas Research Insight Series (Chicago, Ill.: Gas Research Institute, 1983), p. 9.

5. Most LPG vehicles are retrofitted gasoline vehicles. In 1981 Ford and International Harvester began selling small numbers of dedicated production-line LPG vehicles in Canada; General Motors and others followed later, also on a limited basis. Some recent studies of LPG vehicles are R. F. Webb Corp., *An Investigation of Propane as a Motor Vehicle Fuel in Canada* (Ottawa, Ontario: Strategic Studies Branch, Ministry of Transport, 1982); Kathy Roxborough, "Propane Carburetion Market Development 1980–1983" (Ottawa, Ontario: Transportation Energy Division, Ministry of Energy, Mines and Resources, 1984); and selected sections in Hayes/Hill Inc., *An Analysis of the Retrofit Market for Compressed Natural Gas Fleet Vehicles* (American Gas Assoc., 1983).

6. Liquid hydrogen has been advocated as a replacement for jet fuel. It is lighter per energy unit than jet fuel, and although it is also bulkier, weight is the more important consideration. See G. Daniel Brewer, "Liquid Hydrogen Fueled Commercial Aircraft," in *Hydrogen for Energy Distribution, Symposium Paper* (Chicago, Ill.: Institute of Gas Technology, 1978), pp. 541–550.

7. Oak Ridge National Laboratory, *Transportation Energy Data Book: Edition 8* (Springfield, Va.: NTIS, 1985, ORNL-6205), pp. 2–8.

8. Comsis Corp., "A Case Study of the Commercialization of the Diesel Engine in the Domestic Light Duty Vehicle Market" (Prepared for U.S. DOE, Office of Vehicle and Engine R&D, Washington, D.C., 1986). (draft).

9. Oak Ridge National Laboratory, *Transportation Energy Data Book: Edition 8*, pp. 1–40 and 1–41.

10. Ibid., pp. 1–29. On a Btu basis diesel fuel was still considerably less expensive.

11. For an analysis of the diesel car experience see Kenneth Kurani and Daniel Sperling, "The Rise and Fall of Diesel Cars: A Consumer Choice Analysis," *Transportation Research Record* (forth coming).

12. See for instance Arnold Bloch, *Alternative Fuels for Buses: Current Assessment and Future Perspectives*, (Washington, D.C.: Urban Mass Transportation Administration, 1984, NY-11-0023).

13. The following data on methanol bus performance are taken from a "methanol bus information package" that was sent by Charles Gray, Office of Mobile Sources, U.S. Environmental Protection Agency, to Regional Air Division Directors, 30 August 1984. See also W. J. Wells, D. G. Maynard, and M. D. Jackson, "Methanol Fuel Programs: Models for Implementation of an Alternative Transportation Fuel," *Proceedings of the Sixth International Symposium on Alcohol Fuels Technology*, Vol. III (Ottawa, Ontario: May 1984), pp. 507–515.

14. C. Gray, "Methanol Bus Information Package," p. 8.

15. Charles Gray and Jeffrey Alson, *Moving America to Methanol* (Ann Arbor, Mich.: University of Michigan Press, 1985).

16. Written response by GM to questions from the Joint Hearing before the Subcommittees on Fossil and Synthetic Fuels and Energy Conservation and Power on April 25, 1984 (cited in ibid., p. 123).

17. Ibid., p. 123.

18. See E. Eugene Euklund, Richard L. Bechtold, Thomas J. Timbario, and

Peter W. McCallum, "State-of-the-Art Report on the Use of Alcohol in Diesel Engines," *SAE* 940118 (1984).

19. In the mid-1980s R&D programs for natural gas in diesel engines were begun or expanded in New Zealand, Canada, and the U.S.

20. Donald Peterson, "New Automotive Fuels: It's Time for Government Action," *Proceedings of the Sixth International Symposium on Alcohol Fuels Technology*, p. 43.

21. Diesel engines have the same cold-start problem, though to a lesser extent. Diesel car owners in cold climates have adjusted by parking indoors and using electric plug-in heaters for the engine block. In Canada electric plugs have been installed at many outside public parking locations.

22. Amos Golovoy and Roberta J. Nichols, "Natural Gas–Powered Vehicles," *Chem. Tech* (June 1983): 359–363.

23. Roberta J. Nichols, "Technical Aspects of the Use of Methanol-Gasoline Blends as Transport Fuels," Presented to the Joint China-U.S. Seminar on Clean Coal Fuels, Taiyuan, Shanxi Province, People's Republic of China, May 13–18, 1985.

24. Ibid.

25. This $300 cost is a "no frills" retrofit; the cost may vary considerably depending on the vehicle and mechanic.

26. See U.S. General Accounting Office, *Removing Barriers to the Market Penetration of Methanol Fuels* (Gaithersburg, Md.: GAO, 1983), p. 33.

27. Ibid., p. 34. Costs are presented in detail in L. Schieler, M. Fisher, D. Dennler, and R. Nettell, "Bank of America's Methanol Fuel Program," Presented at National Science Foundation Workshop on Automotive Use of Methanol-Based Fuels, Washington, D.C., 10 January 1985.

28. The Bank of America has claimed that a group of thirty-five General Motors cars converted to methanol, with no adjustments of the compression ratio, has a 32 to 39 percent fuel efficiency improvement. However, data collection was erratic and inconsistent. The results are not based on sound research method and are probably highly inaccurate.

29. Kenneth Darrow, "Economic Assessment of Compressed Natural Gas Vehicles for Fleet Applications," p. 9; Canadian Gas Association, *Government Funding Available in Canada* (Ontario, Canada: Don Mills, 1983). R. L. Bechtold and T. J. Timbario, *Status of Gaseous Fuels for Use in Highway Transportation* (Springfield, Va.: NTIS, 1983, DE 84 015205), p. 49.

30. M. J. Ellis, "A Cost-Benefit Assessment of Compressed Natural Gas," Report 314, School of Engineering, University of Auckland, New Zealand, 1983.

31. The Aerospace Corp, *Assessment of Methane-Related Fuels for Automotive Fleet Vehicles* (Springfield, Va.: NTIS, 1982, DOE/CE/50179-1).

32. Thomas Browne (UPS), Testimony before U.S. Congress, House of Reps., Comm. on Interstate and Foreign Commerce, Subcomm. on Energy and Power. *Alternative Fuels and Compatible Engine Designs*. Hearings, 96th Cong., 2d sess., 18 Dec. 1980 (Washington, D.C.: GPO, 1981, 96–237), pp. 50–81.

33. Lindsey-Kaufman Co., *Opinions of Vehicle Manufacturers, Engine*

Manufacturers and End-Users on Use of Alcohol Fuels for Motor Vehicles (Springfield, Va.: NTIS, 1980, PB81-166860), pp. II-25 to II-26.

34. Jouke V. D. Weide and Richard Wineland, "Vehicle Operation with Variable Methanol/Gasoline Mixtures," *Proceedings of the Sixth International Symposium on Alcohol Fuels Technology*, pp. 380–386.

35. Roberta J. Nichols, "Update on Ford's Methanol Vehicle Experience," Presented at the European Fuel Oxygenates Association First Annual Conference, Brussels, April 15, 1986.

36. Hirao Osamu, "Potential of Alcohol as an Alternative Fuel for Internal Combusion Engines," *Proceedings of the Fourth International Symposium on Alcohol Fuel Technology*, São Paulo, Brazil, 1980.

37. The Aerospace Corporation, *Assessment of Methane-Related Fuels for Automotive Fleet Vehicles.*

38. M. R. Swain et al., "Methane Fueled Engine Performance and Emissions Characteristics," in *1983 InterSociety Energy Conversion Engineering Conference*, American Institute of Chemical Engineers.

39. Donald Petersen, "New Automotive Fuels," p. 42.

40. William Abernathy, *The Productivity Dilemma: Roadblock to Innovation in the Automobile Industry* (Baltimore, Md.: Johns Hopkins University Press, 1978).

41. Based on development of conventional gasoline engines in the U.S., 1970s and early 1980s data. See Danilo J. Santini, *Direct and Indirect Costs Arising from Implementation of New Engines* (Argonne, Ill.: Argonne National Laboratory, 1985, ANL/CNSV-TM-156), pp. 12–18.

42. Lawrence J. White, *The Automobile Industry Since 1945* (Cambridge, Mass.: Harvard University Press, 1971).

43. U.S. National Alcohol Fuel Commission, *Fuel Alcohol. Final Report* (Washington, D.C.: GPO, 1981), p. 11.

44. D. Santini, *Direct and Indirect Costs Arising from Implementation of New Engines*, p. 18.

45. See Alan Altshuler et al., *The Future of the Automobile* (Cambridge, Mass.: MIT Press, 1984).

46. For review of these research efforts see Jet Propulsion Laboratory, *California Methanol Assessment*, Vol. II (Pasadena, Calif.: 1983, 83-18), pp. 8–70 to 8–72. A major program on methanol disassociation has been ongoing at the U.S. Solar Energy Research Institute.

47. Section 211 (f), Public Law 95-95.

48. *Federal Register* 44, no. 68, p. 20777.

49. U.S. General Accounting Office, *Removing Barriers to the Market Penetration of Methanol Fuels* (Gaithersburg, Md.: GAO, 1983), p. 14.

50. Mueller Assoc., *Alternative Fuels Utilization and the Automotive Emission Certification Process* (Springfield, Va.: NTIS, 1980, DOE/CS/5605-TI), p. 2.

51. U.S. National Alcohol Fuel, *Fuel Alcohol, Appendix* (Washington, D.C.: GPO, 1981), p. 981.

52. Ibid., p. 895.

53. AB 2004, Calif. legislature.

54. Energy Policy and Conservation Act.

13: MARKET NICHES

1. Larry Ronan and William J. Abernathy, *The Development and Introduction of the Automotive Turbocharger* (Springfield, Va.: NTIS, 1979, DOT-TSC-NHTSA-79-19).

2. Starch and sugar materials were fermented into ethanol and wood was converted into methanol via destructive distillation (thus the common name "wood alcohol" for methanol). Methanol was not made from natural gas until the 1920s.

3. For a detailed description of the U.S. methanol industry see Jet Propulsion Laboratory, *California Methanol Assessment*, Vol. II (Pasadena, Calif.: 1983), chap. 7.

4. Estimated in World Bank, *Emerging Energy and Chemical Applications of Methanol: Opportunities for Developing Countries* (Washington, D.C.: 1982).

5. Ibid., p. 14.

6. Chemical ethanol production in the U.S. has consistently exceeded demand, resulting in excess capacity, similar to chemical methanol production. Ethanol production costs are higher than methanol production costs.

7. See E. S. Lipinsky, "Chemicals from Biomass: Petrochemical Substitution Options," *Science* 212 (26 June 1981): 1465–1471.

8. Only three of thirteen methanol plants are not located on the Gulf Coast; one is the Tennessee Eastman coal-to-methanol plant, one is in Florida, and the other is in Delaware.

9. Although in Brazil, where biomass ethanol costs are much lower than in the U.S., ethanol is used as a chemical. It is also exported to the U.S. for use as a chemical because chemicals are not subject to the same tariff as fuel alcohol.

10. Jet Propulsion Laboratory, *California Methanol Assessment*.

11. Ibid., chap. 9, based on tests by Southern California Edison Co., Genernal Electric, and Florida Power Corp. See A. Weir, W. H. von Klein Smid, and E. A. Sanko, "Test and Evaluation of Methanol in a Gas Turbine System," Final report prepared by Southern Calif. Edison Co. for Electric Power Research Institute Proj. 988-1; R. C. Farmer, "Methanol—A New Fuel Source," *Gas Turbine International*, May–June 1975; and R. C. Revty and T. J. Timbario, *Status of Alcohol Fuels Utilization Technology for Stationary Gas Turbines* (Springfield, Va.: NTIS, 1979, HCP/M2098-01).

12. Ibid., chap. 9 (JPL).

13. On March 4, 1985, the EPA ordered oil refiners to reduce the amount of lead in gasoline by 90 percent by the end of 1985. It also proposed that the date for complete elimination of lead use in gasoline be moved from 1995, as previously proposed, to 1988.

14. D. Dickson, "Europe Mirrors U.S. Debate on Car Exhaust," *Science* 228 (12 April 1985): 160 (also see *Science*, 5 April 1985, p. 37).

15. See John Enos, *Petroleum Progress and Profits: A History of Process Innovation* (Cambridge, Mass.: MIT Press, 1962); and Danilo Santini, *Direct and Indirect Costs Arising from Implementation of New Engines*, Argonne National Laboratory (Springfield, Va.: NTIS, 1985, ANL/CNSV-TM-156).

16. Jet Propulsion Laboratory, *California Methanol Assessment*, Vol. II, pp. 8–26 to 8–31.

17. Ibid., Vol. II, pp. 8–24.

18. Ibid., Vol. II, pp. 8–22.

19. World Bank, *Emerging Energy and Chemical Applications of Methanol: Opportunities for Developing Countries* (Washington, D.C.: 1982), p. 55.

20. The increment would typically be about 1 to 3 cents per gallon. If 2.5 percent cosolvent were used per gallon, and if the cosolvent costs twice that of gasoline per unit volume (which is a high estimate), then the cost increment is 2.5 cents per gallon.

21. Agricultural energy consumption data are highly unreliable because of inaccurate and incomplete data collection and reporting. There are large discrepancies between different sources. See Federal Energy Administration and U.S. Dept. of Agriculture, *Energy and U.S. Agriculture: 1974 Data Base*, vol. 2 (Washington, D.C.: GPO, 1977), cited in Gretchen Kulp, "Agricultural Gasoline Consumption," Oak Ridge National Laboratory, undated and unpublished manuscript (about 1980 Contract W-7405-eng-26).

22. Ibid.

23. For data consistency between years, a single source was used to compare 1974 and 1981 data, even though that source is known to underreport gasoline consumption on farms (as demonstrated in the previously cited manuscript by G. Kulp). The 1974 and 1981 data used here come from U.S. Dept. of Transportation (FHWA), *Highway Statistics* (Washington, D.C.: annual), tables MF-21A and MF-24.

24. Although Ford Motor Co. and others in Brazil are designing tractors and other agricultural equipment to run on ethanol, presumably that technology could be transferred or the equipment exported to the U.S. and elsewhere. The first ethanol tractors were reportedly sold in 1985.

25. David W. Jones, *Urban Transit Policy: An Economic and Political History* (Englewood Cliffs, N.J.: Prentice-Hall, 1985), pp. 62–63.

26. Historical data are found in ibid., p. 62; current data on size and fuel consumption of transit fleets are found in Oak Ridge National Laboratory, *Transportation Energy Data Book: Edition 8* (Springfield, Va.: NTIS, 1985, ORNL-6205).

27. See Charles Gray and Jeffrey Alson, *Moving America to Methanol* (Ann Arbor, Mich.: University of Michigan Press, 1985), chap. 9.

28. *Federal Register* 50 (March 15, 1985): 10606.

29. Gray and Alson, *Moving America to Methanol*, p. 115.

30. Ibid., p. 120.

31. Based on unpublished data from survey of transit fleets in southern California conducted for South Coast Air Quality Management District. Some transit agencies never rebuild their engines; others do so as many as five times. The major transit agency in southern California rebuilds the engine three times.

32. Summary data compiled from other sources on fleet vehicles are provided in Oak Ridge National Laboratory, *Transportation Energy Data Book: Edition 8*. Also see earlier editions. Various surveys and data sources provide

conflicting estimates of total light-duty fleet vehicles. Some estimates ignore light trucks, others measure only vehicles in fleets containing some threshold number of vehicles, most underestimate commercial vehicles, and most seem to ignore Post Office vehicles. For best data on commercial vehicles see Mark R. Berg, M. J. Converse, and D. H. Hill, "Electric Vehicles in Commercial Sector Applications: A Study of Market Potential and Vehicle Requirements" (Prepared for Detroit Edison Co. and Electric Power Research Institute by Institute for Social Research, University of Michigan, 1984); for somewhat dated but best comprehensive review of fleet market see survey conducted by Bobit (publishers of a trade magazine, *Automotive World*) under the direction of Brookhaven National Laboratory in 1977 and 1978. Summary data were published in Joseph Wagner, *Fleet Operator Data Book, vol. 1: National Data* (Springfield, Va.: NTIS, 1979, BNL 50904). Data on light trucks in fleets are available, again in a dated report, from a 1977 survey in Bureau of the Census, Truck Inventory and Use Survey, U.S. Dept. of Commerce, Washington, D.C., May 1980, TC77-T-52, as reviewed in Aerospace Corp., *Assessment of Methane-Related Fuels for Automotive Fleet Vehicles* (Springfield, Va.: NTIS, 1982, DOE/CE 50179-1).

33. Medium- and heavy-duty trucks account for about 21 percent of all fleet vehicles in the U.S. Vehicles in fleets of ten or more account for about 7 percent of all automobiles in the U.S. and about 10 percent of all light trucks (fleets with fewer than ten vehicles represent another 3 to 4 percent of total vehicles). Oak Ridge National Laboratory, *Transportation Energy Data Book: Edition 8*, pp. 2–49.

34. Aerospace Corp., *Assessment of Methane-Related Fuels for Automotive Fleet Vehicles.*

35. Oak Ridge National Laboratory, *Transportation Energy Data Book: Edition 7*, pp. 2–37.

36. See Aerospace Corp., *Assessment of Methane-Related Fuels for Automotive Fleet Vehicles*, tables 5–7 and 5–8.

37. Ibid.

38. When the state of California purchased 500 methanol cars in 1981, the California Highway Patrol was to receive 100 of them, but backed out later because of uncertainty about reliability and safety.

39. Mark R. Berg et al., "Electric Vehicles in Commercial Sector Applications," 1984.

40. Aerospace Corp., *Assessment of Methane-Related Fuels.*

41. Jet Propulsion Laboratory, *California Methanol Assessment*, chap. 8.

42. Ibid.

43. 600,000 vehicles × 1 gallon/15 miles × 15,000 miles/year = 600,000,000/gallons/year.

44. See for instance L. Schieler, M. Fisher, D. Dennler, and R. Nettell, "Bank of America's Methanol Fuel Program: An Insurance Policy That is Now a Viable Fuel," *Proceedings of the Sixth International Symposium on Alcohol Fuels Technology*, vol. 2, 21–25 May 1984, Ottawa, Ontario, pp. 367–372.

45. For a description of methanol fleet programs see F. J. Wiens, M. C.

McCormick, R. J. Ernst, R. J. Morris, and R. J. Nichols, "California's Alcohol Fleet Test Program—Final Results," vol. 3, pp. 367–375; and K. Koyama, K. Chorn, L. Vann, and R. Nichols, "California's Alcohol Fuels," *Proceedings of the Seventh International Symposium on Alcohol Fuels Technology*, Vol. 2, Paris, France, 20–23 October 1986.

14: HOUSEHOLD VEHICLE FUELS MARKET

1. Marianne Millar, M. Singh, A. Vyas, and L. Johnson, "Transportation Energy Outlook Under Conditions of Persistently Low Petroleum Prices," *Transportation Research Record* (forthcoming).

2. Ibid.

3. Garth Harris, L. Arnoux, and P. Phillips, *CNG in Auckland: Survey and Analysis* (Auckland, N.Z.: New Zealand Energy Research and Development Committee, University of Auckland, 1980, Pub. P44).

4. Garth Harris, P. Phillips, L. Richards, and L. Arnoux, *CNG Market Development Study* (Auckland, N.Z.: New Zealand Energy Research and Development Committee, University of Auckland, 1984, Pub. P86).

5. Daniel Sperling and K. Kurani, "Refueling and the Vehicle Purchase Decision: The Diesel Car Case," *SAE* 870644 (1987).

6. Ibid.

7. Daniel Sperling and R. Kitamura, "Refueling and New Fuels: An Exploratory Analysis," *Transportation Research* 20A, no. 1 (1986): 15–23.

8. Canadian Facts, *Management Summary: Natural Gas for Vehicles—Conversion Motivation Study* (Burnaby, B.C.: British Columbia Hydro Gas Operations, 1986).

9. Ibid., p. 3.

10. W. J. Walsh and J. B. Rajan, "Advanced Batteries for Electric Vehicles," *Transportation Research Record* (forthcoming).

11. See Harold L. Walters, "Necessary Incentives for a Methanol Fuel Market," *SAE* 861572 (1986).

12. COMSIS Corp., "*A Case Study of the Commercialization of the Diesel Engine in the Domestic Light Duty Vehicle Market*" (Prepared for U.S. DOE, Office of Vehicle and Engine R&D, Washington, D.C., 1986); and Kenneth Kurani and D. Sperling, "The Rise and Fall of Diesel Cars: A Consumer Choice Analysis," *Transportation Research Record* (forth coming).

13. John Lowe, "On the 'Techno-Economics' of CNG Utilization for Vehicle Transport," Presented at conference on Gaseous Fuels for Transportation, Vancouver, 9 August 1986.

14. The *Lundberg Letter* estimated that 20.3 percent of all gasoline sold in 1986 was premium unleaded. An additional 0.2 percent was premium leaded. Octane ratings are $(R+M) \div 2$.

15. Oak Ridge National Laboratory, *Transportation Energy Data Book: Edition 9* (Springfield, Va.: NTIS, 1987, ORNL-6325).

PART VI: ENVIRONMENTAL DEGRADATION

15: AIR QUALITY

1. Ayub U. Hira, Joseph A. Mulloney, and Gregory J. D'Alessio, "Alcohol Fuels from Biomass," *Envir. Sci. Tech.* 17, no. 5 (1983): 202A–213A.

2. James Antizzo, ed., "Background Material for Workshop on Health and Environmental Effects of Coal Gasification and Liquefaction Technologies," Mitre Corp., 1978, cited in Jonathan Lash and Laura King, eds., *The Synfuels Manual* (New York: Natural Resources Defense Council, 1983), p. 74.

3. Lash and King, eds., *The Synfuels Manual*; Dept. of Home Affairs and Environment, *The Environmental Implications of Three Synthetic Liquid Fuel Technologies—Coal Liquefaction, Shale Oil Production and Biomass Conversion* (Canberra, Australia: Australian Government Publishing Service, 1983).

4. U.S. Department of Energy, *Final Environmental Impact Statement, Solvent Refined Coal II Demonstration Project* (Washington, D.C.: DOE, 1981, DOE/EIS-0069), p. C-58.

5. See John Saw, "Environmental Aspects of Coal Liquefaction in Australia," in Australian Dept. of Home Affairs and Environment, *The Environmental Implications of Three Synthetic Liquid Fuel Technologies*, pp. 1–63.

6. Jonathan Lash and Laura King, *The Synfuels Manual*, p. 121.

7. Ibid., p. 121.

8. Ibid., pp. 106–110.

9. Michael A. Chartock, Michael D. Devine, Stephen C. Ballard, Martin R. Cines, and Martha W. Gillil, "Environmental Issues of Synthetic Transportation Fuels from Coal." Background Paper 4 for U.S. Congress, OTA, *Increased Automobile Fuel Efficiency and Synthetic Fuels* (Springfield, Va.,: NTIS, 1982).

10. Note in table 50 that projected emissions from coal-fired electrical generation plants were much greater than projected emissions from four coal liquids plants. This contrast does not suggest that emissions from liquid fuel production are insignificant, but rather that direct combustion of coal generates unusally large quantities of pollutants.

11. U.S. Department of Energy, *Environmental Analysis of Synthetic Liquid Fuels* (Washington, D.C.: July 1979). This draft report was rewritten when the Reagan Administration took office and published it in an abbreviated form without the siting analysis and other analysis. See U.S. DOE, *Synthetic Fuels and the Environment: An Environmental and Regulatory Impacts Analysis* (Washington, D.C.: DOE, 1980, DOE/EV-0087).

12. For a full analysis of greenhouse effects of transportation fuels, see Mark DeLuchi, Robert Johnston, and Daniel Sperling, "Transportation Fuels and the Greenhouse Effect," *Transportation Research Record* (forthcoming).

13. See ibid. for review of studies.

14. Charles Gray and Jeffrey Alson, *Moving America to Methanol: A Plan to Replace Oil Imports, Reduce Acid Rain, and Revitalize Our Domestic Economy* (Ann Arbor, Mich.: University of Michigan Press, 1985).

15. R. J. Nichols and J. M. Norbeck (Ford Motor Co.), "Assessment of Emissions from Methanol-Fueled Vehicles: Implications for Ozone Air Quality," Paper presented at Annual Meeting of Air Pollution Control Association, Detroit, Michigan, June 1985.

16. See Karen Wilson and M. McCormick, "Methanol as an Ozone Control Strategy," *Proceedings of the Sixth International Symposium on Alcohol Fuel Technology*, Ottawa, Ontario, May 1984, pp. 2–68 to 2–74.

17. Nichols and Norbeck, "Assessment of Emissions from Methanol-Fueled Vehicles."

18. Almost all proposed strategies and controls—e.g., ride sharing—typically gain less than 1 percent reduction in ozone.

19. Larry Berg, Testimony to U.S. Congress, House of Rep., Comm. on Energy and Commerce, Subcomm. on Fossil and Synthetic Fuels and Subcomm. on Energy Cons. and Power, *Methanol as Transportation Fuel*, 98th Cong., 2d sess., 4 and 25 April 1984 (Washington, D.C.: GPO, 1984, 98–145), p. 26.

20. R. J. Nichols (Ford Motor Co.), "Technical Aspects of the Use of Methanol-Gasoline Blends as Transport Fuels," Paper presented to the Joint China-U.S. Seminar on Clean Coal Fuels, Taiyuan, Shanxi Province, The People's Republic of China, May 13–19, 1985.

21. O. M. Bevilacqua, M. J. Bernard, D. Sperling, et al., *An Environmental Assessment of the Use of Alcohol Fuels in Highway Vehicles* (Springfield, Va.: NTIS, 1980, ANL/CNSV-14), tables E.2 and E.3.

22. Ibid., p. 177.

23. Richard L. Bechtold and Thomas J. Timbario, "The Status of Gaseous Fuel for Use in Highway Transportation, A 1983 Perspective" (Prepared for the U.S. Department of Energy, Washington, D.C., November 1983, DOE/CS/56051-10).

24. Methane is actually a potent greenhouse gas—more effective than CO_2—and over the last 400 years methane levels in the atmosphere have doubled, causing a 0.23° C warming worldwide (W. C. Chameides, "Increasing Atmospheric Methane," *Science* 30, no. 17: 568). However, we have calculated elsewhere that natural gas vehicles will not have a significant effect on methane emissions in the U.S. or in the world: Mark DeLuchi, Robert Johnston and Daniel Sperling, "Transportation Fuels and the Greenhouse Effect," *Transportation Research Record* (forthcoming).

25. Tests conducted by DOE at EPA's motor vehicle emissions testing laboratory in Ann Arbor, Michigan. See Aerospace Corp., *Assessment of Methane-Related Fuels for Automotive Fleet Vehicles* (Springfield, Va.: NTIS, 1982, DOE/CE/50179-1).

26. Emission data from laboratory tests, theoretical studies, fleet experience, and field tests are reported and evaluated in Margaret Singh, *State of Knowledge of Environmental Concerns Related to Natural Gas–Fueled Vehicles* (Springfield, Va.: NTIS, 1984, ANL/CNSV-TM-138), pp. 24–27. See also Richard L. Bechtold and Thomas J. Timbario, "The Status of Gaseous Fuels for Use in Highway Transportation, A 1983 Perspective."

27. M. R. Swain et al., "Methane Fueled Engine Performance and Emissions Characteristics," in *1983 Intersociety Energy Conversion Engineering Confer-*

ence Proceedings (New York: American Institute of Chemical Engineers, 1983), pp. 626–629.

28. M. C. Watson et al., "An Australian Hydrogen Car," in *Hydrogen Energy Progress V, Proceedings of the Fifth World Hydrogen Energy Conference* (New York: Pergamon Press, 1987), pp. 1549–1561.

29. The Aerospace Corporation, "Assessment of Methane-Related Fuels for Automotive Fleet Vehicles, Executive Summary" (Prepared for the U.S. Department of Energy, Washington, D.C., 1982, DOE/CE/50179-1.)

30. Charles Gray and Jeffrey Alson, *Moving America to Methanol*, p. 124.

31. Chazi A. Karim. "Some Consideration of the Use of Natural Gas in Diesel Engines," *Nonpetroleum Vehicular Fuels III (Symposium Papers)* (Chicago, Ill.: Institute of Gas Technology, 1983), pp. 337–353.

32. That, in any case, was the claim made at that time. Ibid., p. 1.

33. The first public hearings on the proposal were held on November 7, 1983. As of late 1987 debate was continuing but no specific actions had yet been taken.

16: IMPACTS ON WATER, LAND, AND HUMAN SETTLEMENTS

1. A. U. Hira, J. A. Mulloney, and G. J. D'Alessio, "Alcohol Fuels from Biomass," *Envir. Sci. Technol.* 17, no. 5 (1983).

2. C. S. Barnes and E. J. Hobert, "Alcohol Manufacture—Wastewater Treatment," *Water* 6, no. 4 (December 1979): 22–23.

3. I. P. Willington and G. G. Marten, "Options for Handling Stillage Waste from Sugar-Based Fuel Ethanol Production," *Resources and Conservation* (1982).

4. Argonne National Laboratory, *Environmental and Economic Evaluation of Energy Recovery from Agricultural and Forestry Residues* (Argonne, Ill.: 1979, ANL/EES-TM-58).

5. U.S. Cong., Office of Technology Assessment, *Increased Automobile Fuel Efficiency and Synthetic Fuels* (Washington, D.C.: GPO, 1982); and Jonathan Lash and Laura King, *The Synfuels Manual* (Washington, D.C.: Natural Resources Defense Council, 1983).

6. J. Lash and L. King, *The Synfuels Manual*, 1983; U.S. Department of Energy, *Synthetic Fuels and the Environment* (Washington, D.C.: 1980, DOE/EV-0087).

7. U.S. Cong., OTA, *Increased Automobile Efficiency and Synthetic Fuels.*

8. Ronald Probstein and Harris Gold, *Water in Synthetic Fuel Production* (Cambridge, Mass.: MIT Press, 1978), p. 124.

9. U.S. Cong., OTA, *Increased Automobile Efficiency and Synthetic Fuels*, p. 280.

10. Ibid., p. 280.

11. R. Probstein and H. Gold, *Water in Synthetic Fuel Production.*

12. Another effect would be higher prices for other users, including farmers. The political power of farmers in the arid farming areas of the western U.S. and

the historical percedent of subsidized water for farmers suggests the likelihood of strong political opposition to transfer of water rights to fuel producers.

13. See U.S. Cong., OTA, *An Assessment of Oil Shale Technologies* (Washington, D.C.: 1980); U.S. Environmental Protection Agency, *Energy from the West, Policy Analysis Report* (Washington, D.C.: 1979, 600/7-79-083).

14. J. Lash and L. King, *The Synfuels Manual*, p. 91.

15. See ibid.; and U.S. Cong., OTA, *An Assessment of Oil Shale Technologies.*

16. U.S. Council on Environmental Quality, *Desertification of the U.S.* (Washington, D.C.: GPO, 1981), p. 66.

17. U.S. DOE, *Synthetic Fuels and the Environment.*

18. U.S. Cong., OTA, *An Assessment of Oil Shale Technologies*, pp. 334–335.

19. Ibid.; and U.S. Environmental Protection Agency, *Environmental Perspective on the Emerging Oil Shale Industry* (Cincinnati, Ohio: 1980, EPA-600/w-80-205a), p. 86.

20. M. A. Chartock, M. D. Devine, S.C. Ballard, M. R. Cines, and M. W. Gillil, "Environmental Issues of Synthetic Transportation Fuels from Coal," Background Paper 4 prepared for U.S. Cong., OTA, *Increased Automobile Efficiency and Synthetic Fuels.*

21. U.S. Cong., OTA, *An Assessment of Oil Shale Technologies.*

22. John M. Fowler, *Energy and the Environment* (New York: McGraw-Hill, 1984), p. 201.

23. Soil Conservation Service, *National Erosion Inventory Estimate* (Washington D.C.: U.S. Department of Agriculture, 1978).

24. W. E. Larson, "Crop Residues: Energy Production or Erosion Control?" *J. Soil and Water Conservation* 34 (1979): 74–76.

25. Aerospace Corp., *Environmental Control Perspective for Ethanol Production from Biomass* (Prepared for U.S. DOE, Germantown, Maryland, 1980), p. 20.

26. Cited in U.S. Cong., OTA, *Energy from Biological Processes*, Vol. II, (Washington, D.C.: GPO, 1980), p. 71.

27. This is acknowledged in D. Pimentel, M. A. Moran, S. Fast, et al., "Biomass Energy from Crop and Forest Residues," *Science* 212 (5 June 1981): 1115.

28. Ibid., 1111.

29. Council on Environmental Quality, *The President's Environmental Program* (Washington, D.C.: GPO, 1979), cited in Pimentel et al., "Biomass Energy from Crop and Forest Residues."

30. See Schnittker Assoc., "Ethanol: Farm and Fuel Issues," in U.S. National Alcohol Fuel Commission, *Fuel Alcohol, Appendix* (Washington, D.C.: 1980), p. 553.

31. For various positions on the issue see ibid.; H. D. Braunstein, P. Kanciruk, R. D. Roop, F. E. Sharples, J. S. Tatum, and K. M. Oakes, *Biomass Energy Systems and the Environment* (New York: Pergamon, 1981); U.S.

Cong., OTA, *Energy from Biological Processes*; Aerospace Corp., *Environmental Control Perspective for Ethanol Production from Biomass* (Germantown, Maryland, 1980).

32. U.S. Cong., OTA, *Energy from Biological Processes*, Vol. II, p. 78.

33. Ibid.

34. H. M. Braunstein et al., *Biomass Energy Systems and the Environment*.

35. Ibid., p. 117.

36. U.S. Cong., OTA, *Energy from Biological Processes*, Vol. II, p. 27.

37. U.S. Dept. of Energy, *Environmental Readiness Document, Wood Commercialization* (Washington, D.C.: 1979).

38. U.S. Environmental Protection Agency, *Environmental Implications of Trends in Agriculture and Silviculture*, Vol. 1: "Trend Identification and Evaluation" (Washington, D.C.: 1977, EPA-600/3-77-121).

39. U.S. Cong., OTA, *Energy from Biological Processes*.

40. Ibid., Vol. II, p. 75.

41. Ibid., Vol. I, p. 11.

42. U.S. DOE, *Environmental Analysis of Synthetic Fuels*.

43. Statement of Robert King (Tosco Corp.) to U.S. Cong., House of Rep., Comm. on Science and Technology, Subcomm. on Energy Development and Applications. *The Socioeconomic Impacts of Synthetic Fuels*, 97th Cong., 2d sess., 28 April 1982, (Washington D.C.: GPO, 1982), pp. 51–61.

44. *Wall Street Journal*, 12 August 1981, p. 1.

45. Statement of Allen C. Randle (Union Oil) before U.S. Cong., *The Socioeconomic Impacts of Synthetic Fuels*, p. 19.

46. William N. Rom and Jeffrey Lee, "Energy Alternatives: What Are Their Possible Health Effects?" *Envir. Sci. Technol.* 17, no. 3: 133A–144A.

47. R. J. Sexton, C. S. Weil, and N. I. Condra, "Archives of Environmental Health," Vol. I, 1960, pp. 181–231, cited in Lash and King, *The Synfuels Manual*, p. 101.

48. H. T. Butlin, "Cancer of the Scrotum in Chimney Sweeps and Others," *British Medical Journal* 1 (1892): 1341–1346; 2 (1892): 1–6, 66–71; cited in Diane Y. Sauter, *Synthetic Fuels and Cancer* (New York: Scientists' Institute for Public Information, New York, 1975).

49. A. D. Little Assoc., "A Modal Economic and Safety Analysis of the Transportation of Hazardous Substances in Bulk," U.S. Maritime Administration Dept. of Transportation, Washington, D.C., 1974.

50. National Transportation Safety Board, *Fatality Rates for Surface Freight Transportation* (Washington, D.C.: 1971).

51. Ralph L. Keeney, Ram B. Hulkarni, and Keshavan Nair, "Assessing the Risk of an LNG Terminal," *Technology Review* (October 1978): 64–72.

52. See also the following papers in *Proceedings of the International Symposium on Alcohol Fuel Technology* (Springfield, Va.: 1978, CONF-7775): A. J. Moriarty, "Toxicological Aspects of Alcohol Fuels Utilization"; P. N. D'Eliscu, "Biological Effects of Methanol Spills into Marine, Estuarine, and Freshwater Habitats"; and G. Hagey, A. Parker, and D. Raley, "Methanol and Ethanol—Environmental and Safety Issues."

53. O. M. Bevilacqua, M. J. Bernard, D. Sperling, et al., *An Environmental Assessment of the Use of Alcohol Fuels in Highway Vehicles* (Springfield, Va.: NTIS, 1980, ANL/CNSV-14), p. 7; U.S. Cong., OTA, *Increased Automobile Efficiency and Synthetic Fuels*, p. 266.

54. Ibid., O. M. Bevilacqua et al., p. 6.

55. Italy has the greatest experience with CNG, but unfortunately safety data are not available. The following review of safety hazards of CNG vehicles are primarily based on Aerospace Corp., *Assessment of Methane-Related Fuels for Automotive Fleet Vehicles* (Springfield, Va.: 1982, DOE/CE/50179-1); Margaret M. Singh (Argonne National Laboratory), *State of Knowledge of Environmental Concerns Related to Natural Gas Fueled Vehicles* (Springfield, Va.: NTIS, 1984, ANL/CNSV-TM-138); and M. C. Krupka et al., *Gaseous Fuel Safety Assessment for Light-Duty Automobile Vehicles* (Prepared for U.S. DOE, Livermore, California: Los Alamos National Laboratory, 1983, LA 9829-MS).

56. H. Eichert and M. Fischer, "Hydrogen Safety in Energy Applications Compared with Natural Gas," in *Hydrogen Energy Progress V*, Proceedings of the Fifth World Hydrogen Energy Conference, Toronto, Canada, 15–20 July, ed. T. N. Veziroglu and J. B. Taylor (New York: Pergamon Press, 1984), pp. 1869–1880.

57. An apt analogy is the large number of motorists in the U.S. who, in flagrant violation of federal and state laws, disconnected emission control devices in their vehicles and intentionally "misfueled" their vehicles with less expensive leaded gasoline to save money even though it destroyed the catalysts. Estimates are that 10 to 20 percent of autos were misfueled in the late 1970s and early 1980s.

PART VII: GETTING FROM HERE TO THERE

17: THE CHICKEN-OR-EGG STASIS

1. One of the earliest discussions of this metaphor was in September 1981 at a workshop on methanol as an automotive fuel sponsored by the Mellon Institute at which most participants were from industry. The chicken-or-egg metaphor was addressed as a key consideration (but not analytically treated) in U.S. Congress, House of Representatives, "Methanol as an Automotive Fuel," Staff report prepared by the Subcommittee on Fossil and Synthetic Fuels of the Commission on Energy and Commerce, 98th Congress, 2d session (Washington, D.C.: GPO, 1984, 98-5); and U.S. General Accounting Office, *Removing Barriers to the Market Penetration of Methanol Fuels* (Gaithersburg, Md.: GAO, 1983). The first detailed discussion was in an earlier draft of this chapter in Daniel Sperling, "Alcohol Fuels and the Chicken and Egg Syndrome," *Proceedings of the International Symposium on Alcohol Fuels Technology*, Vol. II (Ottawa, Ontario: May 1984), pp. 393–399.

18: TRANSITION STRATEGIES: SHORTCUTS, DETOURS, AND DEAD ENDS

1. Other organizations have doubtless also developed strategic plans to introduce alternative transportation fuels, but those plans are not publicly available. The U.S. Synthetic Fuel Corporation, as required by the law that brought it into existence, completed a "comprehensive strategy report" in June 1985. It is a self-justifying document that contains only the vaguest guidelines and strategies. The Electric Power Research Institute (Palo Alto, California), research arm of the electric utility industry, has formed a development corporation to promote electric vehicles. They have targeted limited-range fleet vehicles in what may be characterized as an end-use market segmentation strategy.

2. See K. D. Smith, A. Edwards, M. McCormack, and D. Kondoleon, "Alcohol Fuels for California: Establishing the Market," *Proceedings of the Fifth International Symposium on Alcohol Fuel Technology* (Auckland, New Zealand, 1982); and K. D. Smith, D. Fong, D. Kondoleon, and L. Sullivan, "The California Methanol Program: Creating a Market," *Proceedings of the Sixth International Symposium on Alcohol Fuel Technology*. Ottawa, Canada, May, 1984.

3. Paul Wuebben (South Coast Air Quality Management District), "Methanol as a Regional Air Quality Strategy," Paper presented at Transportation Energy Symposium, Institute of Transportation Studies, University of California, Berkeley, 26 April 1985.

4. See Southwest Research Institute, *Properties and Economics of Mechanol-Gasoline Blends with High Methanol Content* (Springfield, Va.: NTIS, 1985); Robert G. Jackson, "The Use of Medium Methanol Blends and On-Site Blending to Implement the Introduction of Methanol to the Motor Fuel Market," *Proceedings of the Sixth International Symposium on Alcohol Fuels Technology*, Vol. II, (Ottawa, Ontario: May 1984), pp. 301–306.

5. The estimate that 1 to 2 percent of VMT by California vehicles is outside California is based on surveys and traffic counts taken at California borders from 1979–1980, and various metropolitan travel surveys. Interstate travel data were provided by Ron White, California Department of Transportation, 15 June 1984.

6. Cost increases would barely be noticeable since alcohol would replace gasoline on a physical, not an energy, basis. Thus one gallon of methanol that might cost $1.00 per gallon ($2 per gasoline-equivalent gallon) would replace gasoline which costs about the same amount on a physical basis. Since only 5% to 15% of the gasoline is replaced by alcohol, the cost difference per gallon and per mile traveled are negligible.

7. Daniel Sperling and Jennifer Dill, "Unleaded Gasoline in the U.S.: A Successful Model of System Innovation," *Transportation Research Record* (forthcoming).

8. Barry McNutt and E. Eugene Ecklund, "Is There a Government Role in Methanol Market Development," *SAE* Paper 861571 (1986).

9. I am now conducting a major study of consumer purchase behavior with respect to nonpetroleum fuels and vehicles with funding from government and industry. Research results will be forthcoming in 1988–1989.

10. S curves are widely used to describe innovation diffusion processes: see, for instance, E. M. Rogers, *Diffusion of Innovations*, 3d ed. (New York: Free Press, 1982).

11. That investment in 1,000 cars or so is an example of a tendency by political bodies to spread funds around as much as possible, even though in many cases they do not contribute to programmatic or policy goals. For an early description of that demonstration program see U.S. Dept. of Energy, *Project Plan for Fleet Operations of Federal Methanol-Fueled Vehicles* (Springfield, Va.: NTIS, 1985, DOE/CE-0126).

PART VIII: MAKING CHOICES

19: A PRICE-DETERMINING VIEW OF ENERGY CHOICES

1. William E. Simon, *A Time for Truth* (New York: McGraw-Hill, 1978), p. 85.

2. Edward L. Morse, "After the Fall: The Politics of Oil," *Foreign Affairs* 64, no. 4 (Spring 1986): 793.

3. Herbert Simon, *Sciences of the Artificial* (Cambridge, Mass.: MIT Press, 1981).

4. Herman Daly, "On Thinking about Future Energy Requirements," Department of Economics, Louisiana State University, Baton Rouge, La., 1976. Cited in Amory Lovins, *Soft Energy Paths: Toward a Durable Peace* (New York: Harper & Row, 1977), p. 68. Originally published by Ballinger.

5. See Walter A. Rosenbaum, *Energy, Politics, and Public Policy* (Washington, D.C.: Congressional Quarterly Press, 1981).

6. The lesson that the cost of defying basic economic principles of supply, demand, and market clearing prices is very large was learned in the 1970s with experiences in energy price regulation. See Richard H. K. Vietor, *Energy Policy in America Since 1945: A Study of Business-Government Relations* (New York: Cambridge University Press, 1984).

7. National Science Foundation, *An Analysis of Federal R&D Funding by Function* (Washington, D.C.: NSF, 1977, 77–236).

8. For a discussion of the role and performance of government in the energy marketplace, see for instance Robert M. Solow, "The Economics of Resources or the Resources of Economics," *American Economic Association* 64, no. 2 (May 1974): 1–14; Richard Schmalensee, "Appropriate Government Policy toward Commercialization of New Energy Supply Technologies," *The Energy Journal* 1, no. 2 (1980): 1–40; and Walter A. Rosenbaum, *Energy, Politics, and Public Policy.*

9. H. R. Heede, R. E. Morgan, and Scott Ridley. "The Hidden Costs of

Energy," Center for Renewable Resources, Suite 638, 1001 Connecticut Ave., N.W., Washington, D.C., 1985.

10. H. R. Heede and A. Lovins, "Hiding the True Costs of Energy Choices," *Wall Street Journal*, 17 September 1985, p. 28.

11. Colin Norman, "Uranium Enrichment: Heading for a Cliff?" *Science* (22 May 1987): 906.

12. *Battelle* Memorial Institute (Pacific Northwest Laboratory), *Analysis of Federal Incentives Used to Stimulate Energy Production* (Richland, Wash.: 1980, PNL-2410 REV II, UC-59). Earlier version published in 1978.

13. Herbert Dennenburg, Insurance Commissioner, State of Pennsylvania, cited in Heede et al., "The Hidden Costs of Energy," p. 7.

14. *Battelle* Memorial Institute, *An Analysis of Federal Incentives Used to Stimulate Energy Production.*

15. U.S. General Accounting Office, *Importance and Impact of Federal Alcohol Fuel Tax Incentives* (Gaithersburg, Md.: GAO, 1984, GAO/RCED-84-1), pp. 49–54.

16. U.S. General Accounting Office, *Removing Barriers to the Market Penetration of Methanol Fuels* (Gaithersburg, Md.: GAO, 1983, GAO/RCED-84-36), pp. 23–24.

17. See Mike Paparian, "Should California Consumers Be Burdened by Utility Project Failures: LNG, A Case Study," Sierra Club, Sacramento, California, undated pamphlet (1982?).

18. See Alan Altschuler, M. Anderson, D. Jones, D. Roos, and J. Womack, *The Future of the Automobile* (Cambridge, Mass.: MIT Press, 1984).

19. As incomes and sensitivity to the inherent inefficiency of standardized, multipurpose vehicles grow, there has been a subtle trend toward somewhat more specialization of vehicles, but old habits and large sunk costs in roadways are powerful conservative forces.

20. Langdon Winner, "Energy Regimes and the Ideology of Efficiency," in George H. Daniels and Mark H. Rose, eds., *Energy and Transport: Historical Perspectives on Policy Issues* (Beverly Hills, California.: Sage Publications, 1982), p. 271.

21. David W. Orr, "U.S. Energy Policy and the Political Economy of Participation," *Journal of Politics* 41, no. 1 (Nov. 1979): 1053.

20: HARD AND SOFT CHOICES?

1. Amory B. Lovins, "Energy Strategy: The Road Not Taken," *Foreign Affairs* 55 (1976): 65–96; see also SRI, *Solar Energy in America's Future* (Washington, D.C.: U.S. Department of Energy, 1977), chap. 5 ("The Broader Issues"), pp. 81–99.

2. See for example H. Nash, ed., *The Energy Controversy: Soft Path Questions and Answers by Amory Lovins and His Critics* (San Francisco: Friends of the Earth, 1979); W. A. Rosenbaum, *Energy, Politics, and Public Policy* (Washington, D.C.: Congressional Quarterly Press, 1981); U.S. Con-

gress, *Alternative Long-Range Energy Strategies*, Joint Hearing before the Select Committee on Small Business and the Committee on Interior and Insular Affairs, U.S. Senate, Dec. 9, 1976 (Washington, D.C.: GPO, 1977), 2 vols; J. B. Robinson, "Apples and Horned Toads: On the Framework-Determined Nature of the Energy Debate," *Policy Science* 15 (1982): 23–45.

3. SRI, *Solar Energy in America's Future*, p. 90.

4. See H. Nash, *The Energy Controversy*, p. 249.

5. See I. Barbour, H. Brooks, S. Lakoff, and J. Opie, *Energy and American Values* (Research Triangle Park, N.C.: Praeger, 1982).

6. Ibid., p. 58.

7. See A. B. Lovins, *Soft Energy Paths: Toward a Durable Peace* (New York: Harper & Row, 1977). Originally published by Ballinger.

8. Ibid., p. 55.

9. Ibid.

10. Ibid., pp. 42–44.

11. Ibid.; and A. B. Lovins and H. Lovins, *Brittle Power: Energy Strategy for National Security* (Andover, Mass.: Brick House Publishing Co., 1982).

12. A. B. Lovins, *Soft Energy Paths*, pp. 38–39.

13. Ibid.; see especially pp. 50–51, 86–103.

14. Office of Technology Assessment, *Energy from Biological Processes*, Vol. I. (Washington, D.C.: GPO, 1980), p. 11.

15. A. B. Lovins, *Soft Energy Paths*, p. 94; see also Lovins and Lovins, *Brittle Power*.

16. U.S. Dept. of Agriculture, *Agricultural Statistics* (Washington, D.C.: GPO, 1982).

17. U.S. Congress, Office of Technology Assessment, *Energy from Biological Processes*, Vol. II (Washington, D.C.: GPO, 1980), pp. 21–22.

18. "The 500 Largest U.S. Industrial Corporations, Ranked by Sales," *Fortune*, 29 April 1985, p. 266.

19. See Lovins and Lovins, *Brittle Power*.

21: FIVE PATHS

1. The possibility of unmarketable surpluses is a real possibility as U.S. grain production continues to expand far beyond the nation's demand for it and beyond what other countries are willing to purchase. Typical annual grain surpluses after exports are over 100 million tons, enough to produce 200 to 300 million barrels of ethanol. If such large surpluses are not converted to alcohol or otherwise sold, the government through various price support and surplus management programs (including grain storage) would incur huge costs, estimated by one economist at $4 billion for 1983 (Abner Womack, cited in "U.S. Farmers This Year Have Reason to Dread a Bountiful Harvest," *Wall Street Journal*, 20 August 1982, p. 1).

2. See for instance David M. Gordon "Do We Need to Be No. 1?" *Atlantic* 257, no. 4 (1986): 100–108. The basic values and beliefs underlying Amory

Lovins's soft energy path are similar, though in a narrower context, to those underlying the "front yard" strategy.

3. See Danilo Santini, "The Petroleum Problem: Managing the Gap," Center for Transportation Research, Argonne National Laboratory, Argonne, Illinois, 1986.

4. Charles L. Gray and Jeffrey Alson, *Moving America to Methanol* (Ann Arbor, Mich.: University of Michigan Press, 1985).

5. Survey conducted by American Gas Association and reported in its newsletter, *NGV Reporter* 3:6 (Nov.–Dec. 1986). The percentage dropped sharply during 1987 from its 11 percent level when Southern California Edison, which operated about 30 percent of all natural gas vehicles in the country (2,500 vehicles), decided to eliminate its natural gas vehicle program.

22: THE 1980s: WHERE ARE WE? WHERE ARE WE GOING?

1. Some states, most with much corn-based agriculture, have provided large subsidies to ethanol fuel production mostly in the form of partial or total exemption of the gasohol blends from fuel taxes. The result of these subsidies and of local support for "home-grown" fuel in those states has been high market penetration levels. In 1984, gasohol (10 percent ethanol and 90 percent gasoline) blends accounted for 32.5 percent of gasoline sold in Iowa, 27 percent in Nebraska, and 18 to 20 percent in Kansas, Indiana, and Kentucky. See Oak Ridge National Laboratory, *Transportation Energy Data Book: Edition 8* (Springfield, Va.: NTIS, 1985, ORNL-6205), pp. 1–19.

2. Langdon Winner, *The Whale and the Reactor: A Search for Limits in an Age of High Technology.* (Chicago, Ill.: University of Chicago Press, 1986), p. X.

3. See for instance Colin Norman, "Radical Surgery in Uranium Enrichment," *Science* 228 (21 June 1985): 1407–1409; William Lanouette, "Dream Machine," *Atlantic Monthly* 251 no. 4 (April 1983): 35.

4. U.S. Department of Commerce, *A Competitive Assessment of the U.S. Methanol Industry* (Washington, D.C.: GPO, 1985), p. 11; and Information Resources, Inc. (Washington, D.C.), personal communication, July 1987.

CONCLUSIONS

1. Mark D. Levine, Paul Craig, "A Decade of United States Energy Policy," *Ann. Rev. Energy* 10 (1985): 557–587.

2. Cited in "Back to the Energy Crisis," *Science* 235 (6 February 1987): 627.

Index

Designer: U.C. Press Staff
Compositor: Asco Trade Typesetting Ltd.
Text: 10/13 Sabon
Display: Sabon